The Curious Lore of Precious Stones

BEING A DESCRIPTION OF THEIR SENTIMENTS AND FOLK LORE, SUPERSTITIONS, SYMBOLISM, MYSTICISM, USE IN MEDICINE, PROTECTION, PREVENTION, RELIGION, AND DIVINATION. CRYSTAL GAZING, BIRTHSTONES, LUCKY STONES AND TALISMANS, ASTRAL, ZODIACAL, AND PLANETARY

BY

GEORGE FREDERICK KUNZ

A.M., PH.D., D.SC.

DOVER PUBLICATIONS, INC.
NEW YORK

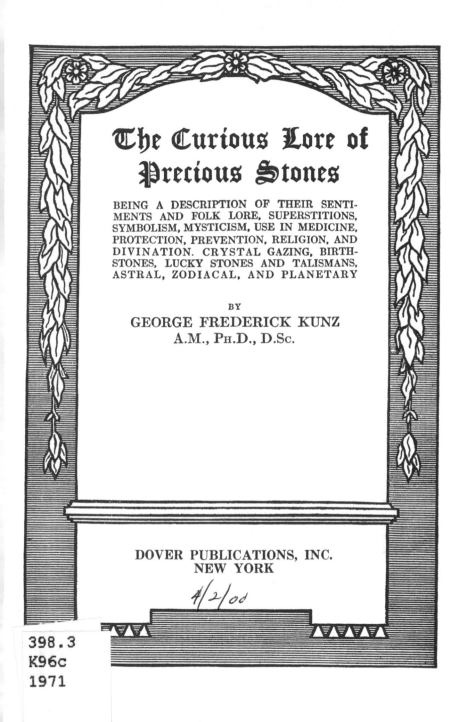

Published in Canada by General Publishing Company, Ltd., 30 Lesmill Road, Don Mills, Toronto, Ontario.
Published in the United Kingdom by Constable and Company, Ltd.

This Dover edition, first published in 1971, is an unabridged republication of the work originally published in 1913 by the J. B. Lippincott Company in Philadelphia. This edition is published by special arrangement with the J. B. Lippincott Company.

International Standard Book Number: 0-486-22227-6
Library of Congress Catalog Card Number: 79-156815

Manufactured in the United States of America
Dover Publications, Inc.
180 Varick Street
New York, N. Y. 10014

WITH HEARTFELT APPRECIATION OF THE NOBLE SPIRIT THAT CONCEIVED AND FOUNDED THE MORGAN-TIFFANY COLLECTION OF GEMS AND THE MORGAN-BEMENT COLLECTIONS OF MINERALS AND METEORITES OF THE AMERICAN MUSEUM OF NATURAL HISTORY, AND THE MORGAN COLLECTION OF THE MUSÉE D'HISTOIRE NATURELLE OF PARIS, AND WHOSE KINDLY ADVICE AND ENCOURAGEMENT HAVE DONE SO MUCH FOR THE PRECIOUS STONE ART, THIS VOLUME IS RESPECTFULLY DEDICATED TO THE MEMORY OF THE LATE

J. PIERPONT MORGAN

Preface

THE love of precious stones is deeply implanted in the human heart, and the cause of this must be sought not only in their coloring and brilliancy but also in their durability. All the fair colors of flowers and foliage, and even the blue of the sky and the glory of the sunset clouds, only last for a short time, and are subject to continual change, but the sheen and coloration of precious stones are the same to-day as they were thousands of years ago and will be for thousands of years to come. In a world of change, this permanence has a charm of its own that was early appreciated.

The object of this book is to indicate and illustrate the various ways in which precious stones have been used at different times and among different peoples, and more especially to explain some of the curious ideas and fancies that have gathered around them. Many of these ideas may seem strange enough to us now, and yet when we analyze them we find that they have their roots either in some intrinsic quality of the stones or else in an instinctive appreciation of their symbolical significance. Through manifold transformations this symbolism has persisted to the present day.

The same thing may be said in regard to the various superstitions connected with gems. Our scientific knowledge of cause and effect may prevent us from accepting any of the fanciful notions of the physicians and astrologers of the olden time; nevertheless, the possession of a necklace or a ring adorned with brilliant diamonds, fair pearls, warm, glowing rubies, or celestial-hued sapphires will to-day make a woman's heart beat faster and bring a blush of pleasure to her cheek. Life will seem better worth living to her; and, indeed, this is no delusion, for life is what our thought makes it, and joy is born of gratified desire. Hence nothing that contributes to increasing the sum of innocent pleasures should be disdained; and surely no pleasure can be more innocent and justifiable than that inspired by the possession of beautiful natural objects.

The author, who possesses what is believed to be the most comprehensive private library on this subject, has obtained many references from material which he has been gathering during the past twenty-five years. Many of the types exist in the collection of folk-lore precious stones exhibited at the World's Columbian Exposition in 1893, and now in the Field Museum of Natural History in Chicago. Other types are drawn from the Morgan Collection exhibited at the Paris Expositions of 1889 and 1900, which, with additions, is now in Morgan Hall, in the American Museum of Natural History, New York City.

Other prominent references are the collection of precious stones in the California Midwinter Memorial Museum, in Golden Gate Park, San Francisco; the Tiffany collection of precious stones, exhibited at the Atlanta Exposition of 1894, now in the National Museum in Washington; the collection exhibited at the Pan-American Exposition, and presented to the Musée d'Histoire Naturelle, in Paris, by the late J. Pierpont Morgan; the collection exhibited at the exposition held in Portland, Oregon, in 1905; and the collection of gems and precious stones exhibited at the Jamestown Exposition, 1907. All of these collections, either entirely or very largely, have been formed by the author.

Some references to sentiment connected with precious stones are embodied in the little work, now in its 21st edition, entitled: "Natal Stones, Sentiments and Superstitions Associated with Precious Stones," compiled by the writer, who has examined nearly all the principal collections in the United States, Europe, Mexico, Canada, and Asiatic Russia.

For courtesies, information and illustrations, I am indebted to the following, to whom my thanks are due:

Prof. Taw Sein Ko, Superintendent of the Archæological Survey, of Burma; Dr. T. Wada, of Tokyo, Japan; Dr. G. O. Clerc, President of the Société Ouralienne des Amis des Sciences Naturelles, Ekaterinebourg, Russia; Dr. Charles Braddock, late Medical Inspector to the King of Siam; Sir Charles Hercules Reed, Curator of Archæology, and Dr. Ernest A. Wallis Budge, Egyptologist, British Museum, London; A. W. Feavearyear, Esq., London; Dr. Salomon Reinach, Director of the Archæological Museum of St. Germain-en-Laye, France; Prof. Giuseppe Belucci, of the University of Perugia; Dr. Peter Jessen, Librarian of the Kunstgewerbe Museum, of Berlin; Miss Belle DaCosta Green; Dr. Frederick Hirth, Chinese Professor, Columbia University, New York; Dr. Clark Wissler, Curator of Archæology, Dr. L. P. Gratacap, Curator of Mineralogy, American Museum of Natural History; Dr. Berthold Laufer, Oriental Archæologist, and Dr. Oliver C. Farrington, Curator of Geology and Mineralogy, Field Museum of Natural History, Chicago; Hereward Carrington, Esq., Psychist, New York; Dr. W. Hayes Ward, Archæologist and Babylonian Scholar; Mrs. Henry Draper, New York; H. W. Kent, Esq., Metropolitan Museum of Art, New York City; Consul General Moser, Colombo, Ceylon; W. W. Blake, Mexico City, who has done so much to encourage Mexican archæological investigation; the late A. Damour, of Paris, the great pioneer of mineralogical archæology; the late Dr. A. B.

Meyer, of Dresden, who, more than anyone else, proved that the *Nephritfrage* or the jade question was to be solved by chemical and mineralogical investigation; the late Rajah Sir Sourindro Mohun Tagore, of Calcutta; and Dr. A. M. Lythgoe, Egyptologist, Metropolitan Museum of Art.

G. F. K.

SEPTEMBER, 1913.

Contents

Illustrations

COLOR PLATES

(Color plates follow page 40.)

DOUBLETONES

LINE CUTS IN TEXT

The Curious Lore of Precious Stones

I

Superstitions and Their Sources

FROM the earliest times in man's history gems and precious stones have been held in great esteem. They have been found in the monuments of prehistoric peoples, and not alone the civilization of the Pharaohs, of the Incas, or of the Montezumas invested these brilliant things from Nature's jewel casket with a significance beyond the mere suggestion of their intrinsic properties.

The magi, the wise men, the seers, the astrologers of the ages gone by found much in the matter of gems that we have nearly come to forgetting. With them each gem possessed certain planetary attractions peculiar to itself, certain affinities with the various virtues, and a zodiacal concordance with the seasons of the year. Moreover, these early sages were firm believers in the influence of gems in one's nativity,—that the evil in the world could be kept from contaminating a child properly protected by wearing the appropriate talismanic, natal, and zodiacal gems. Indeed, folklorists are wont to wonder whether the custom of wearing gems in jewelry did not originate in the talismanic idea instead of in the idea of mere additional adornment.

The influence exerted by precious stones was assumed in medieval times without question, but when the spirit of investigation was aroused in the Renaissance period, an effort was made to find a reason of some sort for the traditional beliefs. Strange as it may seem to us, there was little disposition to doubt that the influence existed; this was taken for granted, and all the mental effort expended was devoted to finding some plausible explanation as to how precious stones became endowed with their strange and mystic virtues, and how these virtues acted in modifying the character, health, or fortunes of the wearer.

When the existence of miracles is acknowledged, there will always be a tendency to regard every singular and unaccountable happening as a miracle; that is to say, as something that occurs outside of, or in spite of, the laws of nature. We even observe this tendency at work in our own time. As regards visual impressions, for instance, if a child of lively imagination enters a half-lighted room and sees a bundle of clothes lying in a corner, the indistinct outline of this mass may be transformed to his mind into the form of a wild animal. The child does not really see an animal, but his fear has given a definite outline and character to the indefinite image printed on the retina.

The writer has always sought to investigate anything strange and apparently unaccountable which has been brought to his notice, but he can truly say he has never found the slightest evidence of anything transcending the acknowledged laws of nature. Still, when we consider the marvellous secrets that have been revealed to us by science and the yet more wonderful things that will be revealed to us in the future, we are tempted to think that there may be something in the old beliefs, some residuum of fact, susceptible indeed of explanation, but very different from what a crass scepticism supposes it to be. Above all, the results of the investigations now pursued in relation to the group of phenomena embraced under the designation of telepathy,—the subconscious influence of one mind over an absent or distant mind,—and the wireless transmission of power in wireless telegraphy and telephony, may go far to make us hesitate before condemning as utterly preposterous many of the tales of enchantment and magical influence. If the unconscious will of one individual can affect the thoughts and feelings of another individual at a great distance and without the intervention of any known means of communication, as is confidently asserted by many competent investigators in the domain of telepathy, their claims being supported by many strange happenings, perhaps the result of coincidence, but possibly due to the operation of some unknown law, does this not give a color of verity to the statements regarding the ancient magicians and their spells?

Auto-suggestion may also afford an explanation of much that is mysterious in the effects attributed to precious stones, for if the wearer be firmly convinced that the gem he is wearing produces certain results, this conviction will impress itself upon his thought and hence upon his very organism. He will really experience the influence, and the effects will manifest themselves just as powerfully as though they were caused by vibrations or emanations from the material body of the stone.

All this may serve to explain the persistence of the belief in magic arts. A few hundred years ago, a Hungarian woman was accused of having murdered two or three hundred young girls, and at her trial she confessed that her object was to use the blood of her victims to renew her youth and beauty, for the blood of innocent virgins was supposed to have wonderful properties. In some parts of England to-day there is a superstitious belief that an article of clothing worn by a person, or anything he has habitually used, absorbs a portion of his individuality. Therefore, it sometimes happens that a handkerchief, for instance, will be stolen and pinned down beneath the waters of a stream on a toad, the pins marking the name of the enemy, the belief being that as this cloth wastes away, so will the body of him who had worn it. In medieval and later times this was the common practice of the sorcerers, although they frequently composed a wax figure rudely resembling the person against whom the spell was directed, and then thrust pins into this figure or allowed it to melt away before a slow fire. The enchantment of the sorcerer was supposed to have caused some essence of the personality to enter into the image, and therefore the living and breathing being felt sympathetically the effects of the ill-treatment inflicted upon its counterfeit.

The persistence of the most cruel and unnatural practices of old-time sorcery is illustrated by the fact that only a few years ago, in the island of Cuba, three women were condemned to death for murdering a white baby so as to use the heart and blood as a cure for diseases. Four other women were sentenced to from fourteen to twenty years' imprisonment as accomplices. When such things happen in Cuba, it is not surprising that in half-civilized Hayti similar crimes are committed. Here the Voodoo priests and priestesses, *papalois* and *mamalois* (papa-kings and mama-queens) require from time to time a human sacrifice to appease their serpent-god. One strange case is related where a stupefying potion, inducing a state of apparent death, was secretly administered to a sick man. When the attending physician pronounced him dead, he was duly interred; but, two days after, the grave was found open and the body had disappeared. The Voodoo worshippers had carried the man away so as to revive him and then sacrifice him at their fearful rites.

In a poem addressed to Marguerite de Valois,—"La Marguerite des Marguerites," as she was called,—by Jean de la Taille de Bondaroy,[1] we read of the diamond that it came from gold and from the sun. But we are told that not only are precious stones endowed with life, they also are subject to disease, old age, and death; "they even take offence if an injury be done to them, and become rough and pale." The sickness of the pearl has been a theme for centuries, and in many cases is only fancied. It is but a subterfuge or deception for a lady to remark that her pearls have sickened; by referring to this sickness, her friends are naturally led to believe that at one time her pearls were fine, perfect ones, when in reality they may never have been so.

The opinion given in 1609, by Anselmus De Boot, court physician to Rudolph II of Germany, regarding the power inherent in certain precious stones,[2] embodies the ideas on this subject held by many of the enlightened minds of that period.

> The supernatural and acting cause is God, the good angel and the evil one; the good by the will of God, and the evil by His permission. . . . What God can do by Himself, He could do also by means of ministers, good and bad angels, who, by special grace of God and for the preservation of men, are enabled to enter precious stones and to guard men from dangers or procure some special grace for them. However, as we may not affirm anything positive touching the presence of angels in gems, to repose trust in them, or to ascribe undue powers to them, is more especially pleasing to the spirit of evil, who transforms

---

[1] Jean de la Taille de Bondaroy, "Le Blason de la Marguerite," Paris, 1574.

[2] De Boot, "Gemmarum et lapidum historia," lib. i, cap. 25, Lug. Bat., 1636, pp. 87, 91.

himself into an angel of light, steals into the substance of the little gem, and works such wonders by it that some people do not place their trust in God but in a gem, and seek to obtain from it what they should ask of God alone. Thus it is perhaps the spirit of evil which exercises its power on us through the turquoise, teaching us, little by little, that safety is not to be sought from God but from a gem.

In the next chapter of his work, De Boot, while extolling the remedial power of a certain group of stones, insists upon the falsity of many of the superstitions regarding these objects.[3]

That gems or stones, when applied to the body, exert an action upon it, is so well proven by the experience of many persons, that any one who doubts this must be called over-bold. We have proof of this power in the carnelian, the hematite, and the jasper, all of which when applied, check hemorrhage. . . . However, it is very necessary to observe that many virtues not possessed by gems are falsely ascribed to them.

Paracelsus, the gifted and brilliant thinker, scientist, and, we must probably add, charlatan of the sixteenth century, whose really extraordinary mental endowment was largely wasted in the effort to impress his followers with the idea that he had a mystic control over supernatural agencies, was the owner of a talismanic jewel which he asserted to be the dwelling-place of a powerful spirit named "Azoth." Some old portraits of the philosopher, or pseudo-philosopher, figure him wearing this jewel, in whose virtues we may fairly doubt that he himself believed, but which furnished part of the paraphernalia be freely employed to gain influence over the credulous.[4]

---

[3] De Boot, "Gemmarum et lapidum historia," lib. i, cap. 26, Lug. Bat., 1636, p. 103.

[4] Mackey, "Memoirs of Extraordinary Popular Delusions," London, n. d., p. 144.

The following passage from the "Faithful Lapidary" of Thomas Nicols,[5] who wrote in the middle of the seventeenth century, illustrates the prevailing opinion in England at that time as to the virtues of precious stones:

Perfectionem effectûs contineri in causa. But it cannot truly be so spoken of gemms and pretious stones, the effects of which, by Lapidists are said to be, the making of men rich and eloquent, to preserve men from thunder and lightning, from plagues and diseases, to move dreams, to procure sleep, to foretell things to come, to make men wise, to strengthen memory, to procure honours, to hinder fascinations and witchcrafts, to hinder slothfulness, to put courage into men, to keep men chaste, to increase friendship, to hinder difference and dissention, and to make men invisible, as is feigned by the Poet concerning Gyges ring, and affirmed by Albertus and others concerning the *ophthalmius lapis,* and many other strange things are affirmed of them and ascribed to them, which are contrary to the nature of gemms, and which they as they are materiall, mixt, inanimate bodies neither know nor can effect, by the properties and faculties of their own constitutions: because they being naturall causes, can produce none other but naturall effects, such as are all the ordinary effects of gemms: that is, such effects as flow from their elementary matter, from their temper, form and essence; such as are the operations of hot and cold, and of all the first qualities, and all such accidents as do arise from the commixtion of the first qualities: such as are hardnesse, heavinesse, thicknesse, colour, and tast. These all are the naturall faculties of gemms, and these are the known effects of the union of their matter, and of the operation of the first qualities one upon another.

The long-continued concentration of vision on an object tends to produce a partial paralysis of certain functions of the brain. This effect may be noted in the helplessness of a bird when its gaze is fixed upon the glittering eyes of a serpent, or in the unwilling obedience yielded by a lion or some other wild animal when forced to look into the intent eyes of its trainer. In the same way those who gaze for a long time and without inter-

[5] Nicols, "Faithful Lapidary," London, 1659, pp. 32, 33.

ruption on a crystal or glass ball, on an opal, a moonstone, a sapphire, or a cat's-eye, may become partially hypnotized or even fall into a profound sleep. The condition induced, whether it be that of semi-trance, of hypnotism, or simply due to the imaginative workings of the brain, is believed to give an insight into the future. This hypnotic effect is probably caused by some gleam or point of light in the stone, attracting and fixing the beholder's gaze. The moonstone, the star sapphire, and the cat's-eye are all gems which possess a moving light, a moving line, or three crossed lines, and they are believed by the Orientals to be gems of good luck. Indeed, it is supposed in the East that a living spirit dwells within these stones, a spirit potent for good.

Superstitious fancies bear the same relation to truth that the shadow of a form does to the form itself. We know that the shadow has no substantial existence, and yet we know equally well that it is cast by some real body; in the same way we may be sure that, however foolish a superstition may appear to be, it has some foundation in fact. Indeed, superstition is associated with the highest attribute of the human mind,—imagination. The realities about us gain much of their charm from sentiment, and all that is great in art and literature owes its being to the transforming energy of pure imagination. Morbid imagination, on the other hand, distorts and degrades the impressions it receives and produces only unlovely or ignoble forms and ideas.

Sentiment may best be expressed as the feeling of one who, on a warm summer's day, is rowing along a shady brook or resting in some sylvan dell, with nothing to interfere with his tranquil mood and nothing to spur him on to action; thus he has only suggestions of hope and indulges in rosy views of life. Reality, on the other hand, may be likened to a crisp winter's morning when one is filled with exhilaration, conscious of the tingle of the cold, but comfortable in the knowledge of wearing a tightly-buttoned garment which will afford protection should the elements become disturbing. Superstition, lastly, can be said to resemble a dark, cold, misty night, when the moon is throwing malevolent shadows which are weird and distorted, while the cold seems to seize one by the throat and arouse a passionate desire to free one's self from its grip in some way, to change a horrible nightmare into a pleasant dream.

In the early part of the last century a series of very interesting experiments designed to demonstrate the effects produced upon a sensitive subject by the touch of precious stones and minerals, were made in the case of the "Seeress of Prevorst," Frederike Hauffe (b. 1801), a woman believed to possess remarkable clairvoyant powers.[6] When pieces of granite, porphyry, or flint were placed in her hand, she was not affected in any way. The finest qualities of fluorspar, on the other hand, had a marked action, relaxing the muscles, causing diarrhœa, and producing a sour taste in the mouth; occasionally a somnambulistic state was induced. This latter condition was also produced by Iceland spar and by the sapphire. While the substances so far noted depressed the vital energy, sulphate of barium stimulated the muscles, produced an agreeable warmth of the body, and made the subject feel as though she could fly through the air. If the application of this material was long continued, the pleasurable sensation found expression in laughter. In the case of witherite, a carbonate of barium,

---

[6] Görres, "Die christliche Mystik," Regensburg, 1840, vol. iii, pp. 190 sqq.

this effect was produced to an even greater degree, for if water in which this mineral had been dipped were swallowed, spasms of laughter resulted.

Rock-crystal also was found to possess a strongly stimulating influence, for if put in the hand, it aroused the subject from a half-slumber, and if placed on the pit of the stomach, it had the power to awaken the seeress from a somnambulistic trance, while at the same time an aromatic odor was diffused around. When, however, the application was continued for some time, the muscles stiffened, until finally an epileptic state ensued. Indeed, the rigidity produced was so great that the limbs resisted all attempts to bend them. The same effect, but in a much less degree, was caused by glass, even by looking at it, or by the tones emitted by a glass object when struck. All colorless silicates, the diamond, and even gypsum, had a similar effect, as did also heliotrope and basalt, either of which caused a bitter taste in the mouth.

The most powerful action was that exerted by hematite, the oxide of iron in this substance inducing a kind of paralysis, with a sensation of inner chill; this condition could only be relieved by the application of a piece of witherite. Octahedrons of magnetite (loadstone) caused a sensation of heaviness and convulsive movements of the limbs, even when the material, wrapped up in paper, was brought near the subject. Spinel, in whose composition oxide of chromium enters, caused the same symptoms as loadstone, except that in this case the force seemed to exert itself from the hand upward along the arm, while with the loadstone the action was downward along the arm to the hand, owing to the attractive quality of this magnetic iron. Ruby called forth a sensation of coldness in the tongue, and rendered this member so heavy that only incoherent sounds could be emitted;

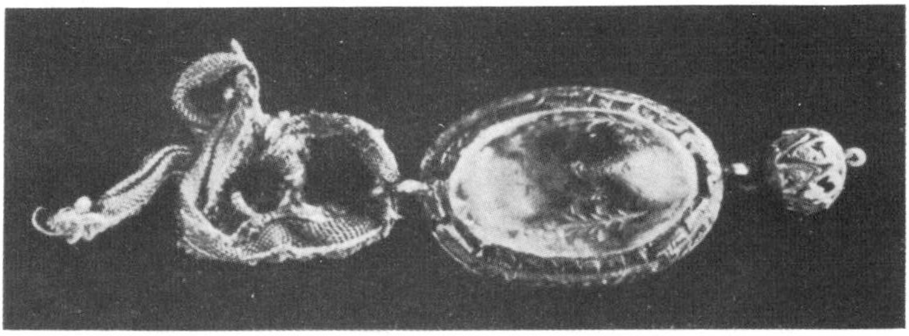

ROCK-CRYSTAL AMULET SET IN SILVER.
Bohemian, tenth century. Field Museum of Natural History.

ROCK-CRYSTAL PLACQUE, ANCIENT MEXICAN.
Field Museum Collection, Chicago.

the fingers and toes also became cold, and the body was agitated by a violent shivering; but to all these bad symptoms succeeded a sense of elasticity and well-being, not, however, without a vague fear that the stone might cause a renewal of the physical depression. When chrysoprase was used, chills and shivering resulted, beginning at the breast and spreading thence over the whole body.

We have touched upon the hypnotic influence exercised by gems, but there can be no doubt that the subject has not been as carefully studied as it deserves to be. That the hypnotic state can be induced by gazing fixedly upon a bright object held just above the eyes is a well-known fact, but quite probably a similar though not so pronounced effect results from gazing on a bright object just before the gazer's eyes. In the case of colored precious stones, the effects of the various color-rays combine with the light effects and strengthen the impression upon the optic nerve. All this, however, concerns only the purely physical impression, but we know that very often the hypnotic state is produced by a mental impression, by the belief, or the fear, that the state will supervene. With precious stones as hypnotizing agents, the mental impression is widely different, for here the physical impression is heightened by the consciousness of the value and rarity of the material. The fascination that a fine set of jewels, with all their sparkle and color, exercises upon the mind of a woman who sees them in their glorious radiance on the neck, the arms, and the head of another woman, is not only due to the beauty of the spectacle, but is largely owing to the consciousness that they are rare and valuable objects and are perhaps eloquent witnesses of the power of love. A dash of envy sometimes serves to render the emotion more complex.

The names of precious stones and semi-precious

stones are frequently used as adjectives, and when so employed convey something more to the mind than do the corresponding adjectives of color. We may instance the following expressions: the "Emerald Isle" and "emerald meadows"; "sapphire seas" and "sapphire eyes"; "ruby wine," "ruby lips," and, in Shakespeare, "the natural ruby of your cheeks"; "coral lips" and "coral ears"; "pearly teeth" and "pearly skin"; "turquoise skies"; "amethystine locks" and, in Roman times, "amber hair." In all these cases the name of the precious mineral is really used as a superlative of the adjective, suggesting the choicest variety of the color or shade. The phrases "hard as adamant" and "clear as crystal" show a similar use of the name of a precious or ornamental stone to express the highest grade of a given quality.

Before the introduction of the "point" system in typography three of the grades of type bore the names of precious stones,—namely, "diamond type," "agate type," and "emerald type"; this latter designation is employed only in England, where "agate type" is called "ruby type." Another size was denominated "pearl type."

A fanciful tale written not long ago treats of the practical inconveniences which would result, could such metaphorical expressions find a realization in fact.[7] At the birth-feast of a certain princess, one of the fairies was not invited; she, nevertheless, made her appearance. After the other fairies had endowed the child with many good qualities, the neglected fairy said, "I will give her vanity, and her vanity shall change her beauty to the things it is said to resemble." However, a friendly fairy

[7] Virna Sheard, "The Jewelled Princess," in Canadian Magazine.

came to the rescue, saying, "I will give her unselfishness, and by it she shall turn her beauty back to what she wishes it to be."

The result can easily be imagined. As the little princess grew up, those who wished to flatter her vanity spoke of her "teeth of pearl," of her "golden hair," of her "coral lips," and of her "sapphire eyes." Upon this her teeth changed to pearls, her hair to spun gold, her lips to coral, and her eyes to two magnificent sapphires. However, beautiful as these were, they did not grant the power of sight, so that the unhappy princess became blind. Not long after this a revolution deprived the king and queen of their throne and they were reduced to great poverty. In these straits the daughter sacrificed her "gold-hair" to relieve their wants, and immediately the spell was dissolved and she regained all her natural beauty.

Shelley, who saw the world illumined by the rainbow hues of poetic fancy, wrote of "diamond eyes," "an emerald sky," "the emerald heaven of trees," "the sapphire ocean," "sapphire-tinted skies," "the sapphire floods of interstellar air," and "the chrysolite of sunrise." For some reason, he does not use the ruby, a favorite stone with many poets, and psychologists might find in this a proof that red appeals less strongly to the idealist than do the other colors.

The principal literary sources for the talismanic and therapeutic virtues attributed to ornamental stones may be divided into several groups, at first more or less independent of each other, but combined to a greater or lesser extent by later writers. Pliny gives, sometimes rather grudgingly, a number of superstitions current in his time, but the Alexandrian literature of the second, third, and fourth Christian centuries provides a much richer field for these superstitions, as shown in the Orphic poem "Lithica," the "Cyrianides," attributed to Hermes Trismegistus, the little treatise "On Rivers," which bore the name of Plutarch, and last, but not least, in the work by Damigeron, which purported to be written by an Arab king named Evax, and sent by him to Tiberius or Nero. The influence exerted by the legends surrounding the stones of the high priest's breastplate, and those chosen as foundation stones for the New Jerusalem, will be treated of elsewhere.

In the seventh, eighth, and ninth centuries, a new literature on this subject made its appearance, probably in Asia Minor. Some of the works were originally written in Syriac and later translated into Arabic. Others were composed in the latter language. This source was drawn upon for the production of the Lapidarium of Alfonso X, of Castile. This compilation, although dating in its present Spanish form from the thirteenth century, is based upon a much older original in "Chaldee" (Syriac?). There can be little doubt that many Hindu superstitions, no longer preserved for us in the literature of India, are reproduced in these Syrio-Arabic works, wherein we have also much that is of Alexandrian origin. This indeed is easily explained by history, for the Arabs, through their widely extended conquests, were led to absorb and amalgamate the date they secured, directly or indirectly, from the East and the West.

While this literature was developing in the Mohammedan world, the tradition of Pliny and Solinus was transmitted to the Christian world of the seventh and succeeding centuries by Isidorus of Seville. This brings us to the remarkable poetical treatise on the virtues of precious stones by Marbodus, Bishop of Rennes, a work written at the end of the eleventh century, and often quoted as that of Evax; indeed, it purports to be by him and really contains a good part of the material composing the treatise of Damigeron or Evax. At the same time Marbodus drew freely upon Pliny, either directly or through Isidorus. For the Middle Ages this poem of Marbodus, already translated into Old French in the twelfth century, became known as the "Lapidario" *par excellence,* and furnished a great part of their material to medieval authors on this subject. Soon, however, extracts from the Arabic sources became available, and the whole mass of heterogeneous material was worked over and recombined in a variety of ways.

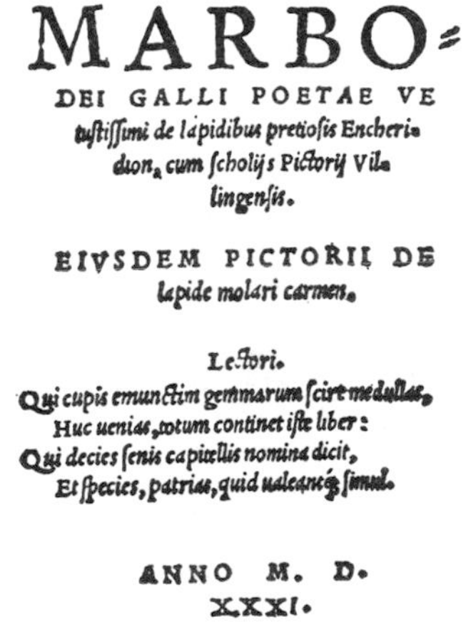
MARBO=
DEI GALLI POETAE VE
tustissimi de lapidibus pretiosis Encheri-
dion, cum scholijs Pictorij Vil-
lingensis.

EIVSDEM PICTORII DE
lapide molari carmen.

Lectori.
Qui cupis emunctim gemmarum scire medullas,
Huc uenias, totum continet iste liber:
Qui decies senis capitellis nomina dicit,
Et species, patrias, quid ualeantque simul.

ANNO M. D.
XXXI.

Title page of the first edition of the poetical treatise on precious stones by Marbodus, Bishop of Rennes, printed in Friburg, 1531.

This complex origin of the traditions explains their almost incomprehensible contradictions regarding the virtues assigned to the different stones, and also the fact that the qualities of one stone are frequently attributed to another one, so that, in the later works on this subject, it becomes quite impossible to present a satisfactory view of the distinguishing qualities and virtues of the separate stones. The habit of copying, without discrimination or criticism, whatever came to hand, and the aim to utilize as much of the borrowed material as possible, is scarcely less a characteristic of the seventeenth and eighteenth century writers than it is of those of a later date. This is in part an excusable and even an unavoidable defect, but it should be minimized as much as possible.

The treatise known under the title "Cyrianides" was, as we have noted, a product of the Alexandrian school. It was asserted to be the work of Hermes Trismegistus, the name given by the Greeks to the Egyptian god Thoth. Here we have a specimen of the species of magic known as litteromancy, or divination by means of the letters of the alphabet, since a stone, a bird, a plant, and a fish, each beginning with the same letter and signifying the four elements, are given for each of the twenty-four letters of the Greek alphabet. These four objects were to be grouped together to form a talisman, the bird being usually engraved on the stone, while a portion of the fish and of the plant was placed in the bezel of the ring in which the stone was to be set.[8] Another, almost contemporary work, is the exceedingly curious and interesting treatise by St. Epiphanius, Bishop of Constantia, on the twelve gems on the "Breastplate of Judgment" of the high priest (Ex., xxviii, 15–21). This unique production is in the form of a letter addressed to Diodorus, Bishop of Tyre, and it is peculiarly valuable as the

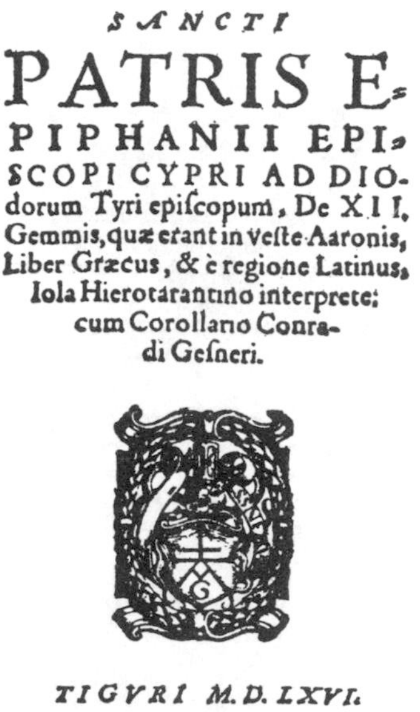
SANCTI
PATRIS E-
PIPHANII EPI-
SCOPI CYPRI AD DIO-
dorum Tyri episcopum, De XII.
Gemmis, quæ erant in veste Aaronis,
Liber Græcus, & è regione Latinus,
Iola Hierotarantino interprete:
cum Corollario Conra-
di Gesneri.

TIGVRI M.D.LXVI.

Title page of the first edition of the Greek treatise by St. Epiphanius on the Gems of the Breastplate, with a Latin version. Edited and issued at Zürich in 1566 (1565) by Conrad Gesner.

[8] De Mély, "Les lapidaires de l'antiquité et du moyen-âge," vol. ii, "Les lapidaires grecs," Paris, 1898, pp. 1–50.

first of a long series of attempts to elucidate the question as to the identity of the twelve stones. The special virtues of each stone are also given, and this treatise may be regarded as the prototype of all the Christian writings on the symbolism of stones.

A most interesting medieval treatise on the virtues of precious stones forms part of the *De rerum natura* of Thomas de Cantimpré (1201–1270), who was a pupil of Albertus Magnus and composed his work between 1230 and 1244. The Latin text has never been printed, but the book was translated into German by Konrad von Megenberg about 1350. Strange to say, the translator did not know the name of the writer and supposed when he began to translate the book that it was by Albertus Magnus. In many cases Thomas de Cantimpré merely copies the statements of older authors, but occasionally he gives us new material, or at least a new version of his originals.

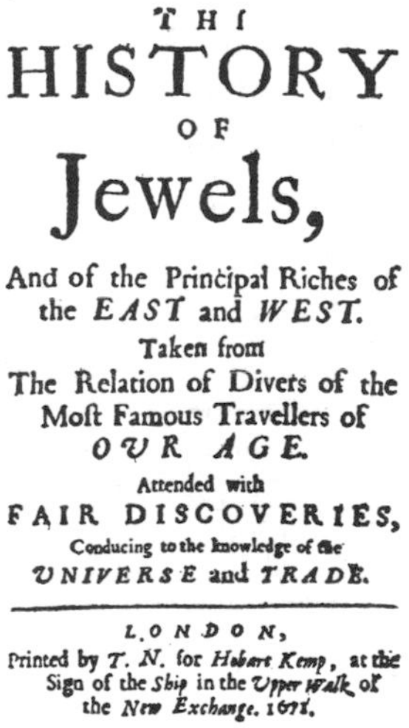
THE
HISTORY
OF
Jewels,
And of the Principal Riches of the *EAST* and *WEST*.
Taken from
The Relation of Divers of the Most Famous Travellers of
OUR AGE.
Attended with
FAIR DISCOVERIES,
Conducing to the knowledge of the
UNIVERSE and *TRADE*.

LONDON,
Printed by *T. N.* for *Hobart Kemp*, at the Sign of the *Ship* in the *Upper Walk* of the *New Exchange*. 1671.

Title page of one of the earliest treatises on precious stones published in England.

The renowned medieval philosopher and theologian, Albertus Magnus (1193–1280), for a short time Bishop of Ratisbon, and who later taught theology in the University of Paris and had the great St. Thomas Aquinas for a pupil, was not altogether free from the superstitious notions of his time, traces of which appear in certain of his numerous writings. Many years after his death some of this material was extracted from his works and, amplified by additions from other sources, was published under the title "Secrets des vertus des Herbes, Pierres et Bestes." Of this there are two versions, one being an epitome of the other and termed respectively "Le Grand Albert" and "Le Petit Albert." These little books were often reprinted and widely circulated, and eventually enjoyed great popularity among the French peasants. Indeed, even to the present day they may still be met with in out-of-the-way parts of rural France.

Among literary deceptions one of the boldest was that practised in the early part of the seventeenth century by Ludovico Dolce. This writer made, in 1565, a literal translation into Italian of the "Speculum lapidum" of Camillo Leonardo, printed in Venice in 1502, and he had the courage to issue it as his own work, under the title "Trattato delle gemme chè produce la natura." In view of the general familiarity with Latin among the better classes at that period, and the numerous fine libraries existing in Venice at the time, it seems most extraordinary that Dolce should have been successful in palming off this work as his own, but even to-day citations are made from Dolce's "Trattato delle gemme" and from Leonardo's "Speculum lapidum," as though these were distinct works.

II

On the Use of Precious and Semi-Precious Stones as Talismans and Amulets

THE use of precious stones in early times as amulets and talismans is shown in many ancient records, and several scholars have assumed that the belief in the magic efficacy of stones gave rise to their use as objects of personal adornment. It is, of course, very difficult either to prove or to disprove such a theory, for, even in the case of the oldest texts, we must bear in mind that they do not in the least represent primitive conditions, and that many thousands of years must have elapsed before a people could attain the grade of civilization necessary for the production of even the simplest literature. For this reason, certain investigators have preferred to seek for a solution of this problem in the customs and habits of the so-called uncivilized peoples of our own time; but we must not forget that conditions which seem to us very rudimentary are, nevertheless, the result of a long process of development. Even if this development was arrested many centuries or millenniums ago, it must have required a very considerable period of time to evolve such usages and conventions as are found even among the lowest races. Indeed, many uncivilized peoples have very complicated rules and observances, testifying to considerable thought and reflection.

Fetichism in all its forms depends upon an imperfect conception of what constitutes life and conscious being, so that will and thought are attributed to inanimate objects. We can observe this in the case of animals and very young children, who regard any moving object as endowed with life. In the case of stones, however, it seems probable that those supposed to be the abode of spirits, good or evil, were selected because their natural form suggested that of some animal or of some portion of the human body. On the other hand, the wearing of what we call precious stones is more likely to have been due to the attraction exercised by bright colors upon the eye of the beholder and to the desire to display some distinguishing mark that would command attention and admiration for the wearer. This tendency runs through the higher animal kingdom, and its workings have served as a foundation for the theory of natural selection.

It seems likely that we have here the true explanation of the motive for the gathering, preserving, and wearing of precious stones. Since these objects are motionless, they can scarcely have impressed the mind of primitive man with the idea that they were alive; they were not imposing by their mass, as were large stones, and their crystalline form scarcely figured any known living shape. Hence their chief, we may even say their only attraction was their color and brilliancy. What effect these qualities had upon the visual sense of primitive man may be safely inferred from the effect such objects produce upon infants. The baby has no fear in regard to a small and brilliantly colored object which is shown to it, but will eagerly put out its hand to seize, hold, and gaze upon a bright-colored stone. As the object is quite passive and easily handled, there is nothing to suggest any lurking power to harm, and therefore there is nothing to interfere with the pleasurable sensation aroused in the optic nerve by the play of color. In this naïve admiration of what is brilliant and colored, the infant undoubt-

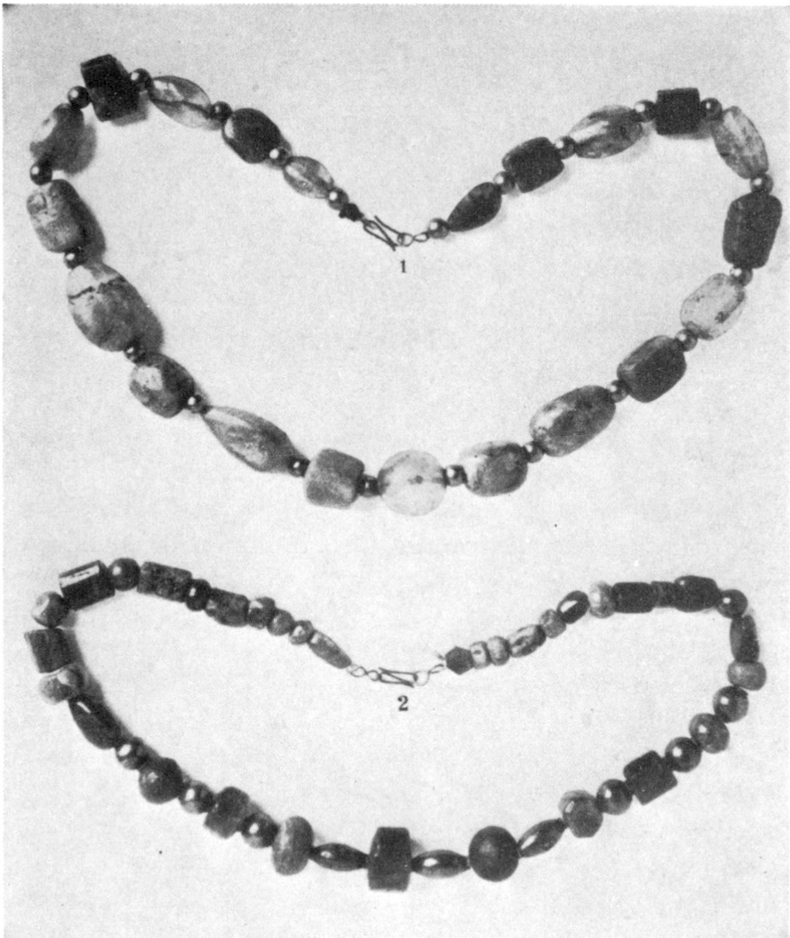

1. Necklace of rock-crystal and amethyst beads, transparent and translucent; very pale; from Egypt. First century.
2. Necklace of antique emeralds with gold beads and amazon stones; from Egypt. First century A.D.

edly represents for us the mental attitude of primitive man.

Probably the first objects chosen for personal adornment were those easily strung or bound together,—for instance, certain perforated shells and brilliant seeds; the softer stones, wherein holes could be easily bored by the help of the simplest tools, probably came next, while the harder gems must have been hoarded as pretty toys long before they could be adjusted for use as ornaments.

Unquestionably, when these objects had once been worn, there was a disposition to attribute certain happenings to their influence and power, and in this way there arose a belief in their efficacy, and, finally, the conviction that they were the abodes of powerful spirits. In this, as in many other things, man's first and instinctive appreciation was the truest, and it has required centuries of enlightenment to bring us back to this love of precious stones for their esthetic beauty alone. Indeed, even to-day, we can see the power of superstitious belief in the case of the opal, which some timid people still fear to wear, although until three or four centuries ago this stone was thought to combine all the virtues of the various colored gems, the hues of which are united in its sparkling light.

A proof that bright and colored objects were attractive in themselves, and were first gathered up and preserved by primitive man for this reason alone, may be found in the fact that certain birds, notable the *Chlamydera* of Australia, related to our ravens, after constructing for themselves pretty arbors, strew the floors with variegated pebbles, so arranged as to suggest a mosaic pavement. At the entrance of the arbors are heaped up pieces of bone, shells, feathers, and stones, which have often been brought from a considerable distance, this giving evidence that the birds have not selected these objects at random. It is strange that the attraction exercised upon the sense of sight by anything brilliant and colored, which is at the same time easily portable and can be handled or worn, should be overlooked by those who are disposed to assert that all ornaments of this kind were originally selected and preserved solely or principally because of their supposed talismanic qualities.

The theory that colored and brilliant stones were first collected by men because of their beauty rather than because of their talismanic virtues, is corroborated by the statement made that seals select with considerable care the stones they swallow, and observers on the fishing grounds have noted this and believe that pebbles of chalcedony and serpentine found there have been brought by the seals.[1]

The popular derivation of the word "amulet" from an Arabic word *hamalât*, signifying something suspended or worn, is not accepted by the best Arabic scholars, and it seems probable that the name is of Latin origin, in spite of the fact that no very satisfactory etymology can be given. Pliny's use of *amuletum* shows that with him the word did not always denote an object that was worn on the person, although this later became its meaning. The old etymology given by Varro (118–29 B.C.), who derived *amuletum* from the verb *amoliri*, "to remove," "to drive away," may not be quite in accord with modern philology, but still has something to recommend it as far as the sense goes, for the amulet was certainly believed to hold dangers aloof, or even to remove them. Talis-

---

[1] Lucas, "The Swallowing Stones by Seals," Science, N. S., vol. xx, No. 512, pp. 537, 538; Report of Fur Seal Investigation, vol. iii, p. 68.

man, however, a word not used in classical times, undoubtedly comes from the Arabic *tilsam,* this being in turn derived from τέλεσμα, used in late Greek to signify an initiation, or an incantation.

It has been remarked that in the earliest Stone Age there is no trace of either idols or images; the art of this period being entirely profane. In the later Stone Age, however, entirely different ideas seem to have gained the ascendancy, for a majority of the objects of plastic art so far discovered have a religious significance. This has evidently proceeded from the conception that every image of a living object absorbs something of the essence of the object itself, and this conception, while a primitive one, still presupposes a certain degree of development. This rule applies more especially to amulets, which were therefore fashioned as beautifully as primitive art permitted, that they might become fitting abodes for the benevolent spirits believed to animate them and render them efficacious.[2]

A curious idol or talisman from Houaïlou, New Caledonia, is in the collection of Signor Giglioli. This is a stone bearing naturally a rude resemblance to the human form.[3] We can easily understand that such an object was looked upon as the abode of some spirit, for similar strange natural formations have been regarded with a species of superstitious awe by peoples much more civilized than the natives of New Caledonia.

For the Middle Ages and even down to the seventeenth century, the talismanic virtues of precious stones were believed in by high and low, by princes and peas-

---

[2] Hoernes, "Urgeschichte der bildenden Kunst," Wien, 1898, p. 108.

[3] Giglioli, "Materiale per lo studio della 'Età della Pietra,'" Archivio per l'Antropologia e l'Etnologia, vol. xxxi, p. 83, Firenze, 1901.

ants, by the learned as well as by the ignorant. Here and there, however, a note of scepticism was sometimes apparent, as in the famous reply of the court jester of Emperor Charles V, to the question, "What is the property of the turquoise?" "Why," replied he, "if you should happen to fall from a high tower whilst you were wearing a turquoise on your finger, the turquoise would remain unbroken."

The doctrine of sympathy and antipathy found expression in the belief that the very substance of certain stones was liable to modification by the condition of health or even by the thoughts of the wearer. In case of sickness or approaching death the lustre of the stones was dimmed, or else their bright colors were darkened, and unfaithfulness or perjury produced similar phenomena. Concerning the turquoise, the prosaic explanation can be offered that this stone is affected to a certain extent by the secretions of the skin; but popular superstition saw the same phenomena in the ruby, the diamond, and other stones not possessing the sensitiveness of the turquoise. Hence the true explanation is to be found in the prevailing idea that an occult sympathy existed between stone and wearer. The sentiment underlying the conception is well expressed by Emerson in the following lines from "The Amulet":

> Give me an amulet
>   That keeps intelligence with you,—
> Red when you love, and rosier red,
>   And when you love not, pale and blue.

A Persian legend of the origin of diamonds and precious stones shows that in the East these beautiful objects were looked upon as the source of much sin and sorrow. We are told that when God created the world he made no useless things, such as gold, silver, precious stones, and diamonds; but Satan, who is always eager to bring evil among men, kept a close watch to spy out the appetites and passions of the human mind. To his great satisfaction he noted that Eve passionately loved the many-colored flowers that decked the Garden of Eden; he therefore undertook to imitate their brightness and color out of earth, and in this way were produced colored precious stones and diamonds. These in after time so strongly appealed to the greed and covetousness of mankind that they have been the cause of much crime and wretchedness.[4]

The present age could afford us nearly as many examples of faith in talismans and amulets as any epoch in the past, if people were willing to confess their real beliefs. However, they are half-ashamed of their fondness for such objects, and fail to see that, back of all the folly and superstition that may find expression in this way, there is a deeper meaning in these talismans than we at first perceive. We may be disposed to smile when we are told that many of the soldiers in the Austro-Prussian War of 1866 carried amulets of some kind upon their persons, and that the great Marshal Canrobert trusted to the protection of an amulet in the Crimean campaign. Of course the Russian army, during the Russo-Japanese War, was amply provided with amulets, religious medals or pictures to which a special virtue had been given by a priestly blessing.

In all these cases, however, it is not the object itself, but the idea for which it stands and which it incorporates, that gives confidence to the wearer, and in this sense the wearing of a talisman is no more a proof of blind super-

---

[4] Rose, "Handleiding tot de Kennis van diamanten," etc., Amsterdam, 1891, p. 110.

stition than is the devotion to a flag, in itself only a few square feet of silk or bunting, but, nevertheless, the symbol of the noblest ideas and feelings, of patriotic devotion to one's native land and to one's fellow-countrymen. The tendency to give a substantial visible form to an abstract idea is so deeply rooted in humanity that it must be looked upon as responding to a human necessity. It is only very rarely that purely intellectual conceptions can satisfy us; they must be given some external, palpable and visible form to exert their greater influences.

Although it may bear a certain superficial likeness to fetichism, this use of signs and symbols is something entirely and radically different, for the idea is never lost sight of, it is only strengthened and vivified by the contemplation of the symbol. Hence, while we know quite well that the symbol is nothing in itself, we know just as well that it has a real power in its relation to the idea it typifies, and we can no more be indifferent to its injury or destruction than we could be indifferent to the injury or destruction of a cherished memento of one whom we have loved and lost.

What super-subtle sense is it that enables some women to endow their gems with a certain individuality, and leads them to feel that these cold, inanimate objects partake of human emotion? A French writer, Mme. Catulle Mendès, gives expression to this when she says that she always wears as many of her rings as possible, because her gems feel slighted when she leaves them unworn. She continues:

> I have a ruby which grows dull, two turquoises which become pale as death, aquamarines which look like siren's eyes filled with tears, when I forget them too long. How sad I should feel if precious stones did not love to rest upon me!

MOSAICS OF TURQUOISE AND ENAMELLED CARNELIAN BEADS, FROM THIBET.

Field Museum, Chicago.

A very beautiful and curious object was found in the Australian opal-fields in 1909. This is a reptilean skeleton resembling a small serpent that has become opalized by natural processes. Perfect in all its details, which are rendered more striking by the splendid play of color, this specimen of Nature's handiwork possesses a beauty and an interest exceeding those to be found in any work of man. As an amulet it certainly is *sui generis,* and in ancient times would have been valued at an immense sum, for the figure of a serpent was a favorite symbol of medical science; even to-day there is little doubt that this strange object will be eagerly sought for by collectors, and will appeal more especially to all who are interested in occult science, and to all who appreciate the poetic and perhaps mystic significance of form, sign, and symbol.

It is impossible to over-estimate the effect of color in determining the supposed influence of gems upon the fortunes or health of the wearers. When we gaze upon the beautiful play of light emitted by a fine ruby or sapphire, we are all conscious of the æsthetic effect produced; but in earlier times, when scientific ideas were not yet prevalent, many other considerations combined to give a peculiar significance to these brilliant gems. Rare and costly as they were, they were supposed to possess mystic and occult powers and were thought to be the abode of spirits, sometimes benevolent and sometimes malevolent, but always endowed with the power to influence human destinies for weal or woe. Coupled with this was the instinctive appreciation of the essential qualities of certain rays of light, and modern science, far from doing away with these ideas, has rather seemed to find a good reason for them. We all know the therapeutic value of the ultra-violet rays, and when the uninstructed mind saw therein the embodiment of purity and chastity, it perhaps realized this health-giving and beneficent function. In the same way the idea of passion was associated with the red and radiant ruby, another concept the relative truth of which has been demonstrated by spectrum analysis, since the red rays are heat-giving and vivifying. But this was not the only source of these primitive ideas in regard to color; the therapeutic effect was often sought and found in some fancied analogy between the color of the gem and the character of the malady or infirmity to be cured. Thus, yellow stones were supposed to be especially efficacious in cases of jaundice, an instance of instinctive homœopathy, based on the dictum *similia similibus curantur.* Following out this train of thought, the red stones were endowed with the power of checking the flow of blood; especially the so-called bloodstone was prescribed for this use, and it was supposed that by its mere touch it could stop the most violent hemorrhages. Green was regarded as the color most beneficial for the sight, and to the emerald and other green stones was ascribed great curative power in this respect. Here, however, the simple influence of the color was later combined with its symbolical significance. In heathen mythology this showed itself in the ascription of the emerald to Venus, as the exponent of the reproductive energies of nature, while in the Christian conception these stones became typical of the resurrection, of the birth into a new and purer life. Nowhere can we find a better illustration of the transforming effect of distinct and diametrically opposite concepts upon the impressions made by natural objects. The pure and colorless and yet brilliant stones, such as the diamond and all other white stones, were naturally brought into connection with the moon, although the diamond, because of its superior qualities and exceptional brilliance and value, was frequently looked upon as the gem of the sun. All gems associated with the moon partook of its enigmatic character. Illuminating the witching hour of the night, when malevolent and treacherous spirits were supposed to hold sway, the moon was sometimes regarded as baleful, as may be seen in the idea that associated lunacy with exposure to the bright rays of the moon; at other times it was supposed to have the power to conjure these evil influences and to drive off the powers of darkness.

The symbolical significance of the colors of precious stones is treated at considerable length by Giacinto Gimma,[5] who has gathered together a great quantity of material on the subject.

Yellow worn by a man denoted secrecy, and was appropriate for the silent lover; worn by a woman it indicated generosity. Golden yellow was, of course, the symbol of the sun and of Sunday. The precious stone was the chrysolite or the yellow jacinth. The animal connected with the color was the lion, doubtless from the association of the zodiacal sign Leo with the midsummer sun. Of the seven ages of man yellow typified adolescence. Roman matrons covered their heads with a yellow veil to show their hope of offspring and happiness. Because garments of this color were a sign of grandeur and nobility, a golden vestment is assigned to the Queen of Heaven as a sign of her pre-eminence, as we read in Psalm xlv, 9: "Upon thy right hand did stand the queen in gold of Ophir." Gimma's explanation of

[5] "Della storia naturale delle Gemme," Napoli, 1730, Vol. I, pp. 131–137.

this as referring to the Virgin Mary is in accord with the Catholic exegesis of his time.

White signified for men friendship, religion, integrity; for women, contemplation, affability, and purity. It was associated with the moon and with Monday and was represented by the pearl. The animal having an affinity with white was quite naturally the ermine. The mystic number was seven, and white was the color of infancy. Among the ancients white was a sign of mourning and sadness, and the Greek matrons attired themselves in white on the death of their husbands. Gimma states that in his time, in Rome, widows used to wear white as mourning for their husbands, while throughout Italy a white band worn around the head was a sign of widowhood.

Red garments on a man indicated command, nobility, lordship, and vengeance; on a woman, pride, obstinacy, and haughtiness. This was the color of the planet Mars and of Tuesday; it was represented by the ruby. Why the lynx should have been selected as the animal for red is rather difficult to understand, but, as the most vivid color, the choice of red as a type of full manhood need not surprise us. Its number was the potent nine, three multiplied by itself. The ancients covered with a red cloth the biers of those who had died valiantly in battle, as Homer[6] shows when he relates that the brothers and companions of Hector covered the urn containing the hero's ashes with soft purple (scarlet) robes. Plutarch asserts that the Lacedemonians clothed their soldiers in red to strike terror into the hearts of their enemies and to manifest a thirst for blood. We might perhaps say much the same of the English "red-coats" to-day. The

[6] Il., xxiv, 795, 796.

Italian code of criminal laws known as the "Digesto Nuovo" was bound in red, to signify that a bloody death awaited thieves and murderers.

Blue on a man's dress indicated wisdom and high and magnanimous thoughts; on a woman's dress, jealousy in love, politeness, and vigilance. Friday and Venus were represented by blue, and the celestial-hued sapphire was the stone in which this color appeared in all its beauty. Blue was a fit symbol of the age of childhood, but it is less easy to understand the choice of the goat as the animal associated with the color. The significant number was six. Natural science, the contemplation of the heavens and of the heavenly bodies, and the study of stellar influences were all typified by blue.

Green signified for men joyousness, transitory hope, and the decline of friendship; for women, unfounded ambition, childish delight, and change. The early verdure of spring might be regarded as at once a symbol of hope and of eventual disappointment, for it must soon pass away. Mercury, and Wednesday, the day of Mercury, were both typified by green, the sly fox being selected as the animal is sympathy with the wily god. The typical green stone is the emerald, youth is the age of man represented by the color, and five the magic number expressing it. In ancient times green was used in the case of those who died in the flower of youth, an emerald being sometimes placed on the index-finger of the corpse, as a sign that the light of hope was spent, for the lower part of the torches used in religious ceremonies was marked with green. Fulvius Pellegrinus relates that, in the tomb of Tullia, the dearly-beloved daughter of Cicero, there was found an emerald, the most beautiful that had ever been seen. This passed into the hands of the Marchesana di Mantova, Isabella Gonzaga da Este. In Italy the graves of young virgins and of children were covered with green branches. When the Codex Justinianus was rediscovered and added to the other Pandects, it was bound in green to signify that these laws were rejuvenated.

Black for men means gravity, good sense, constancy, and fortitude; for young women, fickleness and foolishness, but for married women, constant love and perseverance. The planet Saturn and Saturday are denoted by black. Strange to say, the diamond, the white gem *par excellence,* was selected to represent this sombre hue. Perhaps to offset this the animal chosen was the hog. As black was a mourning color, we need not be surprised that it typified decrepitude. The number eight, the double square, was supposed to have some affinity with black. Black is a symbol of envy, for the thoughts which aim at another's injury cloud the soul and afflict the body. The book of laws treating of dispositions made in view of death was bound in black. The sinister significance of black is well illustrated by what is told of the ruthless Tartar Tamerlane. When he attacked a city, he caused a white tent to be pitched for himself on the first day of the siege, as a sign that mercy would be shown to the inhabitants if they immediately surrendered; on the second day a red tent was substituted, signifying that if the city yielded, all the leaders would be put to death; on the third day, however, a black tent was raised, an ominous signal that no mercy would be shown and that all the inhabitants would be slaughtered.

Violet for a man denoted sober judgment, industry, and gravity; for a woman, high thoughts and religious love. It was the color of the planet Jupiter and of Thursday. As with blue, the sapphire was conceived to present violet most attractively. That the bull should be selected as the animal represented by this color probably arose from some mythological connection with Jupiter, possibly the myth of Europa and the bull. Violet was the color of old age and was associated with the number three.

The influence of color upon the nerves has been noted by some of the leading authorities on hypnotism. For example, Dr. Paul Ferez, finding that red light is stimulating and blue-violet calming, suggests that those who treat patients by means of hypnotism should have two rooms for their reception. In one of these rooms the curtains, wall-paper, chair-coverings, etc., would be red, while in the other they would be of a violet-blue hue. Those suffering from a lack of will-power or from lassitude and depression are to be received in the red room, and those who are a prey to over-excitability are introduced into the blue room. Moreover, according to Dr. Ferez, the sedative qualities of the violet-blue can be utilized in inducing the hypnotic state. For this purpose he recommends a violet-blue disk, which is to be rotated rapidly before the eyes of the patient, the movement serving to attract and hold his gaze better than any immovable object would do.[7]

Red stones such as rubies, carbuncles, and garnets, whose color suggested that of blood, were not only believed to confer invulnerability from wounds, but some Asiatic tribes have used garnets as bullets, upon the contrary principle that this blood-colored stone would inflict a more deadly wound than would a leaden bullet. Such bullets were used by the rebellious Hanzas, in

---

[7] Paper by Dr. Paul Ferez in the Revue de l'Hypnotisme, Paris, No. 10, April, 1906, p. 306.

1892, during their hostilities with the British troops on the Kashmir frontier, and many of these precious missiles were preserved as curiosities.

In his "Colloquy on Pilgrimages," Erasmus makes one of the speakers ask, "Dost thou not see how the artificer Nature delights to represent all things by colors and forms, but more especially in gems?" He then proceeds to enumerate the various images of natural objects in stones. In the ceraunia appeared the thunder-bolt; in the pyrope, living fire; the chalazia (rock-crystal) preserved the form and coldness of the hailstone even if cast into the fire. In the emerald were shown the deep and translucent waves of the sea; the carcinia imitated the form of crabs; the echites, of vipers; the hieracites, of hawks; the geranites, of cranes. The ætites offered the image of an eagle with a white tail; the taos had the form of a peacock; the chelonites, of an asp; while the myrmecites bore within the figure of an ant.[8] The stones bearing this latter name were probably specimens of amber containing ants.

The Greek names of these stones enumerated by Erasmus signify their real or supposed resemblance to certain natural objects, or to something characteristic of such objects. Many of them were fossils, preserving the form of some living organism; a few were entirely fabulous; still others owed their names to some legend or myth illustrating their fancied therapeutic virtues, as in the case of the ætites (eagle-stone) said to be found in the eagle's nest. Evidently this was a quartz pebble.

The oldest magic formulas that have been preserved for us are those of the Sumerians, the founders of the

---

[8] Erasmi, "Colloquia," Lipsiæ, 1713, pp. 597–8. Suggested by Pliny, lib. xxxvii, cap. 71–73.

ancient civilization of Babylonia. Some of them contain references to the use of precious stones as amulets, as appears in the following specimen:

Cords of light-colored wool,
Offered (?) with a pure hand,
For jaundice of the eye,
Bind on the right side (of the patient).
A lululti ring, with sparkling stones
Brought from his own land,
For inflammation of the eye,
On the little finger
Of his left (hand), place.[9]

A curious Babylonian mythological text represents the solar diety Ninib, the son of Bel, as determining the fate of various stones by pronouncing a blessing or a curse upon them. For instance, the dolomite was blessed and declared to be fit material for the statues of kings, while a substance called the *elu* stone was cursed, proclaimed to be unfit for working, and doomed to disintegration. Alabaster was favored by the god, but chalcedony aroused his anger and was condemned.[10]

In these Sumero-Assyrian inscriptions, there is also mention of two stones, the *aban râme* and the *aban la râme*, the "Stone of Love" and the "Stone of Hate" (lit. "non-love").[11] Evidently these stones were believed to excite one or other of these contradictory passions in the hearts of the wearers, and they may be compared with the stones of memory and forgetfulness in the "Gesta Romanorum."

In an ancient Egyptian burial-place at Shêch Abd el-

---

[9] Morris Jastrow, "Die Religion Babyloniens und Assyriens," vol. i, Giessen, 1905, p. 374.

[10] Morris Jastrow, l. c., p. 462.

[11] Delitzsch, "Assyrisches Wörterbuch," Leipzig, 1896, p. 604.

Qurna, excavated by Passalaqua, was found the mummy of a young woman. Not only was it evident from the rich ornaments adorning the body that she had been of noble birth, but it was also apparent that she must have been exceedingly beautiful in form and feature, and must have died in the flower of her age. The hair was artistically braided and adorned with twenty bronze hairpins. About her neck was a remarkably beautiful necklace composed of four rows of beads with numerous pendants representing divinities and sacred symbols. There were also two smaller necklaces with beads of gold, lapis-lazuli, and carnelian; two large jewelled earrings hung from her ears, and on the index-finger of her right hand was a ring set with a scarab; a gold belt garnished with lapis-lazuli and carnelians was bound about her waist and a gold bracelet adorned with semiprecious stones encircled her left wrist. In the sarcophagus was a beautiful mirror of golden-yellow bronze, and three alabaster vases, one still containing some balm or perfume, and another some galena (native lead sulphide) to be used as a cosmetic for the eyes, as well as a little ebony pencil for its application. All these objects are now in the Egyptian collection of the Berlin Museum, and they probably belong to the period of the XVIII Dynasty, about 1500 B.C.

The principal necklace was undoubtedly regarded by the fair Egyptian as an amulet of great power, but it failed to protect her from an untimely end; perhaps, however, its virtues may have aided her soul in its passage through the trials and tests imposed in the underworld. Of the numerous pendants which lent to the necklace its peculiar quality as an amulet, three, in carnelian, figure the god Bes; seven, also in carnelian, the hippo-

1. A necklace of rock crystal, emeralds, hexagonal crystals, and amazon stones; from Egypt.
2. A necklace of onyx and gold beads with the "Lucky Eye" agates; from Egypt. Carnelian, sard; blue and white, and black and white glass beads.

potamus-goddess Toeris, of whom there are besides two representations in lapis-lazuli; then we have a heart of lapis-lazuli; a cat of lapis-lazuli; four falcons of carnelian; one crocodile of carnelian and two of lapis-lazuli; four fish of carnelian, as well as two others of a blackish-white and of a green stone, respectively, and two scorpions of carnelian, and seven flower-forms of the same stone. The greater part of the beads in this necklace are of annular form, of gold, electrum, ivory, or lapis-lazuli; there are a few larger annular or spherical beads of carnelian, chrysoprase, and malachite, and measuring up to 3.5 cm. in diameter.[12]

A necklace, from the time of the Old Empire (c. 3500 B.C.), and having for its chief adornment a turquoise pendant rudely fashioned into the form of an ibex, was found by the German Orient-Gesellschaft at Abusîr el-Meleq in 1905. This necklace, the parts of which were found about the neck of a body, presumably that of a young man, was composed of rounded and annular beads of carnelian and shell, as well as of flat, perforated fragments of turquoise and almandine garnet and an approximately lozenge-shaped bead of amethyst 1.7 cm. long and 1.4 cm. broad. The chief ornament was the turquoise ibex 1.7 cm. in length and 0.9 cm. high.[13] This figure suggests a comparison with the animal and bird forms fashioned out of turquoise that have been found in Indian graves in Arizona and New Mexico, and it probably had the quality of a fetich, or at least of a

[12] "Aegyptische Goldschmiedearbeit," ed. by Heinrich Schäffer, Berlin, 1910, pp. 25–32; necklace figured on Pl. V, other objects on Pls. V–VII.

[13] Ibid., p. 14, Pl. II, figs. 3a, 3b.

talisman, intended to guard the wearer of the necklace from harm.

That there was in Egypt a strong inclination to use a certain particular stone for a given amulet, will be noted in the case of those inscribed with special chapters of the Book of the Dead. This is also true of amulets of certain forms. For instance, the head-rest amulet is usually of hematite as is also the carpenter's square. Of the heart amulets, numbering 47 in the rich collections of the Cairo Museum, nine are of carnelian, four of hematite, two of lapis-lazuli, and two each of green porphyry and green jasper, carnelian being thus the most favored among the more precious materials. Amulets of animal form are plentifully represented in this collection, figuring a large variety of members of the animal kingdom such as the hippopotamus, crocodile, lion, bull, cow, hare, dog-headed ape, cat, dog (somewhat doubtful), jackal, hedgehog, frog, hawk, cobra and fishes, to which list may be added a four-headed ram and a ram-headed sphinx.[14]

One of the special uses of amulets was for seafaring people, for, in ancient times especially, all who went down to the sea in ships were greatly in need of protection from the fury of the elements when they embarked in their small sailing-vessels. A fragment of a Greek Lapidary,[15] probably written in the third or fourth century of our era, gives a list of seven amulets peculiarly adapted for this purpose. The number might suggest a connection with the days of the week, and the amulets

---

[14] See Reisner, "Catalogue générale des antiquités égyptiennes du Musée du Caire: Amulets" Le Caire, 1907.

[15] Pitra, "Specilegium Solesmense," Parisiis, 1855, vol. iii, p. 393.

were perhaps regarded as most efficacious when used on the respective days.

In the first were set a carbuncle and a chalcedony; this amulet protected sailors from drowning. The second had for its gem either of two varieties of the adamas,—one, the Macedonian, being likened to ice (this was probably rock-crystal), while the other, the Indian, of a silvery hue, may possibly have been our corundum; however, the Macedonian stone was regarded as the better. The third amulet bore the beryl, "transparent, brilliant, and of a sea-green hue," evidently the aquamarine beryl; this banished fear. The fourth had for its gem the *druops*, "white in the centre," probably the variety of agate so much favored as a protector against the spell of the Evil Eye. A coral was placed in the fifth amulet, and this was to be attached to the prow of the ship with strips of seal-skin; it guarded the vessel from winds and waves in all waters. For the sixth amulet the *ophiokiolus* stone was selected, most probably a kind of banded agate, for it is said to have been girdled with stripes like the body of a snake; whoever wore this had no need to fear the surging ocean. The seventh and last of these nautical amulets bore a stone called *opsianos*, apparently a resinous or bituminous material, possibly a kind of jet; this came from Phrygia and Galatia, and the amulet wherein it was set was a great protection for all who journeyed by sea or by river.

The ancient treatises on the magic art show that the use of amulets was considered to be indispensable for those who dared to evoke the dark spirits of the netherworld, for without the protection afforded by his amulet the magician ran the risk of being attacked by these spirits. One of these texts gives directions for preparing an amulet, or *phylacterion*, for the "undertaking";

for this a "sweet-smelling" loadstone should be chosen, and should be cut heart-shaped and engraved with the figure of Hecate.[16]

A costly Chinese amulet consists of the diamond, the ruby, and the emerald, to which are added the pearl and coral; Oriental sapphire and topaz are classed with the ruby. An amulet containing these five substances is thought to combine the protecting influences of the different deities presiding over them, and is supposed to lengthen the wearer's life. Sometimes these five princely gems are wrapped up in a paper bearing the names of the respective divinities, to which is added the name of the moon, and those of the twenty-seven constellations, or houses of the moon. Such an amulet, suspended at the entrance of a house, is believed to afford protection to the inmates.[17]

In the language of the ancient Mexicans blood was called *chalchiuhatl,* or "water of precious stones," as the quintessence of what were regarded as the most costly things.[18] Although such poetic designations are in modern times mere figures of speech, among primitive peoples they are more significant, and it is highly probable that with the Aztecs, as with other peoples, the wearing of precious stones was believed to enrich the blood and thus to promote health and vigor, for "the blood is the life."

That gems had sex is asserted by the earliest writers

---

[16] Kropatschek, "De amuletorum apud antiquos usu," Gryphiæ, 1907, p. 24 (Paris papyrus, 2630).

[17] Surindro Mohun Tagore, "Mani Málá," Pt. II, Calcutta, 1881, p. 943.

[18] Seler, "Codex Borgia: Eine altmexicanische Bilderschrift," Berlin, 1904, vol. i, p. 16.

COLOR PLATES

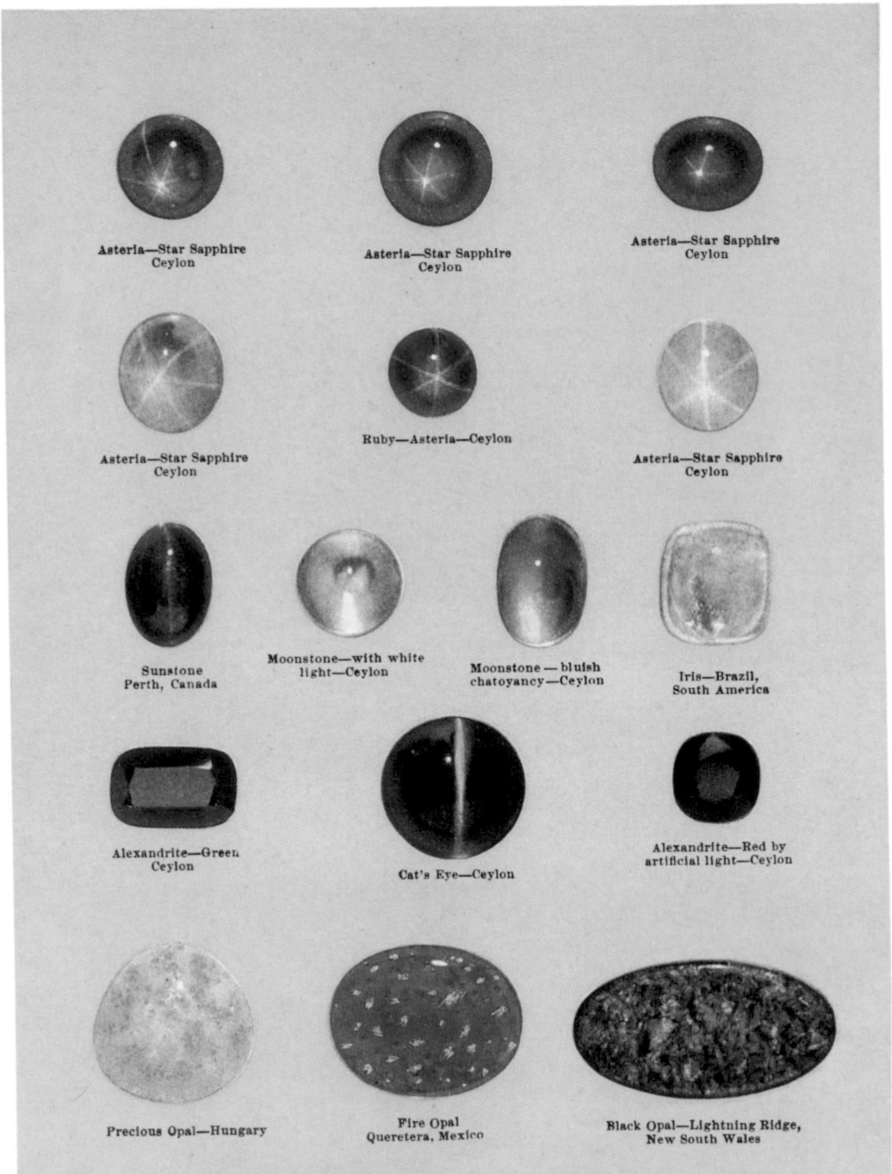

PHENOMENAL GEMS (GEMS EXHIBITING PHENOMENA)

From the J. P. Morgan Collection, American Museum of Natural History, New York

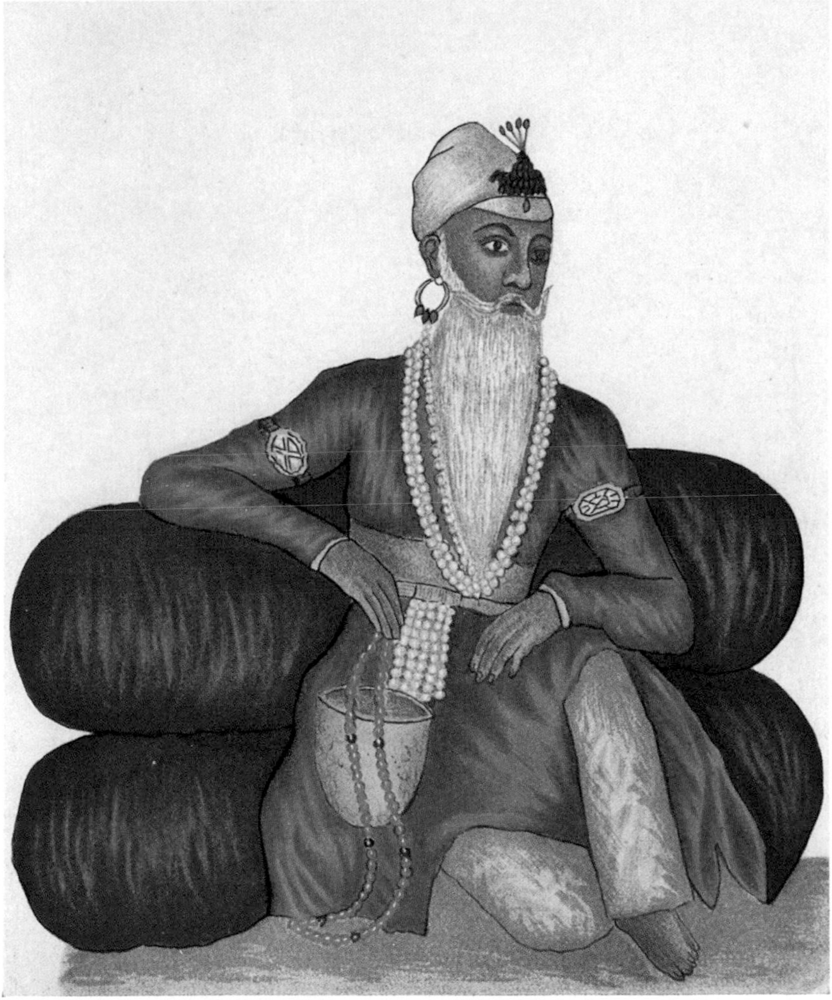

By permission of W. Griggs & Sons, Ltd., London.

MAHARAJA RUNJIT SINGH, RULER OF THE PUNJAB, 1791 TO 1839.

He holds a "rosary" of emeralds, stones prized in the Orient as antidotes to poison. From a portrait by Jiwan Ram, taken at Rupar in 1831. From the Journal of Indian Art and Industry.

SAPPHIRE RING WITH ENCIRCLING DIAMONDS.
Presented to John Cardinal Farley on the occasion of his elevation to the cardinalate.

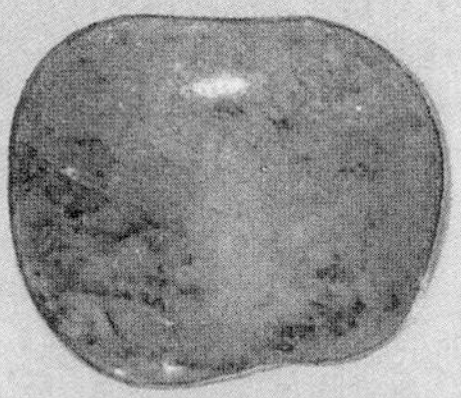

Rubellite, from the Shan Mountains, China. Used as idol's eye in India

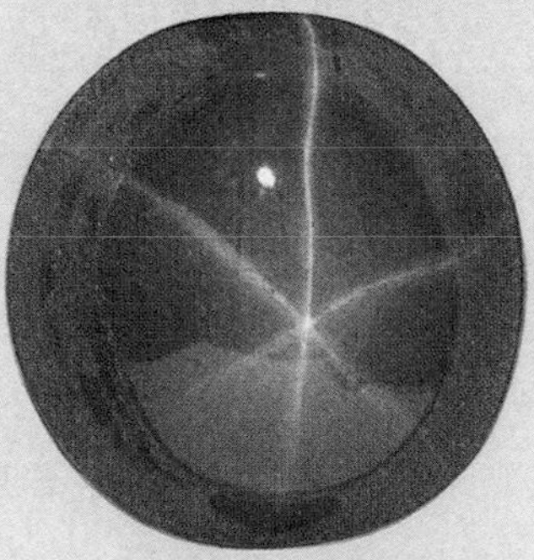

Star of India—Star Sapphire

Engraved Emerald—East Indian Carving—17th Century

GEMS FROM THE MORGAN-TIFFANY COLLECTION.

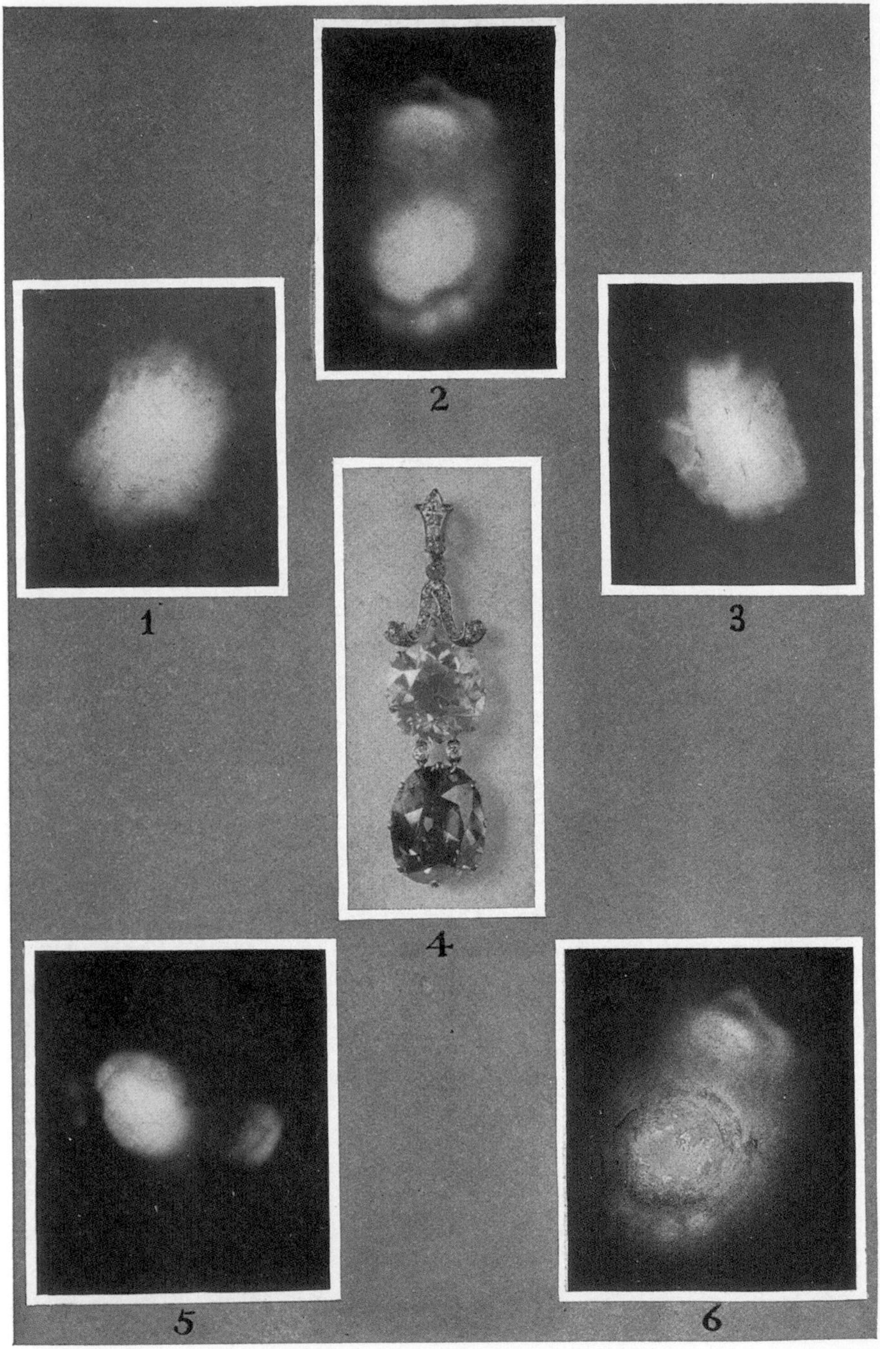

1. Self-print of upper diamond of No. 4 by phosphorescence, produced by rubbing briskly with stick covered by woolen cloth. Exposure one-half minute.

2. Self-print, both diamonds, after one minute's exposure to ultra-violet light, electric action eliminated.

3. Self-print, upper diamond. Exposure one-fourth minute.

4. Upper: blue-white Tiffanyite diamond, 14.86 carats; Bagagem Mine, Brazil. Lower: purple-black diamond, 13.35 carats; Brazil.

5. Self-print, both diamonds; different position.

6. Aspect of both diamonds (No. 4), one minute's exposure, ultra-violet light; blue-white phosphorescing white, purple-black having red glow.

CROSS, ATTACHED AS PENDANT TO THE CROWN OF THE GOTHIC KING RECCESVINTHUS (649–672 A.D.).

Forming part of the treasure discovered in 1858 at Guarrazar in Spain. Now in Musée Cluny, Paris. The cross proper is set with fine sapphires cut en cabochon and eight large pearls. Natural size.

as well as by many of those of a later date. While this must usually be understood as a poetic way of indicating a difference in shade, the darker varieties being regarded as male and the lighter ones as female, Theophrastus, the earliest Greek writer on precious stones, clearly shows that this sexual distinction was sometimes seriously made, for he declares that, wonderful as it might seem, certain gems were capable of producing offspring.

This strange idea was still prevalent in the sixteenth century, and ingenious explanations were sometimes given of the cause of this phenomenon, as appears in the following account by Rueus of germinating diamonds:[19]

> It has recently been related to me by a lady worthy of credence, that a noblewoman, descended from the illustrious house of Luxemburg, had in her possession two diamonds which she had inherited, and which produced others in such miraculous wise, that whoever examined them at stated intervals judged that they had engendered progeny like themselves. The cause of this (if it be permissible to philosophize regarding such a strange matter) would seem to be that the celestial energy in the parent stones, qualified by some one as "*vis adamantifica*," first changes the surrounding air into water, or some similar substance, and then condenses and hardens this into the diamond gem.

The pearl-fishers of Borneo are said to preserve carefully every ninth pearl they find, and place them in a bottle with two grains of rice for each pearl, believing, in spite of all evidence to the contrary, that these particular pearls have the power to engender and breed others. Custom and superstition require that each bottle shall have the finger of a dead man as a stopper.

Talismanic influences are taken into account in the

---

[19] Francisci Ruei, "De gemmis," Tiguri, 1566, f. 4.

wearing of jewelry by Orientals, two bracelets being frequently worn lest one member should become jealous of the other, thus disturbing the equilibrium of the whole organism. The piercing of the ears for ear-rings has been attributed to a desire to chastise the ear for its indiscretion in hearing secrets not intended to be heard, while costly and ornamental ear-rings are set in the ears to console those parts of our anatomy for the suffering caused by the operation of piercing. In the case of necklaces of brilliant metal, adorned with pendants of glittering stones, the talismanic purpose is to attract the beholder's gaze and thus ward off the mysterious and dangerous emanations set forth by the Evil Eye; the necklace, or its ornaments, are supposed to perform a similar service to that rendered by the lightning-rod in diverting the electric discharge.

PEARL DEALER.

From the "Hortus Sanitatis" of Johannis de Cuba [Strassburg, Jean Pryss, ca. 1483]: De lapidibus, cap. lxxviii. Author's library.

At an early date the Christian Church registered its opposition to the practice of wearing amulets. At the Council of Laodicea, held in 355 A.D., it was decreed, in the thirty-fourth canon, that priests and clerks must be neither enchanters, mathematicians, nor astrologers, and that they must not make "what are called amulets," for these were fetters of the soul, and all who wore them should be cast out of the church.[20] This emphatic condemnation of the prevailing usage was not so much a protest against superstition *per se* as against pagan superstition, for almost if not all the amulets in use in the early centuries of our era bore heathen or heretical symbols or inscriptions. In later times the invincible tendency to wear objects of this character found expression in the use of those associated with Christian belief, such, for instance, as relics of the saints, medallions blessed by the priest, etc.

The amulets of the Jews differed in many respects from those used by Christians. The Mosaic prohibition of representations of human or animal forms imposed great restrictions upon the employment of engraved gems, and the Jew was only permitted to wear or carry those bearing merely characters of mystic or symbolic significance. In talmudic times amulets were sometimes hidden in a hollow staff, and they were believed to have more power when concealed from view in this way. They were like concealed weapons, and it was said that, as a father might give such an amulet to a son, so God had given the Law to Israel for its protection.[21]

In the Old French didactic poem, the *Roman de la Rose,* composed in the twelfth century, appear traces of the belief in the magic properties of precious stones. Chaucer translated this poem into English in the fourteenth century and we quote the following lines from his version. They describe the costume of the symbolical figure, Riches.

---

[20] Histoire critique des pratiques superstitieuses; par un prêtre de l'Oratoire," Paris, 1702, p. 320.

[21] Blum, "Das altjüdische Zauberwesen," Strassburg, 1898, p. 91.

Richesse a girdle hadde upon
The bokel of it was of a stoon
Of Vertue greet, and mochel of might.

That stoon was greetly for to love,
And til a riche mannes bihove
Worth al the gold in Rome and Fryse.

The mordaunt[22] wrought in noble wyse,
Was of a stoon full precious,
That was so fyn and vertuous,
That hool a man it coude make
Of palasye and of tooth-ake.[23]

At the trial, in 1232, of Hubert de Burgh, chief justiciar, one of the charges brought against him was that he had surreptitiously removed from the English treasury an exceedingly valuable stone, possessing the virtue of rendering the wearer invincible in battle, and had given it to Llewellyn, King of Wales, the enemy of his own sovereign, Henry III of England (1207–1272).[24] This must have taken place about 1228, when Henry was engaged in a war with the Welsh.

That precious stones could, under certain circumstances, lose the powers inherent in them was firmly believed in medieval times. If handled or even gazed upon by impure persons and sinners, some of the virtues of the stones departed from them. Indeed, there were those who held that precious stones, in common with all created things, were corrupted by the sin of Adam. Therefore, in order to restore their pristine virtue it might become

---

[22] A projection serving to fasten down the belt.

[23] Compleat Works of Geoffrey Chaucer, ed. Skeat, Oxford, 1849, vol. i, p. 139.

[24] Matthæi Paris, "Historia major," London, 1684, p. 318.

necessary to sanctify and consecrate them, and a kind of ritual serving this purpose has been preserved in several old treatises. The subject is sufficiently curious to warrant here the repetition of one of these forms. The stones which required consecration were to be wrapped in a perfectly clean linen cloth and placed on the altar. Then three masses were to be said over them, and the priest who celebrated the third mass, clad in his sacred vestments, was to pronounce the following benediction:[25]

> The Lord be with us. And with thy spirit. Let us pray. Almighty God and Father, who manifestedst thy virtue to Elias by certain senseless creatures, who orderedst Moses, Thy servant, that, among the sacerdotal vestments, he should adorn the Rational of Judgment with twelve precious stones, and showedst to John, the evangelist, the famous city of Jerusalem, essentially constituted by the same stones, and who hadst the power to raise up sons to Abraham from stones, we humbly beseech Thy majesty since Thou hast elected one of the stones to be a dwelling-place for the majesty of Thy heart, that Thou wilt deign to bless and sanctify these stones by the sanctification and incarnation of Thy name, so that they may be sanctified, blessed, and consecrated, and may receive from Thee the effect of the virtues Thou hast granted to them, according to their kinds, and which the experience of the learned has shown to have been given by Thee; so that whoever may wear them on him may feel the presence of Thy power and may be worthy to receive the gift of Thy grace and the protection of Thy power. Through Jesus Christ, Thy Son, in whom dwells all sanctification, benediction, and consecration; who lives with Thee and reigns as God for all eternity, Amen. Thanks be to God.

Konrad of Megenburg also gives this benediction in his "Buch der Natur."

Luther tells the following humorous tale of a Jew who was a vender of amulets:

---

[25] "Le Grand Lapidaire" of Jean de Mandeville, Vienna, 1862, pp. 126–128.

There is sorcery among the Jews and their sorcerers think: "If we succeed, it is well for us; if we fail, a Christian is the sufferer; what care we for that?" . . . But Duke Albert of Saxony acted shrewdly. When a Jew offered him a button, inscribed with curious characters and signs, and asserted that this button gave protection from cuts, thrusts, and shots, the Duke answered: "I will test that upon thyself, O Jew." Hereupon he led the man to the gate, hung the button at his neck, drew his own sword, and thrust the fellow through the body. "The same fate would have happened to me," said the Duke, "as has happened to thee."[26]

Ruskin, with his keen poetic insight into the working of natural laws, saw in the formation of crystals the action of both "force of heart" and "steadiness of purpose." He thus found himself, consciously or unconsciously, in agreement with the old fancies which attributed a species of personality to precious stones. Just as the Hindu regarded an imperfectly shaped crystal as a bringer of ill luck to the owner, so Ruskin sees in such a crystal the signs of an innate "immorality," if we may use this expression. Of a crystal aggregation of this type he writes as follows:[27]

Opaque, rough-surfaced, jagged on the edge, distorted in the spine, it exhibits a quite human image of decrepitude and dishonour; but the worst of all signs of its decay and helplessness is, that half-way up, a parasite crystal, smaller, but just as sickly, has rooted itself in the side of the larger one, eating out a cavity round its root, and then growing backwards, or downwards, contrary to the direction of the main crystal. Yet I cannot trace the least difference in purity of substance between the first most noble stone, and this ignoble and dissolute one. The impurity of the last is in its will or want of will.

There is established a very pretty custom of assigning to the various masculine and feminine Christian

---

[26] Güdermann, "Das jüdische Unterrichtswesen," Wien, 1873, p. 225.

[27] "Ethics of the Dust," New York, 1886, p. 96.

names a particular gem, and such name-gems are often set together with natal and talismanic gems and with gems of one's patron saint. It is considered an exceedingly good omen when it happens that all three gems are of the same sort.

GEMS FOR FEMININE NAMES.

Adelaide	Andalusite
Agnes	Agate
Alice	Alexandrite
Anne	Amber
Beatrice	Basalt
Belle	Bloodstone
Bertha	Beryl
Caroline	Chalcedony
Catherine	Cat's-eye
Charlotte	Carbuncle
Clara	Carnelian
Constance	Crystal
Dorcas	Diamond
Dorothy	Diaspore
Edith	Eye-agate
Eleanor	Elæolite
Elizabeth	Emerald
Ellen	Essonite
Emily	Euclase
Emma	Epidote
Florence	Fluorite
Frances	Fire-opal
Gertrude	Garnet
Gladys	Golden Beryl
Grace	Grossularite
Hannah	Heliotrope
Helen	Hyacinth

Irene	Iolite
Jane	Jacinth
Jessie	Jasper
Josephine	Jadeite
Julia	Jade
Louise	Lapis-lazuli
Lucy	Lepidolite
Margaret	Moss-agate
Martha	Malachite
Marie	Moldavite
Mary	Moonstone
Olive	Olivine
Pauline	Pearl
Rose	Ruby
Sarah	Spodumene
Susan	Sapphire
Therese	Turquoise

GEMS FOR MASCULINE NAMES.

Abraham	Aragonite
Adolphus	Albite
Adrian	Andalusite
Albert	Agate
Alexander	Alexandrite
Alfred	Almandine
Ambrose	Amber
Andrew	Aventurine
Archibald	Axinite
Arnold	Aquamarine
Arthur	Amethyst
Augustus	Agalmatolite
Benjamin	Bloodstone
Bernard	Beryl

Charles .... Chalcedony
Christian .... Crystal
Claude .... Cyanite
Clement .... Chrysolite
Conrad .... Crocidolite
Constantine .... Chrysoberyl
Cornelius .... Cat's-eye

Dennis .... Demantoid
Dorian .... Diamond

Edmund .... Emerald
Edward .... Epidote
Ernest .... Euclase
Eugene .... Essonite

Ferdinand .... Feldspar
Francis .... Fire-opal
Frederick .... Fluorite

George .... Garnet
Gilbert .... Gadolinite
Godfrey .... Gagates
Gregory .... Grossularite
Gustavus .... Galactides
Guy .... Gold quartz

Henry .... Heliolite
Herbert .... Hyacinth
Horace .... Harlequin opal
Hubert .... Heliotrope
Hugh .... Heliodor
Humphrey .... Hypersthene

James .... Jade
Jasper .... Jasper
Jerome .... Jadeite
John .... Jacinth
Joseph .... Jargoon
Julius .... Jet

Lambert	Labradorite
Lawrence	Lapis-lazuli
Leo	Lepidolite
Leonard	Loadstone
Mark	Malachite
Matthew	Moonstone
Maurice	Moss-agate
Michael	Microcline
Nathan	Natrolite
Nicholas	Nephrite
Oliver	Onyx
Osborne	Orthoclase
Osmond	Opal
Oswald	Obsidian
Patrick	Pyrope
Paul	Pearl
Peter	Porphyry
Philip	Prase
Ralph	Rubellite
Raymond	Rose-quartz
Richard	Rutile
Robert	Rock-crystal
Roger	Rhodonite
Roland	Ruby
Stephen	Sapphire
Theodore	Tourmaline
Thomas	Topaz
Valentine	Vesuvianite
Vincent	Verd-antique
Walter	Wood-opal
William	Willemite

III

On the Talismanic Use of Special Stones[1]

Agate

THE author of "Lithica" celebrates the merits of the agate in the following lines:[2]

> Adorned with this, thou woman's heart shall gain,
> And by persuasion thy desire obtain;
> And if of men thou aught demand, shalt come
> With all thy wish fulfilled rejoicing home.

This idea is elaborated by Marbodus, Bishop of Rennes, in the eleventh century, who declares that agates make the wearers agreeable and persuasive and also give them the favor of God.[3] Still other virtues are recounted by Camillo Leonardo, who claims that these stones give victory and strength to their owners and avert tempests and lightning.[4]

The agate possessed some wonderful virtues, for its wearer was guarded from all dangers, was enabled to vanquish all terrestrial obstacles and was endowed with a bold heart; this latter prerogative was presumably the

---

[1] See also the writer's pamphlet: "The Folk-Lore of Precious Stones," Chicago, 1894; a paper read before the Folk-Lore Congress held at the World's Columbian Exhibition, and describing the Kunz Collection exhibited in the Anthropological Building there. This collection is now in the Field Museum, Chicago.

[2] King's version in his "Natural History of Precious Stones," London, 1865, p. 392.

[3] Marbodei, "De lapidibus," Friburgi, 1531, fol. 10.

[4] Camilli Leonardi, "Speculum lapidum," Venetia, 1502, fol. 22.

secret of his success. Some of these wonder-working agates were black with white veins, while others again were entirely white.[5]

The wearing of agate ornaments was even believed to be a cure for insomnia and was thought to insure pleasant dreams. In spite of these supposed advantages, Cardano asserts that while wearing this stone he had many misfortunes which he could not trace to any fault or error of his own. He, therefore, abandoned its use; although he states that it made the wearer more prudent in his actions.[6] Indeed, Cardano appears to have tested the talismanic worth of gems according to a plan of his own,—namely, by wearing them in turn and noting the degree of good or ill fortune he experienced. By this method he apparently arrived at positive results based on actual experience; but he quite failed to appreciate the fact that no real connection of any kind existed between the stones and their supposed effects. In another treatise this author takes a somewhat more favorable view of the agate, and proclaims that all varieties render those who wear them "temperate, continent, and cautious; therefore they are all useful for acquiring riches.[7]

According to the text accompanying a curious print published in Vienna in 1709, the attractive qualities of the so-called coral-agate were to be utilized in an air-ship, the invention of a Brazilian priest. Over the head of the aviator, as he sat in the air-ship, there was a network of iron to which large coral-agates were attached.

[5] Albertus Magnus, "Le Grand Albert des secretz des vertus des Herbes, Pierres et Bestes. Et aultre livre des Merveilles du Monde, d'aulcuns effetz causez daulcunes bestes," Turin, Bernard du mont du Chat (c. 1515). Liv. ii, fol. 8 recto.

[6] Cardani, "De subtilitate," Basileæ, 1560, p. 460.

[7] Cardani, "De gemmis," Basileæ, 1585, p. 323.

These were expected to help in drawing up the ship, when, through the heat of the sun's rays, the stones had acquired magnetic power. The main lifting force was

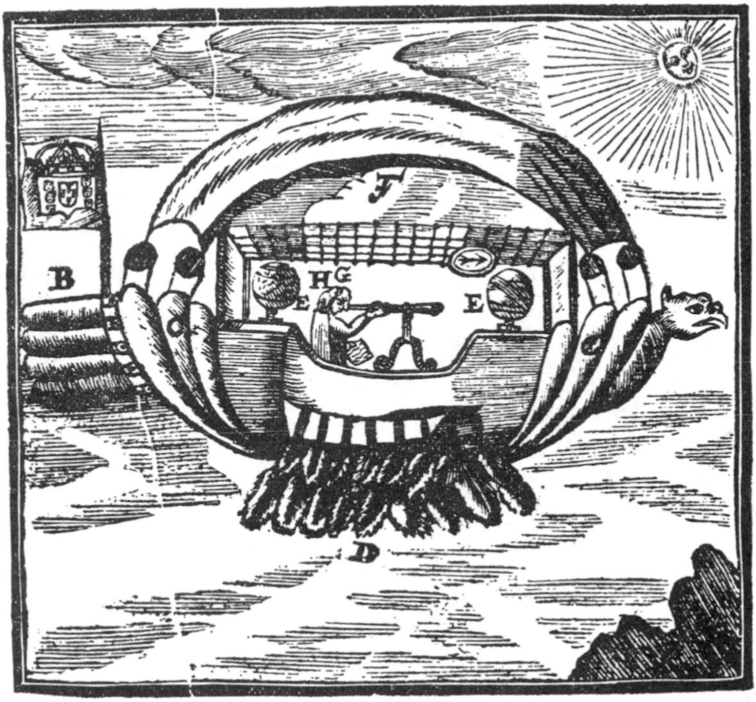

AN AIR-SHIP OF 1709.

In the network above the figure were to be set coral-agates, supposed to possess such magnetic powers as to keep the craft aloft. From Valentini, "Museum Museorum," Pt. III, Franckfurt am Mayn, 1714, p. 35. Author's library.

provided by powerful magnets enclosed in two metal spheres; how the magnets themselves were to be raised is not explained.[8]

[8] Valentini, "Museum museorum oder die vollständige Schau-Bühne," Franckfurt am Mayn, 1714, vol. ii, pt. 3, p. 34; figure of air-ship on p. 35.

About the middle of the past century, the demand for agate amulets was so great in the Soudan that the extensive agate-cutting establishments at Idar and Oberstein in Germany were almost exclusively busied with filling orders for this trade. Brown or black agates having a white ring in the centre were chiefly used for the fabrication of these amulets, the white ring being regarded as a symbol of the eye. Hence the amulets were supposed to neutralize the power of the Evil Eye, or else to be emblematic of the watchfulness of a guardian spirit. The demand for these amulets has fallen off greatly, but when it was at its height single firms exported them to the value of 40,000 thalers ($30,000) annually, the total export amounting to hundreds of thousands of thalers. Even at present a considerable trade in these objects is still carried on. That there is a fashion in amulets is shown by the fact that, while red, white, and green amulets are in demand on the west coast of Africa, only white stones are favored for this use in Northern Africa.

Alexandrite

There are a few talismanic stones which have gained their repute in our time, notably the alexandrite, a variety of chrysoberyl found in Russia, in the emerald mines on the Takowaya, in the Ural region. The discovery of this variety is stated to have been made in 1831 on the day Alexander II (then heir-apparent) reached his majority, and it was therefore named alexandrite, by Nordenskjöld, the mineralogist. The stone as found in gem form rarely weighs over from one to three carats, and is characterized by a marked pleochroism of a splendid green changing to a beautiful columbine red. But in Ceylon much larger gems are found, some few weighing

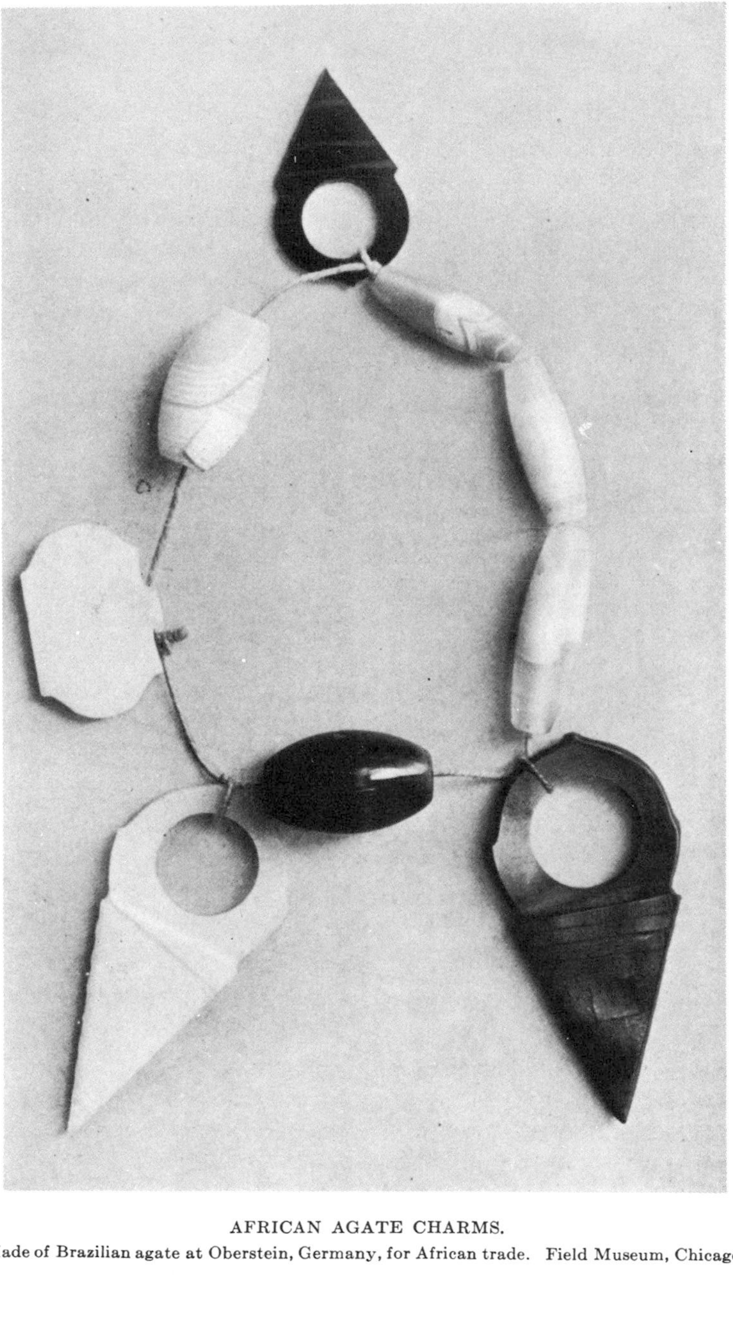

AFRICAN AGATE CHARMS.
Made of Brazilian agate at Oberstein, Germany, for African trade. Field Museum, Chicago.

60 carats each, although rarely of more than one or two carats. The color is of a darker and more bottle-like green, and the change by night renders them darker and more granitized than the Russian stones, which are extremely rare. As red and green are the Russian national colors, the alexandrite has become a great favorite with the Russians, and is looked upon as a stone of good omen in that country. Such, however, is its beauty as a gem that its fame is by no means confined to Russia, and it is eagerly sought in other lands as well.

Amber

Amber was one of the first substances used by man for decoration, and it was also employed at a very early period for amulets and for medicinal purposes. More or less shapeless pieces of rough amber, marked with circular depressions, have been found in Prussia, Schleswig-Holstein, and Denmark, in deposits of the Stone Age. These depressions are sometimes regularly disposed and at other times irregularly, and seem intended to imitate similar depressions found in large stones and rocks, often the work of man's hand, but occasionally the result of natural causes. In Hoernes' opinion they marked the resting place of the spirit or spirits believed to animate the stone, and hence it is probable that the amber fragments were used as talismans or amulets.[9]

For the ancient Greek poets, the grains of amber were the tears annually shed over the death of their brother Phaëthon by the Heliades after grief had meta-

[9] Hoernes, "Urgeschichte der bildenden Kunst," Vienna, 1898, p. 376. Figured in S. Muller's "Ordn. af Danm. Olds.," i, Pl. XV, Figs. 252 sq.

morphosed them into poplars growing on the banks of the Eridanus (the modern river Po).[10] In a lost tragedy of Sophocles, he saw the origin of amber in the tears shed over the death of Meleager by certain Indian birds. For Nicias it was the "juice" or essence of the brilliant rays of the setting sun, congealed in the sea and then cast up upon the shore. A more prosaic explanation likened amber to resin, and regarded it as being an exudation from the trunks of certain trees. Indeed, the poetic fancy we have just noted is the same idea clothed in a metaphorical or mythological form. Another fancy represented amber to be the solidified urine of the lynx, hence one of its names, *lyncurius*.[11]

THE TREE THAT EXUDES AMBER.

From the "Hortus Sanitatis," of Johannis de Cuba [Strassburg, Jean Pryss, ca. 1483]; De lapidibus, cap. lxx. Author's library.

The brilliant and beautiful yellow of certain ambers and the fact that this material was very easily worked served to make its use more general, and it soon became a favorite object of trade and barter between the peoples of the Baltic Coast and the more civilized peoples to the

[10] Ovidii, "Metamorphoses," lib. ii, 11. 340 sqq. Some have proposed to read Redanus instead of Eridanus and have seen in the former name the designation of a stream flowing into the Vistula.

[11] Plinii, "Naturalis Historia," lib. xxxvii, cap. 7.

south. Schliemann found considerable amber from the Baltic in the graves of Mycenæ, and the frequent allusions to it in the works of Latin authors of the first and succeeding centuries testify to its popularity in the Roman world.

Probably the very earliest allusion in literature to the ornamental use of amber appears in Homer's Odyssey,[12] where we read:

> Eurymachus
> Received a golden necklace, richly wrought,
> And set with amber beads, that glowed as if
> With sunshine. To Eurydamas there came
> A pair of ear-rings, each a triple gem,
> Daintily fashioned and of exquisite grace.
> Two servants bore them.

Amber ingeniously carved into animal forms has been discovered in tumuli at Indersoen, Norway.[13] These curious objects were worn as amulets, and the peculiar forms were supposed to enhance the power of the material, giving it special virtues and rendering it of greater value and efficacy.

Pieces of amber with singular natural markings were greatly esteemed, especially when these markings suggested the initials of the name of some prominent person. Thus, we are told that Friedrich Wilhelm I of Prussia paid to a dealer a high price for a piece of amber on which appeared his initials. The same dealer had another piece on which he read the initials of Charles XII of Sweden. When he received the news of this king's death, he bitterly lamented having lost the opportunity of selling him amber for a high price. But he was cleverly consoled by Nathaniel Sendal, the relator of the

---

[12] Bk. xviii, 11, 295–298, trans. of William Cullen Bryant.

[13] Du Chaillu, "The Viking Age," New York, 1889, vol. ii, p. 314. (Figs. 1210, 1211, 1212.)

story, who easily persuaded the dealer that the markings could just as well signify the initials of some other name. Sendal adduces this as a proof that the letters read on such pieces of amber were as much the product of the observer's imagination as of the markings on the material.[14] Those who secured amber so mysteriously marked by Nature's hand probably felt that they had obtained a talisman of great power, especially destined for their use.

Amethyst

While the special and traditional virtue of the amethyst was the cure of drunkenness, many other qualities were attributed to this stone in the fifteenth century. For Leonardo,[15] it had the power to control evil thoughts, to quicken the intelligence, and to render men shrewd in business matters. An amethyst worn on the person had a sobering effect, not only upon those who had partaken too freely of the cup that intoxicates, but also upon those over-excited by the love-passion. Lastly, it preserved soldiers from harm and gave them victory over their enemies, and was of great assistance to hunters in the capture of wild animals. The amethyst shared with many other stones the power to preserve the wearer from contagion.[16]

A pretty legend in regard to the amethyst has been happily treated in French verse. The god Bacchus, offended at some neglect that he had suffered, was determined to avenge himself, and declared that the first person he should meet, when he and his train passed along, should be devoured by his tigers. Fate willed it that this

[14] Sendelii, "Electrologiæ," Elbingæ, 1725, Pt. I, p. 12, note.

[15] Camilli Leonardi, "Speculum lapidum," Venetia, 1502, fol. 22.

[16] Johannis de Cuba, "Hortus Sanitatis," [Strassburg, 1483] tractatus de lapibus, cap. vii.

1. Amber ornament, perforated, from Assyrian grave.
2. Amber ring ornament from Pompeii.
3. Large annular bead of amber from Mexico. Aztec work.
4. Amber wedding necklace. Eighteenth century. Baltic Provinces.
5. Amber beads. Worn by African natives.

luckless mortal was a beautiful and pure maiden named Amethyst, who was on her way to worship at the shrine of Diana. As the ferocious beasts sprang toward her, she sought the protection of the goddess, and was saved from a worse fate by being turned into a pure white stone. Recognizing the miracle and repenting of his cruelty, Bacchus poured the juice of the grape as a libation over the petrified body of the maiden, thus giving to the stone the beautiful violet hue that so charms the beholder's eye.[17]

From the various descriptions of this stone given by ancient writers, it appears that one of the varieties was probably the purple almandine or Indian garnet, and it is not improbable that we have here the reason for the name amethyst and for the supposed virtue of the stone in preserving from drunkenness. For if water were poured into a vessel made of a reddish stone, the liquid would appear like wine, and could nevertheless be drunk with impunity.

Beryl

Arnoldus Saxo, writing about 1220, after reciting the virtues of the beryl as given by Marbodus, after Evax and Isidorus, reports in addition that the stone gave help against foes in battle or in litigation; the wearer was rendered unconquerable and at the same time amiable, while his intellect was quickened and he was cured of laziness.[18] In the old German translation of Thomas de Cantimpré's "De Proprietatibus Rerum," we read that

---

[17] Belleau, "Œuvres poétiques," ed. Marty-Laveaux, Paris, 1878, vol. ii, pp. 172 sqq. The poem in which this tale occurs is the "Amours et nouveaux eschanges des pierres précieuses," written in 1576 and dedicated to Henri III.

[18] Rose, "Aristotles de lapidibus und Arnoldus Saxo," in Zeitschr. für D. Alt., New Series, vol. vi, p. 431.

the beryl reawakens the love of married people (er hat auch die art daz er der elaut lieb wiederpringt).[19]

Bloodstone

The heliotrope or bloodstone was supposed to impart a reddish hue to the water in which it was placed, so that when the rays of the sun fell upon the water they gave forth red reflections. From this fancy was developed the strange exaggeration that this stone had the power to turn the sun itself a blood-red, and to cause thunder, lightning, rain, and tempest. The old treatise of Damigeron relates this of the bloodstone, adding that it announced future events by producing rain and by "audible oracles." Probably the conjurors, before proceeding to use the stone for their incantations, watched the heavens and waited until they noticed the signs of an approaching storm. They then interpreted

Capitulum.xc.

A PRACTICAL TEST OF THE VIRTUES OF THE BLOODSTONE TO PREVENT NOSE-BLEED.

From the "Hortus Sanitatis" of Johannis de Cuba [Strassburg, Jean Pryss, ca. 1483]; De lapidibus, cap. xc. Author's library.

---

[19] Konrad von Megenberg, "Buch der Natur," ed. by Dr. Franz Pfeiffer, Stuttgart, 1861, p. 436.

the sounds of the wind and thunder in various ways, so as to give apt answers to the questions addressed to them touching future events. It is well known that the sighing of the wind, and, indeed, all those natural sounds which constitute the grand symphony of Nature, were interpreted by prophets and seers into articulate speech. Damigeron also declares that the bloodstone preserved the faculties and bodily health of the wearer, brought him consideration and respect, and guarded him from deception.[20]

In the Leyden papyrus the bloodstone is praised as an amulet in the following extravagant terms:

> The world has no greater thing; if any one have this with him he will be given whatever he asks for; it also assuages the wrath of kings and despots, and whatever the wearer says will be believed. Whoever bears this stone, which is a gem, and pronounces the name engraved upon it, will find all doors open, while bonds and stone walls will be rent asunder.[21]

Carbuncle

The carbuncle was recommended as a heart stimulant; indeed, so powerful was its action, that the wearers were rendered angry and passionate and were even warned to be on their guard against attacks of apoplexy.[22] The blood-red hue of the stone also suggested its use as a symbol of the divine sacrifice of Christ on the cross. However, not only in Christianity was this stone used to illustrate religious conceptions, for the Koran affirms that the Fourth Heaven is composed of car-

---

[20] Pitra, "Specilegium Solesmense," Parisiis, 1855, vol. iii, p. 325.

[21] Kropatschek, "De amuletorum apud antiquos usu," Gryphiæ, 1907, p. 16.

[22] Cardani, "Philosophi opera quædam lectu digna," Basileæ, 1585, p. 323. "De gemmis."

buncle. In mythical fancies too this stone played its part, for dragon's eyes were said to be carbuncles.

Rumphius [23] states that in 1687 he was told by a chirurgeon that the latter had seen in the possession of one of the rulers in the island of Amboin a carbuncle said to have been brought by a serpent. The story ran that this ruler, when a child, had been placed by his mother in a hammock attached to two branches of a tree. While there a serpent crept up to him and dropped a stone upon his body. In gratitude for this gift the parents of the child fed and cared for the serpent. The stone is described as having been of a warm yellow hue, verging on red; it shone so brightly at night that a room could be illuminated by it. It eventually passed into the possession of a King of Siam.

Carnelian

Talisman ist Karneol
Gläubigen bringt er Glück und Wohl;
Steht er gar auf Onyx' Grunde,
Küss' ihm mit geweihtem Munde!
Alles Übel treibt er fort,
Schützet dich und schützt den Ort;
Wenn das eingegrabene Wort
Allah's Namen rein verkündet;
Dich zu Lieb' und Tat entzündet;
Und besonders werden Frauen
Sich am Talisman erbauen! [24]

Carnelian is a talisman,
It brings good luck to child and man;
If resting on an onyx ground,
A sacred kiss imprint when found.

---

[23] Rumphius, "Amboinsche Rariteitkamer," Amsterdam, 1741, p. 308.

[24] Goethe Westösterlicher Divan I, Segenspfänder.

It drives away all evil things;
To thee and thine protection brings.
The name of Allah, king of kings,
If graven on this stone, indeed,
Will move to love and doughty deed.
From such a gem a woman gains
Sweet hope and comfort in her pains.

The wearing of carnelians is recommended by the Lapidario of Alfonso X [25] to those who have a weak voice or are timid in speech, for the warm-colored stone will give them the courage they lack, so that they will speak both boldly and well. This is in accord with the general belief in the stimulating and animating effects produced by red stones.

On a carnelian is engraved in Arabic characters a prayer to keep away evil and to deliver the wearer from all the tricks of the devil and from the envious. The inscription reads in translation:

In the name of God the Just, the very Just!
I implore you, O God King of the World,
God of the World, deliver us from the devil
Who tries to do harm and evil to us through
Bad people, and from the evil of the envious.

Throughout all the East people are afraid of the envious. They believe that if you envy a person for his health or his wealth or any good thing he may have, he will lose it in a short time, and it is the devil who incites the envy of some people against others. So it is supposed that by wearing this stone, bearing this prayer against the envious, their envy will cease to do you harm.

The popularity of the carnelian as a talismanic stone

---

[25] "Lapidario del Rey D. Alfonso X," codice original, Madrid, 1881, fol. 77, p. 49.

among Mohammedan peoples is said to be due to the fact that the Prophet himself wore, on the little finger of his right hand, a silver ring set with a carnelian engraved for use as a seal. One of the most famous of the imâms, Jafar, lent the weight of his authority to the belief in the virtue of the carnelian, for he declared that all the desires of any man who wore this stone would be gratified. Hence in Persia the name of one of the twelve imâms, comprising Ali and his successors, is frequently engraved on this stone.[26]

CARNELIAN SEAL, WORN BY NAPOLEON I, NAPOLEON III, AND THE PRINCE IMPERIAL.

This most interesting seal is described by the Rev. C. W. King, the writer on Antique Gems. It is carnelian, octagonal-shaped, and upon it is engraved the legend: "The slave Abraham relying upon the Merciful (God)." Napoleon III wore it on his watch-chain. He said about it: "The First Consul picked it up with his own hands during the campaign in Egypt and always carried it about him, as his nephew did later." The Prince Imperial received it with the following message: "As regards my son, I desire that he will keep, as a talisman, the Seal which I used to wear attached to my watch." He carried the seal upon a string fastened about his neck in obedience to the injunction of his father. At the time of his lamentable death it must have been carried off in South Africa by the Zulus, when they stripped his body, and it has never been recovered.

An Armenian writer of the seventeenth century reports that in India the *lâl* or balas-ruby, if powdered and taken in a potion was believed to banish all dark forebodings and to excite joyous emotions. To the carnelian was attributed a virtue somewhat analogous to that ascribed to the turquoise, as anyone wearing a carnelian was proof

[26] Hendley, "Indian Jewellery," London, 1909, p. 158.

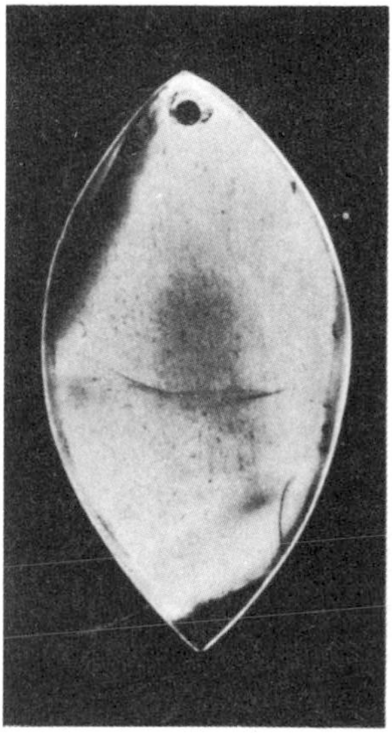

CHALCEDONY VOTIVE CHARM FROM MEXICO.
Aztec. Field Museum, Chicago.

CURIOUS ALTAR OF POWALAWA INDIANS OF ARIZONA.
The ceremonial objects are grouped around a crystal of rock-crystal in the centre. (See page 254.)

against injury from falling houses or walls; the writer emphasizes this by stating that "no man who wore a carnelian was ever found in a collapsed house or beneath a fallen wall."[27]

Chalcedony

An ingenious though far-fetched explanation of the power attributed to chalcedony of driving away phantoms and visions of the night is supplied by Gonelli, writing in 1702. For him the source of this asserted power was to be found in what has been erroneously termed the alkaline quality of the stone. This dissipated the evil humors of the eye, thus removing the diseased condition of that organ which caused the apparitions to be seen.[28] However absurd this explanation may be, it nevertheless shows that the author put little faith in visible ghosts, and rightly enough recognized the purely subjective character of such phenomena.

Chrysoberyl

The cat's-eye variety of chrysoberyl, or precious cat's-eye, is used by the natives of Ceylon as a charm against evil spirits. As a proof of the high value set upon the gem in India, De Boot states that a cat's-eye estimated as worth ninety gold pieces in Lusitania was sold for six hundred in India.[29] Some of the finest specimens come from Ceylon.

---

[27] Arakel, "Livre d'histoire," chap. liii; transl. in Brosset, "Collection d'historiens arméniens," St. Pétersburg, 1874, vol. i, pp. 544, 545.

[28] Josephi Gonelli, "Thesaurus philosophicus, seu de gemmis," Neapoli, 1702, p. 112.

[29] "Gemmarum et lapidum historia," Lug. Bat., 1636, p. 230.

Chrysolite

The "Serpent Isle," in the Red Sea, was stated by Agatharcides to be the source whence came the topaz (chrysolite); here, by the mandate of the Egyptian kings, the inhabitants collected specimens of this stone and delivered them to the gem-cutters for polishing.[30] These simple details are elaborated by Diodorus Siculus into the legend that the island was guarded by jealous watchers who had orders to put to death any unauthorized persons who approached it. Even those who had the right to seek the gem could not see the chrysolite in daytime; only after nightfall was it revealed by its radiance; the seekers then marked well the spot and were able to find the stone on the following day.[31]

From this Egyptian source, and possibly from others exploited by the Egyptians, have come the finest chrysolites (peridots, or olivines), the most magnificent examples of this gem. These found their way into the cathedral treasures of Europe, evidently by loot or trade at the period of the Crusades, and are generally called emeralds. Those most notable are in the Treasury of the Three Magi, in the great "Dom," or Cathedral at Cologne. Some of these gems are nearly two inches long.

In our own land beautiful specimens can be seen in the Morgan collection at the American Museum of Natural History and in the Higinbotham Hall in the Field Museum of Natural History, Chicago, Illinois.

Pliny quotes from Juba the tradition that the topaz (chrysolite) derived its name from the Island of To-

---

[30] Agatharcides, "De Mare Erythræo," §2. The topaz of the ancients was unquestionably the gem commonly called chrysolite at present (olivine, peridot).

[31] Diodorus Siculus, lib. iii, cap. 38.

pazos, in the Red Sea, the first specimen having been brought thence by the procurator Philemon, to Berenice, mother of Ptolemy II, Philadelphus. This monarch is said to have had a statue of his wife Arsinoë made from the stone.[32] If there be any foundation for this latter statement, the precious gift sent by Philemon must have been a mass of fluor-spar, or some similar material. More than three hundred years after Pliny's time, Epiphanius, evidently repeating another version of this tradition, states that the "topaz" was set in the diadem of the "Theban queen."

Chrysolite (olivine, peridot), to exert its full power, required to be set in gold; worn in this way it dispelled the vague terrors of the night. If, however, it were to be used as a protection from the wiles of evil spirits, the stone had to be pierced and strung on the hair of an ass and then attached to the left arm.[33] The belief in the virtue of the chrysolite to dissolve enchantments and to put evil spirits to flight was probably due to the association of the stone with the sun, before whose life-giving rays darkness and all the powers of darkness were driven away.

Chrysoprase

Wonderful things are told of the virtue of the chrysoprase, for Volmar states that, if a thief sentenced to be hanged or beheaded should place this stone in his mouth, he would immediately escape from his executioners.[34] Although we are not informed in what way this fortunate result was attained, it seems likely that the

[32] Plinii, "Naturalis Historia, lib. xxxvii, cap. 32.

[33] Marbodei, "De lapidibus," Friburgi, 1531, fol. 16.

[34] Volmar, Steinbuch, ed. by Hans Lambel, Heilbronn, 1877, p. 22.

stone was believed to make the thief invisible, and thus possessed a virtue often attributed to the opal.

A strange story regarding a magic stone reputed to have been worn by Alexander the Great is related by Albertus Magnus. According to this recital, Alexander, in his battles, wore a "prase" in his girdle. On his return from his Indian campaign, wishing one day to bathe in the Euphrates, he laid aside his girdle, and a serpent bit off the stone and then dropped it into the river.[35] Even Albertus, who is far from critical, admits that the story seems like a fable, and it probably belongs to a comparatively late period. As the term "prase" is used very loosely by early writers, this "victory stone" may have been an emerald or possibly jade.

Coral

The appreciation of coral as an ornament, or for amulets, seems to presuppose a certain development of civilization, for savage tribes greatly prefer glass ornaments. Many attempts have been made to introduce coral beads instead of glass beads among such tribes, but with no success, as the cheaper, but brighter, glass always commands a higher price.[36]

To still tempests and traverse broad rivers in safety was the privilege of one who bore either red or white coral with him. That this also stanched the flow of blood from a wound, cured madness, and gave wisdom, was said to have been experimentally proved.[37]

---

[35] Alberti Magni, "Opera Omnia," ed. Borgnet, Parisiis, 1890, vol. v, p. 43. De mineralibus, lib. ii, tract. 2.

[36] Bauer, "Edelsteinkunde," Leipzig, 1909, p. 750.

[37] Albertus Magnus, "Le Grand Albert des secretz des vertus des Herbes, Pierres et Bestes. Et aultre livre des Merveilles du Monde, d'aulcuns effetz causez daulcunes bestes," Turin, Bernard du mont du Chat (c. 1515). Liv. ii, fol. 9 recto.

KABYLE JEWELRY.

Of Mediterranean coral and pearls. Field Museum, Chicago.

Coral, which for twenty centuries or more was classed among the precious stones, to retain its power as an amulet, must not have been worked, and in Italy only such pieces are valued for this purpose as have been freshly gathered from the sea or have been cast up by the sea on the shore. To exercise all its power against spells, or enchantments, coral must be worn where its brilliant color makes it conspicuous; if, however, it should by accident be broken, the separate pieces have no virtue, and the magic power ceases, as though the spirit dwelling in the coral had fled from its abode. The peasant women are careful to guard the corals they wear for a special purpose from the eyes of their husbands, for the substance is believed to grow pale at certain seasons, regaining its pristine hue after a short interval of time. Indeed, the women believe that the coral shares their indisposition with them. All this serves to show that a kind of vital force is believed to animate the material, gaining or losing in vigor according to certain conditions, and finally disappearing when the form is broken. These beliefs are all clearly traceable to the animistic ideas of primitive man.[38]

Diamond

The diamond is to the pearl as the sun is to the moon, and we might well call one the "king-gem" and the other the "queen-gem." The diamond, like a knight of old,—brilliant and resistant, is the emblem of fearlessness and invincibility; the pearl, like a lady of old, —pure and fair to look upon, is the emblem of modesty and purity. Therefore it does not seem unfitting that

[38] Bellucci, "Il feticismo primitivo in Italia," Perugia, 1907, pp. 22–25.

the diamond should be presented as a token to the pearl, and that pearls should go with the diamond. The virtues ascribed to this stone are almost all directly traceable either to its unconquerable hardness or to its transparency and purity. It was therefore thought to bring victory to the wearer, by endowing him with superior strength, fortitude, and courage. Marbodus [39] tells us it was a magic stone of great power and served to drive away nocturnal spectres; for this purpose it should be set in gold and worn on the left arm. For St. Hildegard the sovereign virtue of the diamond was recognized by the devil, who was a great enemy of this stone because it resisted his power by day and by night.[40] Rueus [41] calls it "a gem of reconciliation," as it enhanced the love of a husband for his wife.

Cardano [42] takes a more pessimistic view of the qualities of the diamond. He says:

> It is believed to make the wearer unhappy; its effects therefore are the same upon the mind as that of the sun upon the eye, for the latter rather dims than strengthens the sight. It indeed renders fearless, but there is nothing that contributes more to our safety than prudence and fear; therefore it is better to fear.

The diamond was often associated with the lightning and was sometimes believed to owe its origin to the thunderbolt, but we do not recall having seen elsewhere the statement made in an anonymous Italian manuscript of the fourteenth century. Here it is expressly

---

[39] "De lapidibus," Friburgi, 1531, f. 8.

[40] St. Hildegardæ, "Opera Omnia," in Pat. Lat. ed Migne, vol. cxcvii, col. 1254.

[41] "De gemmis," Tiguri, 1566, f. 52.

[42] "Philosophi opera quædam lectu digna," Basileæ, 1585, p. 322. "De gemmis."

asserted that the diamond is sometimes consumed or melted when it thunders.[43] Certainly, that the same force that was supposed to have formed the stone should be able to dissolve it, is not an illogical idea. That the diamond can be entirely consumed at a high temperature was a fact not known in Europe in the fourteenth century, and therefore the belief in the destructive effect of the electric current must have arisen from superstitious or poetic fancies, and not from any vague conception of the true nature of the diamond.

In the Talmud we read of a gem, supposed to have been the diamond, which was worn by the high priest.[44] This stone served to show the guilt or innocence of one accused of any crime; if the accused were guilty, the stone would grow dim, but if he were innocent, it would shine more brilliantly than ever. This quality is also alluded to by Sir John Mandeville, who wrote:

> It happens often that the good diamond loses its virtue by sin and for incontinence of him who bears it.

The Hindus classed diamonds according to the four castes. The Brahmin diamond gave power, friends, riches and good luck; the Kshatriya diamond prevented the approach of old age; the Vaisya stone brought success, and the Sudra, all manner of good fortune. On the other hand, in the treatise on gems by Buddhabhatta[45] we read:

> A diamond, a part of which is the color of blood or spotted with red, would quickly bring death to the wearer, even if he were the Master of Death.

---

[43] Anonymous writer in Ital. MS. of the fourteenth century in the author's library; fol. 41 p. verso.

[44] See page 278 for description of this diamond by St. Epiphanius.

[45] Finot, "Les lapidaires indiens," Paris, 1896, p. 9.

The Arabians and Persians, as well as the modern Egyptians, agree in attributing to the diamond a wonderful power to bring good fortune, and Rabbi Benoni, a mystic of the fourteenth century, treating of its magic virtues, asserts that it produces somnambulism, and, as a talisman, so powerfully attracts the planetary influences that it renders the wearer invincible; it was also said to provoke a state of spiritual ecstasy. An alchemist of the same century, Pierre de Boniface, asserted that the diamond made the wearer invisible.

A curious fancy, prevalent in regard to many stones, attributed sex to the diamond, and it is therefore not surprising that these stones were also supposed to possess reproductive powers. In this connection Sir John Mandeville wrote:

> They grow together, male and female, and are nourished by the dew of heaven; and they engender commonly, and bring forth small children that multiply and grow all the year. I have oftentimes tried the experiment that if a man keep them with a little of the rock, and water them with May dew often, they shall grow every year and the small will grow great.

The following lines from a translation of the celebrated Orphic poem, written in the second century, show the high esteem in which the adamas was held at that time:

> The Evil Eye shall have no power to harm
> Him that shall wear the diamond as a charm,
> No monarch shall attempt to thwart his will,
> And e'en the gods his wishes shall fulfil.

This probably refers either to colorless corundum, the so-called "white sapphire," or to quartz. The writer is disinclined to believe that the ancients knew the diamond.

The ancient Hindu gem-treatise of Buddhabhatta asserts that the diamond of the Brahmin should have the whiteness of a shell or of rock-crystal; that of the Kshatriya, the brown color of the eye of a hare; that of the Vaisya, the lovely shade of a petal of the *kadali* flower; that of the Sudra, the sheen of a polished blade. To kings alone the sages assigned two classes of colored diamonds,—namely, those red as coral and those yellow as saffron. These were exclusively royal gems, but diamonds of all other shades could be set in royal jewels.[46]

A typical diamond is thus described in a Hindu gem-treatise:[47]

> A six-pointed diamond, pure, without stain, with pronounced and sharp edges, of a beautiful shade, light, with well-formed facets, without defects, illuminating space with its fire and with the reflection of the rainbow, a diamond of this kind is not easy to find in the earth.

According to a wide-spread superstition, the talismanic power of a diamond was lost if the stone were acquired by purchase; only when received as a gift could its virtues be depended on.[48] The same belief is noted regarding the turquoise. The spirit dwelling in the stone was thought to take offence at the idea of being bought and sold, and was supposed to depart from the stone, leaving it nothing more than a bit of senseless matter. If, however, the diamond (or turquoise) were offered as a pledge of love or friendship, the spirit was quite willing to transfer its good offices from one owner to another.

The Talmud shows us that the Jewish Rabbis some-

---

[46] Finot, "Les lapidaires indiens," Paris, 1896, p. 8.

[47] Finot, l. c., p. 9.

[48] Konrad von Megenberg, "Buch der Natur," ed. by Dr. Franz Pfeiffer, Stuttgart, 1861, p. 433.

times endeavored to enliven their exhaustive discussions of ritual and legal questions by telling "good stories" to each other. One of these may be given as illustrating at once the wild improbability of some of these recitals and the belief in the wonderful magic virtues of the diamond:[49]

R. Jehudah of Mesopatamia used to tell: Once while on board of a ship, I saw a diamond that was encircled by a snake, and a diver went to catch it. The snake then opened its mouth, threatening to swallow the ship. Then a raven came, bit off its head, and all water around turned into blood. Then another snake came, took the diamond, put it in the carcass, and it became alive; and again it opened its mouth, in order to swallow the ship. Another bird then came, bit off its head, took the diamond and threw it on the ship. We had with us salted birds, and we wanted to try whether the diamond would bring them to life, so we placed the gem on them, and they became animated and flew away with the gem.

It is said that the first large diamonds discovered by Europeans in South Africa were found in the leather bag of a sorcerer. Although large stones or fragments of rock are usually the objects of adoration as fetiches in Africa, any small stone that is wrapped in colored rags and worn on the neck may be regarded in the same way.[50] Several competent authorities state that these diamonds were the playthings of some Boer children.

Al Kazwini relates as follows the marvellous tale of the Valley of Diamonds:[51]

---

[49] New edition of the Babylonian Talmud, ed. and trans. by Michael L. Rodkinson, vol. v (xiii), Baba Batra, New York, 1902, p. 207.

[50] Ratzel, "Völkerkunde," Leipzig, 1885, vol. i, p. 36.

[51] Dr. Julius Ruska, "Das Steinbuch aus der Kosmographie des al-Kazwini," Beilage zum Jahresbericht 1894–5 der Oberrealschule Heidelberg, p. 35. See Aristoteles De Lapidibus und Arnoldus Saxo, ed. Rose, Z.f.D.A. New Series VI, pp. 364, 365, 389, 390. The "other writer" is probably Ahmed Teifashi.

"Aristotle[52] says that no one except Alexander ever reached the place where the diamond is produced. This is a valley, connected with the land Hind. The glance cannot penetrate to its greatest depths and serpents are found there, the like of which no man hath seen, and upon which no man can gaze without dying. However, this power endures only as long as the serpents live, for when they die the power leaves them. In this place summer reigns for six months and winter for the same length of time. Now, Alexander ordered that an iron mirror should be brought and placed at the spot where the serpents dwelt. When the serpents approached, their glance fell upon their own image in the mirror, and this caused their death. Hereupon, Alexander wished to bring out the diamonds from the valley, but no one was willing to undertake the descent. Alexander therefore sought counsel of the wise men, and they told him to throw down a piece of flesh into the valley. This he did, the diamonds became attached to the flesh, and the birds of the air seized the flesh and bore it up out of the valley. Then Alexander ordered his people to pursue the birds and to pick up what fell from the flesh."

"Another writer states that the mines are in the mountains of Serendib (Ceylon) in a very deep gorge, in which are deadly serpents. When people wish to take out the diamonds they throw down pieces of flesh, which are seized by vultures and brought up to the brink of the gorge. There such of the diamonds as cling to the flesh are secured; these are of the size of a lentil or a pea. The largest pieces found attain the size of a half-bean."

In his version of the tale, one form of which appears in the seventh voyage of Sindbad the Sailor, Teifashi states that the finest corundum gems were washed down the streams that flowed from Adam's Peak, on the island of Ceylon; in time of drought, however, this source of supply ceased. Now it happened that many eagles built their nests on the top of this mountain, and the gem-seekers used to place large pieces of flesh at the foot of the mountain. The eagles pounced upon these and bore them away to their nests, but were obliged to alight from

[52] The work on precious stones attributed to Aristotle was composed in Arabic probably in the ninth century.

time to time in order to rest, and while the pieces of flesh lay on the rock, some of the corundums became lightly attached to this, so that when the eagles resumed their flight the stones dropped off and rolled down the mountain side.[53]

These oft-repeated tales are explained by Dr. Valentine Ball as originating in the Hindu custom of sacrificing cattle when new mines were opened, and leaving on the spot a certain part of the meat as an offering to the guardian deities. As these pieces of meat were soon carried away by birds of prey, the legend arose that the diamonds were obtained in this way. This custom still prevailed in some parts of India when Dr. Ball wrote.[54]

The effect exercised by Hindu superstition on even the most enlightened Europeans of our day may be recognized in the fact that the gifted prima donna, Mme. Maeterlinck, the wife of the foremost living European poet, has confessed that she wears a diamond suspended on her forehead because her husband believes that this brings good fortune to the wearer. This forehead-jewel is characteristically Hindu and enjoys in India the reputation of being especially auspicious.

Emerald

The emerald was believed to foreshow future events,[55] but we do not learn whether visions were actually seen in the stone, as they were in spheres of rock-crystal or beryl, or whether the emerald endowed the wearer with a supernatural fore-knowledge of what was to come. As

---

[53] Teifashi, "Fior di pensieri sulle pietre preziose," Firenzi, 1818, p. 13.

[54] Proc. of the Royal Irish Academy, 2d Ser., Polite Literature and Antiquities, vol. ii, Dublin, 1879–1888, p. 303.

[55] Epiphanii, "De XII gemmis," Tiguri, 1565, fol. 5.

a revealer of truth, this stone was an enemy of all enchantments and conjurations; hence it was greatly favored by magicians, who found all their arts of no

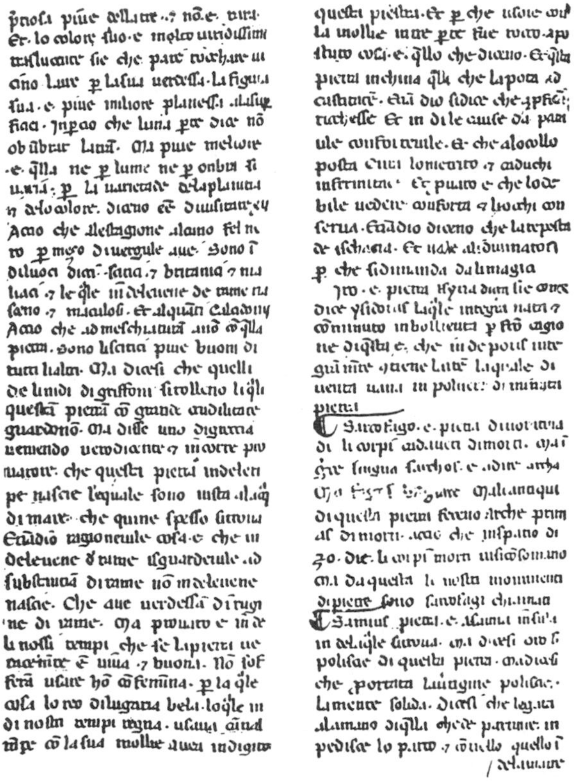

SPECIMEN PAGE OF ITALIAN MANUSCRIPT OF THE FOURTEENTH CENTURY.

Containing an Italian version of the "De Mineralibus" of Albertus Magnus. On this page is the account of the emerald, set in a ring worn by King Bela IV of Hungary (1235–1270), that was fractured when he caressed his wife. Author's library.

avail if an emerald were in their vicinity when they began to weave their spells.[56]

[56] Morales, "De las piedras preciosas," Valladolid, 1604, fol. 101.

To this supernatural power inherent in the stone, enabling it to quicken the prophetic faculty, may be added many other virtues. If any one wished to strengthen his memory or to become an eloquent speaker, he was sure to attain his end by securing possession of a fine emerald.[57] And not only the ambitious, but also those whose hearts had been smitten by the shafts from Cupid's bow found in this stone an invaluable auxiliary, for it revealed the truth or falsity of lover's oaths. Strange to say, however, the emerald, although commonly assigned to Venus, was often regarded as an enemy of sexual passion. So sensitive was the stone believed to be in this respect that Albertus Magnus relates of King Bela of Hungary, who possessed an exceptionally valuable emerald set in a ring, that, when he embraced his wife while wearing this ring on his finger, the stone broke into three parts.[58]

In Rabbinical legend it is related that four precious stones were given by God to King Solomon; one of these was the emerald. The possession of the four stones is said to have endowed the wise king with power over all creation.[59] As these four stones probably typified the four cardinal points, and were very likely of red, blue, yellow, and green color respectively, we might conjecture that the other three stones were the carbuncle, the lapis-lazuli, and the topaz.

After stating that the emerald sharpens the wits and quickens the intelligence, Cardano declares that it therefore made people more honest, for "dishonesty is

---

[57] Marbodei, "De lapidibus, Friburgi, 1531, fol. 48; Camilli Leonardi, "Speculum lapidum," Venetia, 1502, fol. xliii.

[58] Fol. 55 recto of Ital. MS., 14th Century. Reference is to Bela IV (1235–1270). Lo reo dilugaria bela loqale in di nostri tempi regna.

[59] Weil, "Biblische Legenden," p. 225.

nothing but ignorance, stupidity, and ill-nature." The same writer adds that the stone was believed to make men economical and hence to make them rich, but of this he was very sceptical, since the experience of others as well as his own showed that the emerald possessed very little power in this direction.[60]

A talismanic emerald, once the property of the Mogul emperors of Delhi, has recently been shown in Europe. The stone is of a rich deep green, and weighs 78 carats. Around the edge in Persian characters runs the inscription: "He who possesses this charm shall enjoy the special protection of God."

Emerald sharpened the wits, conferred riches and the power to predict future events. To evolve this latter virtue it must be put under the tongue. It also strengthened the memory. The light-colored stones were esteemed the best and legend told that they were brought from the "nests of griffons."[61]

Gypsum

Gypsum when fibrous—the fibres being long and straight—is known as "satin spar." This material is frequently cut rounded, or *en cabochon,* across the fibres; sometimes it is cut in the form of beads, or of pear-shaped drops, which are mounted in earrings, scarf-pins, or necklaces. The material is frequently found in Russia, England, and elsewhere, and is cut in England or Russia. Some of the cut stones are mounted in brass, or gilded

---

[60] Cardani, "Philosophi opera quædam," Basileæ, 1585, p. 328. "De gemmis."

[61] Albertus Magnus, "Le Grand Albert des secrets des vertus des Herbes, Pierres et Bestes. Et aultre livre des Merveilles du Monde, d'aulcuns effetz causez daulcunes bestes," Turin, Bernard du mont du Chat (c. 1515). Liv. ii, fol. 11.

brass, and sold as luck stones at Niagara, the claim being made that the " satin spar " was taken from beneath the Falls at great peril, as occasionally small deposits of this kind of gypsum are found under the Falls.

From time to time small consignments of this material have been sent to Japan, as the Japanese value it possibly on account of its purity, or owing to the fact that it has the effect of the cat's-eye. It is quite cheap, and at the same time very soft, so that it can be scratched with the finger-nail. That found in Russia is of a golden-yellow or salmon color, and is worked into various ornaments, the one popular form being egg-shaped, and, because of their form, such objects are frequently given as Easter gifts. The same material is also known in Egypt, and is cut in the same egg form, the ornaments being called " Pharaoh's eggs," although just which Pharaoh this refers to is not stated. They are also believed to possess qualities of protection and to bring good fortune.

Hematite

The virtues of the hematite were praised in an ancient gem-treatise written by Azchalias of Babylon for Mithridates the Great, King of Pontus (d. 63 B.C.), a sovereign who was passionately fond of precious stones, and possessed a splendid collection of them, both engraved and unengraved. Azchalias, as cited by Pliny[62] taught that human destinies were influenced by the virtues inherent in precious stones, and asserted that the hematite, when used as a talisman, procured for the wearer a favorable hearing of petitions addressed to kings and a fortunate issue of lawsuits and judgments. It is a red oxide of iron, which when abraded shows a red streak; whence the

[62] "Naturalis historia," lib. xxxvii, cap. 60.

name hematite, from the Greek *haima,* "blood." As an iron ore and hence associated with Mars, the god of war, this substance was also considered to be an invaluable help to the warrior on the field of battle if he rubbed his body with it. Probably, like the loadstone, it was believed to confer invulnerability.

The high degree of skill possessed by the Pueblo workers is strikingly shown in a finely inlaid hematite cylinder found in Pueblo Bonito. The inlays are of turquoise and are designed to make the cylinder a conventional representation of a bird. The wings are indicated by turquoise inlays of pyramidal outline, curved so as to follow the curvature of the cylinder, the head being figured by a conical piece of turquoise attached to one end. This conical termination bore a small bird-figure carved in relief.[63] When we consider the difficulties the Indian workers had to overcome in the execution of this artistic task with the tools at their command, we can well realize that this object, probably an amulet, must have been considered very valuable, and was most likely the property of some one of high rank in the tribe or community.

Jacinth

The jacinth was more especially recommended as an amulet for travellers, because of its reputed value as a protection against the plague and against wounds and injuries, the two classes of perils most feared by those who undertook long journeys. Moreover, this stone assured the wearer a cordial reception at any hostelry he

[63] George H. Pepper, "The Exploration of a Burial-room in Pueblo Bonito, New Mexico," Putnam Anniversary Volume, New York, 1909, p. 239; Fig. 5.

visited.[64] It was said to lose its brilliancy and grow pale and dull if the wearer or any one in his immediate neighborhood became ill of the plague. In addition to these qualities the jacinth augmented the riches of the owner, and endowed him with prudence in the conduct of his affairs.[65]

St. Hildegard, the Abbess of Bingen (d. 1179), gives the following details as to the proper use of the *jachant* (jacinth):[66]

> If any one is bewitched by phantoms or by magical spells, so that he has lost his wits, take a hot loaf of pure wheaten bread and cut the upper crust in the form of a cross,—not, however, cutting it quite through,—and then pass the stone along the cutting, reciting these words: "May God, who cast away all precious stones from the devil . . . cast away from thee, N., all phantoms and all magic spells, and free thee from the pain of this madness."

The patient is then to eat of the bread; if, however, his stomach should be too feeble, unleavened bread may be used. All other solid food given to the sick person should be treated in the same manner. We are also told that if any one has a pain in his heart, the pain will be relieved provided the sign of the cross be made over the heart while the above mentioned words are recited.

The wearer of a jacinth was believed to be proof against the lightning, and it was even asserted that wax that had been impressed by an image graven on this stone averted the lightning from one who bore the seal. That the stone really possessed this power was a matter of common report, it being confidently declared that in re-

---

[64] Marbodei, "De lapidibus," Friburgi, 1531, fol. 38.

[65] Cardani, "Philosophi opera quædam," Basileæ, 1585, p. 323. "De gemmis."

[66] S. Hildegardæ, Opera omnia; in Pat. Lat. ed. J. P. Migne, vol. cxcvii, Parisiis, 1855, col. 1251.

gions where many were struck by lightning, none who wore a jacinth were ever harmed. By a like miracle it preserved the wearer from all danger of pestilence even though he lived in an air charged with the disease. A third virtue was to induce sleep. Of this, Cardano states that he was in the habit of wearing rather a large jacinth, and had found that the stone "seemed to dispose somewhat to sleep, but not much." He adds, in explanation of its slight efficacy, that his stone was not bright red, nor of the best sort, but of a golden hue, differing much from the best.[67]

Jade

The name jade includes two distinct minerals, nephrite and jadeite. The former is a silicate of magnesia, of exceedingly tough structure, and ranks 6.5 in the scale of hardness, while jadeite, a silicate of alumina, is more crystalline and not as tough as nephrite and has a hardness of 7. A variety having a rich emerald-green hue is called by the Chinese *feits'ui,* "Kingfisher plumes"; it is also denominated Imperial jade.

The original form of the Chinese character *pao,* signifying "precious," consists of the outline of a house, within which are the symbols of jade beads, shell, and an earthen jar. This shows that at the very early time when these characters were first used, the Chinese already collected jade and employed it for personal adornment.[68] The oldest form of the ideograph for "king", , appears to be the symbol for a string of jade beads, which are even

[67] Cardani, "De subtilitate," Basileæ, 1560, pp. 442–3.

[68] Chalfant, "Early Chinese Writing," Mem. of Carnegie Museum, vol. iv, No. 1, Pittsburg, 1906, p. 10 and Pl. XX, No. 275. See also Pl. X, No. 132; *pei,* "shell," "value," as shells were used as money in very ancient times.

now used in China as insignia for high rank and authority.[69]

Jade amulets of many different forms are popular with the Chinese. One representing two men is called "Two Brothers of Heavenly Love," and is often given to friends. A phœnix of jade is a favorite ornament for young girls and is bestowed upon them when they come of age. To a newly-wedded pair is given the figure of a man riding on a unicorn and holding castanets in his hand; this signifies that an heir will be born in due time.

Such is the fondness of the Chinese for jade that those who can afford the luxury of its possession are wont to carry with them small pieces, so that they may have them always at hand; for they believe that, when handled, something of the secret virtue of the substance is absorbed into the body. When struck, jade is thought to emit a peculiarly melodious sound, which for the Chinese poet resembles the voice of the loved one; indeed, jade is termed the concentrated essence of love.

Fashioned into the form of a butterfly, a piece of jade acquires a special romantic significance in China, because of a Chinese legend which relates that a youth in his eager pursuit of a many-hued butterfly made his way into the garden of a rich mandarin. Instead of being punished for his trespass, the youth's unceremonious visit led to his marriage with the mandarin's daughter. Hence the figure of a butterfly is a symbol of successful love, and Chinese bridegrooms are wont to present jade butterflies to their fiancées.

A Chinese jade ornament constituting a child's amulet assumes a form approximating to that of a padlock. When this is attached to a child's neck, it is supposed to

[69] Chalfante, "Early Chinese Writing," Pl. XXII, No. 299.

bind the little one to life and protect it from all danger in infantile diseases. A jade object of a different kind is sometimes used at nuptial feasts in China. This is a cup having the form of a cock, and both bride and groom drink from it. The form of this vessel is accounted for by a legend to the effect that when a beautiful white cock saw its young mistress, who had often petted it, throw herself into a well in a transport of despair at the loss of her lover, the faithful fowl sought and found death in the same way, so as not to be separated from its mistress.

Among the splendid Chinese jade carvings of the Woodward Collection is a curious symbolic ornament carved out of the rare *fei-ts'ui yü,* or "kingfisher-green jade," a rich emerald green jadeite with translucent green shading. This ornament, executed in the beginning of the eighteenth century and believed to be a product of the Imperial Jade Works in Peking, figures the natural form of a so-called "hand-of-Buddha" citron, the finger-like protuberances of the fruit suggesting this strangely fanciful name. The Chinese regard this as a most felicitous emblem, denoting at once a long life and abundance of riches for its enjoyment. In the present carving the figure of a bat clinging to the foliage enveloping the fruit constitutes an added omen of good fortune, the Chinese character *fu* signifying at once "bat" and "happiness," another proof of what we are prone to call Chinese queerness, for with the superstitious of our race the bat is always looked upon as especially ill-omened.[70]

It is a well-known fact that many analogies have been found between the customs, usages, and products of the more civilized aborigines of the New World and those of

[70] "Catalogue of the Woodward Collection of Jades and other Hard Stones," by John Getz, Privately printed (New York), 1913, p. 11, No. 24.

the ancient Egyptians. Another instance is offered by the custom of placing a piece of *chalchihuitl* (jade?) or of some other green stone in the mouth of a noble, after his death, and calling this his heart. Among the lower classes a *texaxoctli,* a stone of small value, was used for the same purpose. We shall see that, in the Egyptian "Book of the Dead," directions are given for putting a semi-precious stone on or in a mummy, as a symbol, and designating this the heart of the deceased person. For the use of a green stone for this purpose by the ancient Mexicans, Mrs. Zelia Nuttall finds a reason in the two meanings of the Nahuatl word *xoxouhqui-yollotl,* which is used to signify a "free man," the literal meaning being a "fresh or green heart." Hence, the stone was a symbol of the rank of the deceased as well as of his heart.[71] The fact that jade celts have been found cut into several pieces is taken to indicate the high value placed upon this material; for it has been conjectured by Dr. Earle Flint, that a living chief would cut a piece from the jade he wore as a sign of his rank, in order to provide a suitable ornament or amulet for a dead kinsman.

To certain of the Chinese "tomb-jades"—that is, jade amulets deposited with the dead—has been given the name *han-yü,* or "mouth-jade," because these amulets, supposed to afford protection to the dead, were placed in their mouths. The Metropolitan Museum of Art in New York contains a fine collection of 279 specimens of jade from Chinese tombs, found within the past five or six years, and presented to the museum by Mr. Samuel F.

[71] Zelia Nuttall, "The Fundamental Principles of Old and New World Civilization," Cambridge, Mass., 1901, p. 195. Archæological and Ethnographical Papers of the Peabody Museum, Harvard University, vol. ii.

Peters. In color these jades are not especially attractive, for the material has acquired a brownish stain, due to the products of decomposition of the body, and also to the absorption of some of the chemical constituents of the other objects in the tomb, during the long period of time, in many cases a thousand years or more, since the bodies were consigned to their final resting place.

So multifarious are the uses to which jade is put by the Chinese, and so great is their admiration of its qualities, that they regard it as the musical gem *par excellence.* A series of oblong pieces of jade, of the same length and width, usually about 1.8 feet long and 1.35 feet wide, and numbering from 12 to 24, constitute a chime, the difference in the notes emitted by the material when sharply struck depending upon the varying thickness of the separate pieces. What is designated the "stone chime" used in court and religious ceremonials, is composed of 16 undecorated stones, while a series known as the singers' chime consists of from 12 to 24 pieces carved into fantastic shapes. This use of jade for the production of musical sounds dates far back in the Chinese annals. We are told that when Confucius was much troubled at the ill-success of his efforts to reform the Chinese morals of his day, he sought consolation in playing on the "musical stone." A peasant who noted this in passing by, exclaimed, as he heard the sounds: "Full indeed is the heart of him who beats the musical stone like that!"[72]

A jade ornament greatly favored by the Maoris of

---

[72] The Bishop Collection. "Investigations and Studies in Jade," New York, privately printed, 1906, vol. i, pt. iii, "Jade as a Mineral," by George Frederick Kunz, p. 117. Nos. 421 and 646 of the collection are excellent examples of this special jade.

New Zealand bore the name *hei-tiki* ("a carved image for the neck"). The ornaments of this class are very rude and grotesque representations of the human face or form, and were generally regarded as schematically figuring some departed ancestor. The head sometimes slanted right or left, so that the eyes, which were very large and occasionally inlaid with mother-of-pearl, were on an angle of forty-five degrees. These ornaments were prized not only as memorials, but because, having been worn by successive ancestors, they were supposed to communicate something of the very being of those ancestors to such descendants as were privileged to wear the treasured heirloom in their turn. In many cases, when the family was dying out, the last male member would leave directions that his *hei-tiki* should be buried with him, so that it might not fall into the hands of strangers.[73]

So rare was this New Zealand jade, known to the Maoris as *punamu* (green-stone), that the aid of a *tohunga,* or wizard, was regarded as necessary to learn where it could be found. On setting forth on a search for this material, the jade-seekers would take with them a *tohunga,* and when the party reached the region where jade was usually found the *tohunga* would retire to some solitary spot and would fall into a trance. On awaking he would claim that the spirit of some person, dead or living, had appeared to him and had directed to search in a particular place for the jade. He would then conduct

---

[73] The Bishop Collection. "Investigations and Studies in Jade," New York, 1906, vol. i, p. 12. Privately printed and edition limited to 100 copies. For a description of this monumental work see "The Printed Catalogue of the Heber R. Bishop Collection of Jade," by George Frederick Kunz, supplement to the Bulletin of the Metropolitan Museum of Art for May, 1906, Occasional Notes, No. 1..

the party to this place, where a larger or smaller piece of jade was invariably found. Of course the wizard had previously assured himself of the presence of the stone in the place indicated.

To this jade was given the name of the man whose spirit had revealed its location, and in many cases the grotesque form given to the stone was conceived to represent this man. We can easily understand the reverence accorded to the *hei-tikis* when we consider that they were not only prized as heirlooms, which had been handed down by the successive heads of the family, but were also believed to have been originally found in such a mysterious way.

When the head of the family died, his *hei-tiki* was generally buried with his body, but was exhumed after a shorter or longer time by the nearest male relative. As we have noted, if no representative of the family remained, the heirloom was allowed to remain in the grave. The fact that tribal or intertribal feuds sometimes arose in regard to the possession of a *hei-tiki* serves to prove the peculiar virtues ascribed to them.

While there can be little doubt that the heirloom was supposed to represent, in a very general way, the person whose name it bore, the particular form given it was largely determined by the natural shape of the mass, which was slowly and patiently fashioned into the form it eventually acquired. Though this was mainly due to the imperfect means of which the artist disposed, there was probably a conviction that the form of the natural stone was not the result of accident, but was in itself significant and required only to be rendered more clear and definite. The fabrication of the *hei-tikis* of the Maoris is said to have ceased in the early part of the last century. The greater number of those that have been collected in New Zealand appear to have been made from one hundred to one hundred and fifty years ago.[74]

Jasper

The jasper had great repute in ancient times as a rain-bringer, and the fourth century author of "Lithica" celebrates this quality in the following lines:[75]

> The gods propitious hearken to his prayers,
> Whoe'er the polished grass-green jasper wears;
> His parched glebe they'll satiate with rain,
> And send for showers to soak the thirsty plain.

Evidently the green hue of this translucent stone suggested its association with the verdure of the fields in an even closer degree than was the case with transparent green stones such as the emerald, etc. Another early authority, Damigeron, mentions this belief, and states that only when properly consecrated would the jasper do service in this way.[76] Jasper was also credited in the fourth century with the virtue of driving away evil spirits and protecting those who wore it from the bites of venomous creatures.[77] An anonymous German author of the eleventh or twelfth century recommends the use of this stone for the cure of snake bites, and states that if it be placed upon the bitten part the matter will come out

---

[74] See Fischer, "Ueber die Nephritindustrie der Maoris in Neuseeland," Archiv für Anthropologie, vol. xv, Braunschweig, 1884, pp. 463–466.

[75] King's version in his Natural History of Precious Stones, London, 1865, p. 382.

[76] Pitra, "Specilegium Solesmense," Parisiis, 1855, p. 328.

[77] Epiphanius, "De XII gemmis," Tiguri, 1565, fols. 7, 8.

from the wound.[78] Here the cure is operated, not by the absorbent quality of the stone, but by its supposed power to attract poison or venom to itself, thus removing the cause of disease.

A popular etymology of the Greek and Latin name for jasper is reported by Bartolomæus Anglicus, who writes that "in the head of an adder that hyght Aspis is founde a lytyl stone that is called Jaspis." The same authority pronounces this stone to be of "wunder vertue," and says that "it hath as many vertues as dyvers coloures and veines."[79] This is fully in accord with tradition, for, as color was at least as important as chemical composition in determining the talismanic or therapeutic worth of the different stones, the great variety of colors and markings in the different jaspers naturally indicated their use in many different ways.

Jet

Jet has been found among the palæolithic remains in the caves of the "Kesslerloch," near Thayngen, Canton Schaffhausen, Switzerland. The material was evidently derived from the deposits in Würtemberg and was shaped by flint chips. Quite possibly jet, as well as amber, was already regarded as possessing a certain talismanic virtue. Such ornaments, when worn, were believed to become a part of the very body and soul of the wearer, and were therefore to be guarded with jealous care.[80] In the

---

[78] Birlinger, "Kleinere deutsche Sprachdenkmäler," in Germania, vol. viii (1863), p. 302.

[79] Bartolomæi Anglici "De proprietatibus rerum," London, Wynkyn de Worde, 1495, lib. xvi, cap. 51, De Jaspide. Old English version by John of Trevisa.

[80] Hoernes, "Urgeschichte der bildenden Kunst," Wien, 1898, pp. 22, 24.

palæolithic cave-deposits of Belgium also, jet appears, the supply being in this instance derived from northern Lorraine. The fragments had been rounded and pierced through the centre.[81] This indicates their use as parts of a necklace or as pendants. Necklaces, bracelets, and rings were especially favored for the wearing of talismanic gems, since the stones could easily be so set that they would come in direct contact with the skin.

Jet was one of the materials used by the Pueblo Indians for their amulets. An exceptionally well-executed figure of a frog made of this material was found in Pueblo Bonito, in 1896, by Mr. Pepper. The representation is much more realistic than is the case in the other figures of this type from this region. Turquoise eyes have been inserted in the head of the figure and a band of turquoise surrounds the neck.[82]

Lapis-Lazuli

Both in Babylonia and in Egypt, lapis-lazuli was very highly valued, and this is shown by the use of its Assyrian name (*uknu*) in poetic metaphor. Thus, in a hymn to the moon-god Sin, he is addressed as the "strong bull, great of horns, perfect in form, with long flowing beard, bright as lapis-lazuli."[83] This may remind us of the "hyacinthine locks" of classical literature.

Lapis-lazuli, "a blue stone with little golden spots," was a cure for melancholy and for the "quartern fever,"

[81] Dupont, "L'homme pendant les âges de la pierre," Brussels, 1872, pp. 156 sqq.

[82] Pepper, "The Exploration of a Burial-room in Pueblo Bonito," Putnam Anniversary Volume, New York, 1909, p. 237.

[83] Ward, "Seal Cylinders of Western Asia," Washington, D. C., 1910, p. 121; citing Jastrow, "Religion," p. 303.

JASPER PENDANT.

Aztec Mexican. Used to stanch blood.

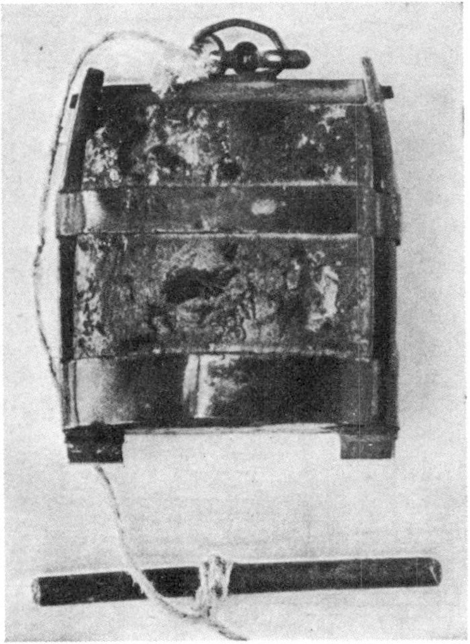

PIECE OF NATURAL LOADSTONE.

Used in sixteenth century for medicinal purposes.

ARAGONITE PENDANT.

Used for votive purposes in Armenia. Field Museum, Chicago.

an intermittent fever returning each third day, or each fourth day counting in the previous attack.[84]

Loadstone

We have the authority of Plato (Ion, 533 D) for the statement that the word *magnetis* was first applied to the loadstone by the tragic poet Euripides (480–405 B.C.), the more usual name being "the Heraclean stone." These designations refer to two places in Lydia, Magnesia and Herakleia, where the mineral was found.[85] Pliny states, on the authority of Nicander, that a certain Magnes, a shepherd, discovered the mineral on Mount Ida, while pasturing his flock, because the nails of his shoes clung to a piece of it.[86]

We are told by Pliny that Ptolemy Philadelphus (309–247 B.C.), planning to erect a temple in honor of his sister and wife Arsinoë, called in the aid of Chirocrates, an Alexandrian architect. The latter engaged to place therein an iron statue of Arsinoë which should appear to hang in mid-air without support. However, both the Egyptian king and his architect died before the design could be realized.[87] This story of an image held in suspense by means of powerful magnets set in the floor and roof, and sometimes also in the walls of a temple, is repeated in a variety of forms by early writers. Of

---

[84] Albertus Magnus, "Le Grand Albert des secretz des vertus des Herbes, Pierres et Bestes. Et aultre livre des Merveilles du Monde, d'aulcuns effetz causez daulcunes bestes," Turin, Bernard du mont du Chat (c. 1515). Liv. ii, fol. 11, recto.

[85] The Timæus of Plato, ed. by R. R. Archer-Hind, London, 1888, p. 302, note.

[86] Plinii, "Historia naturalis," Venetiis, 1507, fol. 269 verso, lib. xxxvi, cap. 16.

[87] Plinii, l. c., fol. 254, verso, lib. xxxiv, cap. 14.

course, there was no real foundation for such tales, as the thing is altogether impracticable.

The Roman poet Claudian (fifth century A.D.) relates that the priests of a certain temple, in order to offer a dramatic spectacle to the eyes of the worshippers, caused two statues to be executed,—one of Mars in iron, and another of Venus in loadstone. At a special festival these statues were placed near to each other, and the loadstone drew the iron to itself. Claudian vividly describes this:

> The priests prepare a marriage feast.
> Behold a marvel! Instant to her arms
> Her eager husband Cythereia charms;
> And ever mindful of her ancient fires,
> With amorous breath his martial breast inspires;
> Lifts the loved weight, close round his helmet twines
> Her loving arms, and close embraces joins,
> Drawn by the mystic influence from afar.
> Flies to the wedded gem the God of War.
> The Magnet weds the Steel: the sacred rites
> Nature attends, and th' heavenly pair unites.[88]

There was current as early as the fourth century a curious belief that a piece of loadstone, if placed beneath the pillow of a sleeping wife, would act as a touchstone of her virtue. This first appears in the Alexandrian poem "Lithica," and it has been thus quaintly Englished by a fourteenth century translator:

> Also magnes is in lyke wyse as adamas; yf it be sett under the heed of a chaste wyfe, it makyth her sodenly to beclyppe [embrace] her husbonde; & yf she be a spowse breker, she shall meve her oute of the bed sodenly by drede of fantasy.[89]

---

[88] King's metrical version in his "Natural History of Gems," London, 1865, p. 226.

[89] John of Travisa's version (made in 1396) of Bartholomæus Anglicus' "De proprietatibus rerum," London, Wynkyn de Worde, 1495, lib. xvi, cap. 43, De Magnete.

The same writer attempts an explanation of the popular fancy that when powdered loadstone was thrown upon coals in the four corners of a house, the inmates would feel as though the house were falling down; of this he says: "That seemynge is by mevynge [moving] that comyth by tornynge of the brayn."[90]

In classical writings the fascination exercised by a very beautiful woman is sometimes likened to the attractive power of the loadstone, as notably by Lucian,[91] who says that if such a woman looks at a man she draws him to her, and leads him whither she will, just as the loadstone draws the iron. To the same idea is probably due the fact that in several languages the name given to the loadstone indicates that its peculiar power was conceived to be a manifestation of the sympathy or love of one mineral substance for another. This is commonly believed to be the sense in which we should understand the French designation *aimant,* namely, as the participle of the verb *aimer,* "to love"; however, some etymologists prefer to derive the word from *adamas,* sometimes used in Low Latin for the loadstone, although properly signifying the diamond. It is certainly worthy of note that in two such dissimilar languages as Sanskrit and Chinese, the influence of this idea appears in the names given to the loadstone. In Sanskrit the word is *chumbaka* or "the kisser," and in Chinese *t' su shi,* or "the loving-stone." Chin T'sang Khi, a Chinese author of the eighth century, wrote that "the loadstone attracts iron just as does a tender mother when she calls her children to her.[92]

---

[90] Bartolomæi Anglici, "De proprietatibus rerum," l. c.

[91] Lucian, Imag. I.

[92] Klaproth, "Lettre à M. le Baron A. de Humboldt sur l'invention de la boussole," Paris, 1834, p. 20.

A rich growth of Mohammedan legends grew up about the exploits of Alexander the Great, a striking example being given on another page, and in one of them it is related that the Greek world-conqueror provided his soldiers with loadstones as a defence against the wiles of the jinns, or evil spirits; the loadstone, as well as magnetized iron, being regarded as a sure defence against enchantments and all the machinations of malignant spirits.[93]

In the East Indies it is said that a king should have a seat of loadstone at his coronation; probably because the magnetic influence of the stone was supposed to attract power, favor, and gifts to the sovereign. But it is not only in the Orient that magnetite is prized for its talismanic powers, for even in some parts of our own land this belief is still prevalent. Large quantities of loadstone are found at Magnet Cove, Arkansas, and it is estimated that from one to three tons are sold annually to the negroes to be used in the Voodoo ceremonies as conjuring stones. The material has been found in land used for farming purposes, and many pieces have been turned up in ploughing for corn; these vary from the size of a pea to masses weighing from ten to twenty pounds. They occur in a reddish-brown, sticky soil; their surface is smooth and brown and they have the appearance of water-worn pebbles. In July, 1887, an interesting case was tried in Macon, Georgia, where a negro woman sued a conjuror to recover five dollars which she had paid him for a piece of loadstone to serve as a charm to bring back her wandering husband. As the market value of this mineral was only seventy-five cents a pound, and the piece

---

[93] From El Kazwini's "Adjâïl el makluquat"; cited in marginal note, vol. i, pp. 310, 311, of El Damu's "Hayat el hayauân," Cairo, 1313 (1895).

was very small, weighing but a few ounces, the judge ordered that the money should be refunded.[94]

Malachite

For some reason not easy to fathom, malachite was considered to be a talisman peculiarly appropriate for children. If a piece of this stone were attached to an infant's cradle, all evil spirits were held aloof and the child slept soundly and peacefully.[95] In some parts of Germany, malachite shared with turquoise the repute of protecting the wearer from danger in falling, and it also gave warning of approaching disaster by breaking into several pieces.[96] This material was well known to the ancient Egyptians, malachite mines having been worked between Suez and Sinai as early as 4000 B.C.

The appropriate design to be engraved upon malachite was the image of the sun. Such a gem became a powerful talisman and protected the wearer from enchantments, from evil spirits, and from the attacks of venomous creatures.[97] The sun, as the source of all light, was generally regarded as the deadly enemy of necromancers, witches, and demons, who delighted in the darkness and feared nothing more than the bright light of day.

Moonstone

The moonstone is believed to bring good fortune and is regarded as a sacred stone in India. It is never displayed for sale there, except on a yellow cloth, as yellow

---

[94] Kunz, "Gems and Precious Stones of North America," New York, 1890, p. 192.

[95] Marbodei, "De lapidibus," Friburgi, 1531, fol. 51; Camilli Leonardi, "Speculum lapidum," Venetia, 1502, fol. xxxviii.

[96] Chiocci, "Museum Calceolarium," Veronæ, 1622, p. 227.

[97] De Boot, "Gemmarum et lapidum historia," Lug. Bat., 1636, p. 264, lib. ii, cap. 113.

is an especially sacred color. As a gift for lovers the moonstone takes a high rank, for it is believed to arouse the tender passion, and to give lovers the power to read in the future the fortune, good or ill, that is in store for them. To gain this knowledge, however, the stone must be placed in the mouth while the moon is full.[98]

Antoine Mizauld[99] tells us of a selenite or moonstone owned by a friend of his, a great traveller. This stone, about the size of the gold piece known as the gold noble, but somewhat thicker, indicated the waxing and waning of the moon by a certain white point or mark which grew larger or smaller as did the moon. Mizauld relates that to convince himself of the truth of this he obtained possession of the stone for one lunar month, during which time he sedulously observed it. The white mark first appeared at the top. It was like a small millet-seed, increasing in size and moving down on the stone, always assuming the form of the moon until, on reaching the middle, it was round like the full moon; then the mark gradually passed up again as the moon diminished. The owner declared that he had "vowed and dedicated this stone to the young king [Edward VI], who was then highly esteemed because he had good judgment in regard to rare and precious things."

Onyx

The onyx, if worn on the neck, was said to cool the ardors of love, and Cardano relates that everywhere in India the stone was worn for this purpose.[100] This belief is closely related to the idea commonly associated with the onyx,—namely, that it provoked discord and separated

[98] Marbodei, "De lapidibus," Friburgi, 1531, fol. 51.

[99] "Les secrets de la Lune," Paris, 1571.

[100] Cardani, "De subtilitate," lib. vii, Basileæ, 1560, p. 464.

OBSIDIAN MASK, FROM THE FAYOUM, EGYPT.

Twelfth Dynasty. Late De Lesseps Collection. Collection of Mrs. Henry Draper. The obsidian is the typical stone of Mexico.

lovers. The close union and yet the strange contrast between the layers of black and white may have suggested this.

Pyrites

Crystals of iron pyrites (pyrite, native iron disulphide) are sometimes used as amulets by the North American Indians, and the belief in their magic power is attested by their presence in the outfit of miscellaneous objects which the medicine-men use in the course of their incantations. Because these gleaming yellow crystals are occasionally mistaken for gold, the name "fool's gold" has been popularly bestowed upon them.[101]

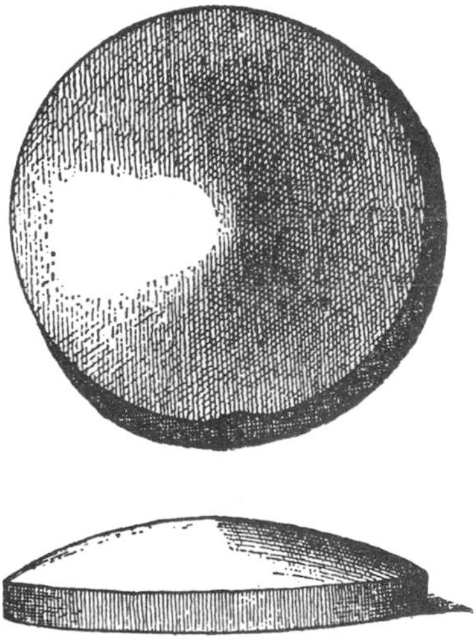

OBSIDIAN MIRROR, FROM OAXACA, MEXICO. NOW IN TROCADÉRO MUSEUM, PARIS.

See "Gems and Precious Stones of North America," by George Frederick Kunz, New York, 1890, p. 299.

Of this material the ancient Mexicans made wonderful mirrors, one side being usually polished flat, while the other side was strongly convex. Frequently this side was curiously carved with some symbolic representation as appears in the case of a pyrite mirror of the Pinard collection in the Trocadéro, Paris.[102]

[101] "Handbook of American Indians North of Mexico," ed. by Frederick Webb Hodge; Smithsonian Inst.; Bur. Am. Ethn., Bull. 30; Washington, 1910, Pt. 2, p. 331.

[102] Kunz, "Gems and Precious Stones of North America," New York, 1890, pp. 299, 300.

Rock-crystal

The popular belief in his time as to the origin of rock-crystal is voiced by St. Jerome, when, using the words of Pliny, although not citing his authority, he says that it was formed by the congelation of water in dark caverns of the mountains, where the temperature was intensely cold, so that, "While a stone to the touch, it seems like

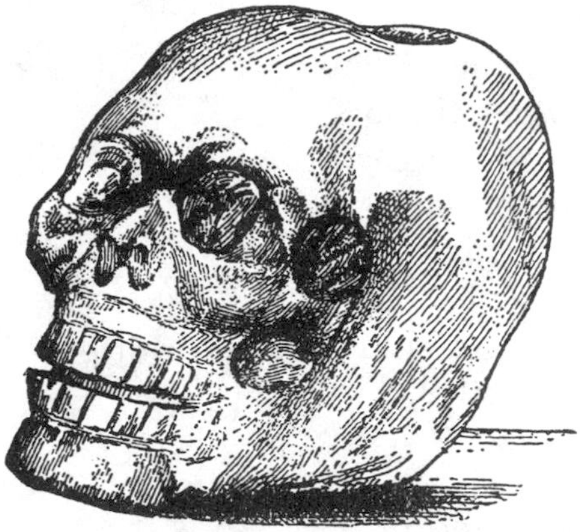

ROCK-CRYSTAL SKULL, ANCIENT MEXICAN

Weighing 475¼ oz. Troy. Now in the British Museum, London. From "Gems and Precious Stones of North America," by George Frederick Kunz, New York, 1890, p. 285.

water to the eye." This belief was evidently due to the fact that rock-crystal was so often found in mountain clefts and caverns. Symbolically, it signified that those within the portals of the Church should keep themselves free from stain and have a pure faith.[103]

The Chinese emperor Wu was devoted to the service of the gods and of the immortal spirits. He built many edifices for religious purposes, and all the doors of these

---

[103] Sancti Eusebii Hieronymi "Opera Omnia," ed. Migne, vol. iv, Parisiis, 1865, col. 545.

buildings were made of white rock-crystal, so that a flood of light poured into the interior. Although the Chinese texts call this material rock-crystal, it is possible that the name was applied to glass when that substance was but recently introduced into China.[104]

Regarding this same "rock-crystal" a humorous tale is related. Muan-fen, a mandarin who had a great terror of draughts, was once received in the palace by one of the Chinese emperors. The doors of the audience chamber were of rock-crystal and were tightly closed, but, because of the transparency of the material, they seemed to be wide open, and the emperor was greatly amused to note that Muan-fen was shivering with cold, although the temperature of the room was quite comfortable.[105]

An exceptionally fine specimen of Aztec work is a skull carved out of rock-crystal. It weighs 475¼ ounces Troy, and measures 8¼ inches in width.

Ruby

The ruby has many names in Sanskrit, some of them clearly showing that it was more valued as a gem by the Hindus than any other. For instance, it is called *ratnaraj,* "king of precious stones," and *ratnanâyaka,* "leader of precious stones;" another name, applied to a particular shade of ruby is *padmarâga,* "red as the lotus."[106]

The glowing hue of the ruby suggested the idea that an inextinguishable flame burned in this stone. From this fancy came the assertion that the inner fire could not

---

[104] Pfizmeier, "Beiträge zur Geschichte der Edelsteinen und des Goldes," Sitzungsbericht d. phil. hist. Kl., Wien, vol. lviii, 1868, p. 200.

[105] Pfizmeier, l. c., p. 201.

[106] Garbe, "Die indische Mineralien; Naharari's Rajanighantu, Varga XIII, Leipzig, 1882, p. 70.

be hidden, as it would shine through the clothing or through any material that might be wrapped around the stone.[107] If cast into the water the ruby communicated its heat to the liquid, causing it to boil. The dark and the star rubies were called "male" stones, the others, more especially, however, those of lighter hue, being considered as "female" stones. All varieties served to preserve the bodily and mental health of the wearer, for they removed evil thoughts, controlled amorous desires, dissipated pestilential vapors, and reconciled disputes.[108]

In the "Lapidaire" of Philippe de Valois, it is said that "the books tell us the beautiful clear and fine ruby is the lord of stones; it is the gem of gems, and surpasses all other precious stones in virtue." In the time of Marbodus (end of the eleventh century A.D.) the same proud place was assigned to the sapphire. The ruby is spoken of in similar terms in the "Lapidaire en Vers," where it is called "the most precious of the twelve stones God created when He created all creatures" By Christ's command the ruby was placed on Aaron's neck, "the ruby, called the lord of gems; the highly prized, the dearly loved ruby, so fair with its gay color." [109]

As with diamonds, rubies also were divided by the Hindus into four castes. The true Oriental ruby was a Brahmin; the rubicelle, a Kshatriya; the spinel, a Vaisya, and lastly, the balas-ruby, a Sudra. The possession of a *padmarâga,* or Brahmin ruby, conferred perfect safety upon the owner, and as long as he owned this precious stone he could dwell without fear in the midst of enemies

---

[107] Epiphanii, "De XII gemmis," Tiguri, 1565, fol. 5.

[108] Camilli Leonardi, "Speculum lapidum," Venetia, 1502, fol. xxvi.

[109] Pannier, "Les lapidaires français," Paris, 1882, pp. 246, 264, 295. Cited in Schofield, "The Pearl," Pub. of Mod. Lang. Asso. of Am., vol. xxiv, Pt. 4, p. 599.

and was shielded from adverse fortune. However, great care had to be taken to preserve this ruby of the first class from contact with inferior specimens, as its virtue would thereby be contaminated, and its power for good correspondingly diminished.[110]

The many talismanic virtues of the ruby are noted in the fourteenth century treatise attributed to Sir John Mandeville.[111] Here the fortunate owner of a brilliant ruby is assured that he will live in peace and concord with all men, that neither his land nor his rank will be taken from him, and that he will be preserved from all perils. The stone would also guard his house, his fruit-trees, and his vineyards from injury by tempests. All the good effects were most surely secured if the ruby, set in ring, bracelet, or brooch, were worn on the left side.

The gorgeous ruby, the favorite gem of Burma, where the finest specimens are found, is not only valued for its beauty, but is also believed to confer invulnerability. To attain this end, however, it is not thought to be sufficient to wear these stones in a ring or other piece of jewelry, but the stone must be inserted in the flesh, and thus become, so to speak, a part of its owner's body. Those who in this way bear about with them a ruby, confidently believe that they cannot be wounded by spear, sword, or gun.[112] As it is often remarked that the most daring and reckless soldiers pass unscathed through all the perils of war, we can understand that this superstition may sometimes appear to be verified.

---

[110] Surindro Mohun Tagore, "Mani Málá," Pt. I, Calcutta, 1879, p. 199.

[111] "Le grand lapidaire de Jean de Mandeville," from the ed. of 1561, ed. by J. S. del Sotto, Vienne, 1862, p. 8.

[112] Taw Sein Ko, communication from his "Burmese Necromancy."

Sapphire

The sapphire is noted as a regal gem by Damigeron, who asserts that kings wore it about their necks as a powerful defence from harm. The stone preserved the wearer from envy and attracted divine favor.[113] For royal use, sapphires were set in bracelets and necklaces, and the sacred character of the stone was attested by the tradition that the Law given to Moses on the Mount was engraved on tablets of sapphire.[114] While we should probably translate here "lapis-lazuli" instead of "sapphire," all such passages were later understood as referring to the true sapphire, which is not found in pieces of the requisite size.

In the twelfth century, the Bishop of Rennes lavishes encomiums upon this beautiful stone. It is quite natural that this writer should lay especial stress upon the use of the sapphire for the adornment of rings, for it was in his time that it was beginning to be regarded as the stone most appropriate for ecclesiastical rings. The sapphire was like the pure sky, and mighty Nature had endowed it with so great a power that it might be called sacred and the gem of gems. Fraud was banished from its presence and necromancers honored it more than any other stone, for it enabled them to hear and to understand the obscurest oracles.[115]

The traditional virtue of the sapphire as an antidote against poison is noted by Bartolomæus Anglicus, who claims to have seen a test of its power, somewhat similar to that recorded by Ahmed Teifashi of the emerald. In

---

[113] Pitra, "Specilegium Solesmense," Parisiis, 1855, vol. iii, p. 328.

[114] Epiphanii, "De XII gemmis," Tiguri, 1565, fol. 6.

[115] Marbodei, "De lapidibus," Friburgi, 1531, fols. 46, 47.

John of Trevisa's version this passage reads as follows:[116]

His vertue is contrary to venym, and quencheth it every deale. And yf you put an attercoppe[117] in a boxe and hold a very saphyre of Inde at the mouth of the boxe ony whyle, by vertue thereof the attercoppe is overcome & dyeth as it were sodenly, as Dyasc. sayth [pseudo Dioscorides]. And this same I have assayed oft in many and dyvers places. His vertue kepeth and savyth the syght, & clearyth eyen of fylthe wythout ony greyf.

Voicing the general belief that the sapphire was endowed with power to influence spirits, Bartolomæus says that this stone was a great favorite with those who practised necromancy, and he adds: "Also wytches love well this stone, for they wene that they may werke certen wondres by vertue of this stone."[118]

There was in the South Kensington Museum, in London, a splendid sapphire of a peculiar tint. In the daylight it shows a beautiful rich blue color, while by artificial light it has a violet hue and resembles an amethyst. In the eighteenth century this stone was in the collection of Count de Walicki, a Polish nobleman, and Mme. de Genlis used it as the theme of one of her stories, entitled "Le Saphire Merveilleux." Here the sapphire is used as a test of female virtue, the change of color indicating unfaithfulness on the part of the wearer. If the owner of the stone wished to prove that the subject of the test was innocent, she was made to wear the sapphire for three hours of daylight; but in the opposite case the test was so timed that it began in daylight and ended when the

---

[116] Bartolomæi Anglici, "De proprietatibus rerum," London, Wynkyn de Worde, 1495, lib. xvi, cap. 86, De Saphiro.

[117] Old English for spider.

[118] Bartolomæus Anglicus, l. c.

candles or lamps had been lighted. This sapphire, still known as the "Saphire Merveilleux," was for a time in the collection of the Duke of Orleans, who bore the name of Philippe Egalité during the French Revolution.

The star sapphire is that variety of sapphire in which, when the stone is cut and rounded off horizontal with the dome of the crystal, the light is condensed across the three lines of crystalline interference. Three cross lines produce a star which moves as a source of light, or as it is moved from the source of light. Star sapphires very rarely possess the deep blue color of the fine blue sapphire; generally the color is somewhat impure, or of a milky-blue, or else a blue-gray, or sometimes almost a pure white. The blue-gray, gray, and white stones frequently show a much more distinct star, possibly from the fact that there are more inclusions between the layers of the crystals than with the darker blue stones, as it is the set of interference bands that produces the peculiar light. Just as the eye agate was used in some countries to preserve against the Evil Eye, so the moving star is believed by the Cingalese to serve as a protection and a guard against witchcraft of all kinds.

The great Oriental traveller, Sir Richard Francis Burton, had a large star sapphire or asteria, as it was called. He referred to it as his talisman, for it always brought him good horses and prompt attention wherever he went; in fact, it was only in those places where he received proper attention that he would show it to the natives, a favor they greatly appreciated, because the sight of the stone was believed to bring good luck. The fame of Burton's asteria travelled ahead of him, and it served him well as a guiding-star. De Boot, writing in the seventeenth century, states that such a stone was called Siegstein (victory-stone) among the Germans.

The remarkable asteria, known as the "Star of India," in the Morgan-Tiffany Collection in the American Museum of Natural History, has a more or less indefinite historic record of some three centuries, but after its many wanderings it has now found a worthy resting-place in the great Museum. Its weight is 543 carats.[119]

The asteria, or star sapphire, might be called a "Stone of Destiny," as the three cross-bars which traverse it are believed to represent Faith, Hope, and Destiny. As the stone is moved, or the light changes, a living star appears. As a guiding gem, warding off ill omen and the Evil Eye, the star-sapphire is worn for the same reasons as were the *oculus mundi* and the *oculus Beli.* One of the most unique of talismanic stones, it is said to be so potent that it continues to exercise its good influence over the first wearer even when it has passed into other hands.

Sard

The sard was regarded as a protection against incantations and sorcery, and was believed to sharpen the wits of the wearer, rendering him fearless, victorious, and happy.[120] The red hue of this stone was supposed to neutralize the malign influence of the dark oynx, driving away the bad dreams caused by the latter and dispelling the melancholy thoughts it inspired.

---

[119] The subject of the origin, development and reform of the carat-weight has been fully treated by the author in the Trans. of the Soc. of Min. Engineers, 1913, pp. 1225–1245, "The New International Metric Diamond Carat of 200 milligrams."

[120] Marbodei, "De lapidibus," Friburgi, 1531, fol. 50, note of Pictor Villengensis.

Serpentine

The Italian peasants of to-day believe that pebbles of green serpentine afford protection from the bites of venomous creatures. These stones are usually green with streaks or veins of white, and the name was derived from their fancied resemblance to a serpent's skin. In addition to their prophylactic powers, if any one has been bitten by such a creature, the stone, when applied to the wound, is supposed to draw out the poison. Here, as in the case of coral, the hand of man must not have shaped the amulet; it should be in its natural state. As a general rule, however, the belief that the touch of any iron instrument, such as the tool of the gem-cutter, destroys the magic efficacy of the substance, is less firmly held in regard to stones than in reference to coral.[121]

Topaz

See Chrysolite.

Turquoise

While there was a tendency to attribute the virtues originally ascribed to one particular stone to others of the same or similar color and appearance, certain stones were regarded as possessing special virtues not commonly attributed to others. A notable instance of this is the quality supposed to inhere in the turquoise. This stone was known in Egypt from a very early period and is later described by Pliny under the name of *callais*. For Pliny, and for all those who derived their information from him or from the sources he used, the turquoise only participated in the virtues assigned to all blue or

---

[121] Bellucci, "Il feticismo primitivo in Italia," Perugia, 1907, pp. 25, 26.

greenish-blue stones; but from the thirteenth century, when the name turquoise was first employed, we read that the stone possessed the power to protect the wearer from injury by falling, more especially from horseback; later, this was extended to cover falls from a building or over a precipice. A fourteenth century authority, the "Lapidaire" of Sir John Mandeville, states that the turquoise protected horses from the ill-effects resulting from drinking cold water when overheated by exertion, and it is said that the Turks often attached these stones to the bridles and frontlets of their horses as amulets. They are also so used in Samarcand and Persia. We might therefore be justified in supposing that the turquoise was originally used in the East as a "horse-amulet," and the belief in its power to protect from falls may have arisen from the idea that it rendered the horse more sure-footed and enduring. As the horse was often regarded as a symbol of the sun in its rapid course through the blue heavens, the celestial hue of the turquoise may have caused it to be associated in some way with the horse. We can only hazard this as a plausible conjecture.

Probably the earliest notice of the peculiar superstition in regard to the turquoise—namely, that it preserves the wearer from injury in case of falling—is contained in Volmar's thirteenth century "Steinbuch," where we read:

> Whoever owns the true turquoise set in gold will not injure any of his limbs when he falls, whether he be riding or walking, so long as he has the stone with him.[122]

Anselmus de Boot, court physician of Emperor Rudolph II, tells a story of a turquoise that, after being thirty years in the possession of a Spaniard, was offered

---

[122] Volmar, "Steinbuch," ed. by Hans Lambel, Heilbronn, 1877, p. 19.

for sale with the rest of the owner's property. Every one was amazed to find it had entirely lost its color; nevertheless De Boot's father bought it for a trifling sum. On his return home, however, ashamed to wear so mean-looking a gem, he gave it to his son, saying, "Son, as the virtues of the turquoise are said to exist only when the stone has been given, I will try its efficacy by bestowing it upon thee." Little appreciating the gift, the recipient had his arms engraved on it as though it had been only a common agate and wore it as a signet. He had scarcely worn it a month, however, before it resumed its pristine beauty and daily seemed to increase in splendor. Could we accept this statement as true we would have here an altogether unique instance of the recovery by a turquoise of the blue color it had lost.

Not long after, the powers of De Boot's turquoise were put to the test. As he was returning to Bohemia from Padua, where he had just taken his degree, he was forced to traverse a narrow and dangerous road at night. Suddenly his horse stumbled and threw him heavily to the ground, but, strange to say, neither horse nor rider was injured by the fall. Next morning, while washing his hands, De Boot remarked that about a quarter of his turquoise had broken away. Nevertheless the stone did not lose its virtue. Some time afterward, when the wearer was lifting a very heavy pole, he felt all at once a sharp pain in his side and heard his ribs crack, so that he feared he had injured himself seriously. However, it turned out that he had not broken any bones but had simply strained himself; but, on looking at his turquoise, he saw that it had again broken into two pieces.[123]

---

[123] De Boot, "Gemmarum et lapidum historia," Lug. Bat., 1636, pp. 266–268.

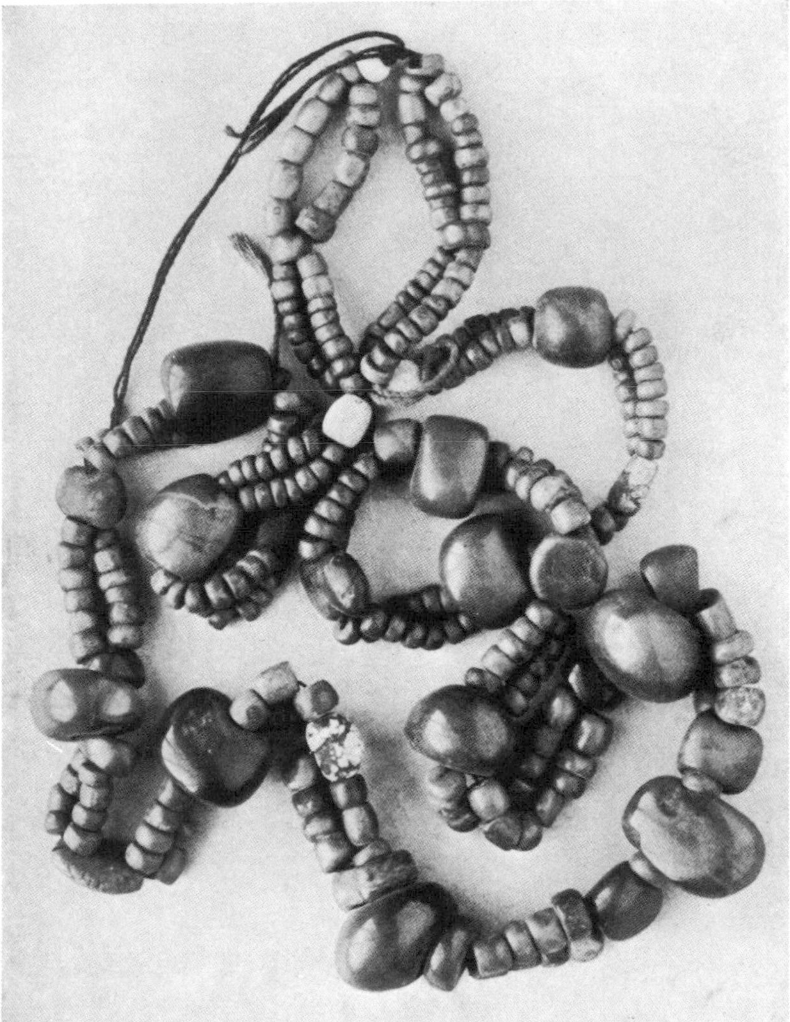

TURQUOISE NECKLACE, THIBET.
Field Museum, Chicago.

A singular virtue ascribed to the turquoise was that of striking the hour correctly, if the stone were suspended from a thread held between the thumb and index-finger in such a way that a slight vibration would make the stone strike against the side of a glass. De Boot states that he made the experiment successfully, but he very sensibly explains the apparent wonder by the unconscious effect of the mind on the body. The expectation that the stone was going to strike a certain number of times induced an involuntary movement of the hand.[124]

The turquoise seems to have been worn almost exclusively by men at the beginning of the seventeenth century, for De Boot, writing in 1609, said that it was so highly regarded by men that no man considered his hand to be well adorned unless he wore a fine turquoise. Women, however, rarely wore this gem.[125] This custom was much in vogue among the Englishmen who travelled in the Orient, until a score of years ago.

The Persians fully appreciate the beauty and power of this, their national stone, and they have a saying that to escape evil and attain good fortune one must see the reflection of the new moon either on the face of a friend, on a copy of the Koran, or on a turquoise,[126] thus ranking this stone with two most precious things, a friend and the source and warrant of religion. Possibly we should take this proverbial saying to indicate that whoever has a true friend, a copy of the sacred volume or a turquoise will be preserved from harm.

The turquoise of the Los Cerillos mines in New Mexico is rudely extracted by building large fires at the

---

[124] De Boot, "Gemmarum et lapidum historia," Lug. Bat., 1636, pp. 169, 170.

[125] De Boot, l. c., p. 270.

[126] Hendley, "Indian Jewelry," London, 1909, p. 158.

base of the rock until it becomes heated, when cold water is dashed over it, the sharp change of temperature splitting up the rock. Some of the fragmentary material thus secured is worked up in the region into heart-shaped ornaments, or amulets, locally called malacates. The religious veneration with which many of the New Mexico Indians still regard the turquoise was noted by Major Hyde, when he explored the region in 1880, for some Pueblo Indians from Santo Domingo, New Mexico, expressed strong disapproval of his action in extracting turquoise from the old mine, as they looked upon this as a sacred stone which should not pass into the possession of those whose Saviour was not a Montezuma.[127]

The ruins called Los Muertos, situated nine miles from Tempe, Arizona, have furnished a peculiarly interesting amulet or fetish of Zuñi workmanship. This is a seashell which has been coated with black pitch, in which are encrusted turquoises and garnets so disposed in mosaic as to represent clearly enough the figure of a toad, the sacred emblem of the Zuñis.[128]

The sacred character with which this stone was invested is shown by the wealth of turquoise ornaments found in some of the burials, notably in those of Pueblo Bonito, unearthed by Mr. George H. Pepper in 1896.[129] This is one of the Chaco Cañon groups of ruins, in the northwestern part of New Mexico. In one case nearly nine thousand beads and pendants of turquoise were found on or about a single skeleton. There was abun-

[127] Kunz, "Gems and Precious Stones of North America," New York, 1890, pp. 61, 62, pl. opposite p. 56.

[128] Kunz, l. c., see pl. 2, fig. A.

[129] Pepper, "The Exploration of a Burial-room in Pueblo Bonito, New Mexico," Putnam Anniversary Volume, New York, 1909, pp. 196–252.

dant evidence in the special care bestowed upon the burial that the deceased must have been a man of high rank, and the condition of the skull plainly indicated that he had met a violent death. The 1980 beads found on the breast of the skeleton are believed to have been strung as a necklace, and the position of other masses of these beads renders it probable that they had been used for bracelets or anklets, the strings having decayed and disappeared in the course of time. The most interesting of the turquoise objects are, however, the pendants worked into various forms designed to favor the entrance of some guardian spirit into the stone. In this single burial were found pendants shaped more or less roughly into the forms of a rabbit, a bird, an insect (?), a human foot and a shoe. Around another burial in the same chamber were strewn nearly six thousand turquoise beads and pendants.[130] In all 24,932 beads were found in these burials.

Another very interesting object from Pueblo Bonito, and one having probably a special ceremonial use and value, is a turquoise basket,—that is to say, a cylindrical basket three inches in diameter and six inches long, originally made of slender splints with a coating of gum in which 1214 small pieces of turquoise have been set. These are very closely set and form a complete mosaic covering for the object. The legends of the Navahos contain allusions to "turquoise jewel baskets," and Mr. Pepper raises the question whether or no this can refer to those made by the Pueblo Indians.[131]

The Apache name for the turquoise is *duklij*, which

[130] Pepper, "The Exploration of a Burial-room in Pueblo Bonito, New Mexico," pp. 223, 224.

[131] Pepper, l. c., p. 227.

signifies either a green or a blue stone, no distinction being made between the two colors. This stone is highly prized for its talismanic virtues. Indeed the possession of a turquoise was indispensable for a medicine-man, as without it he would not receive proper recognition. That some of the powers of the thunder-stone were ascribed to the turquoise by the tribes appears from the fancy that a man who could go to the end of a rainbow after a storm and search in the damp earth would find a turquoise. One of its supposed powers was to aid the warrior or hunter by assuring the accuracy of his aim, for if a turquoise were affixed to a gun or bow the shot sped from the weapon would go straight to the mark.[132]

A lady prominent in the London world is said to possess the power of restoring to their pristine hue turquoises that have grown pale. According to report, this lady is often called upon to use her peculiar gift by friends whose turquoises have faded.[133] While the improvement supposed to be noted may be more imaginary than real in many cases, there is little doubt that this stone is exceptionally sensitive to the action of certain emanations, and may, at times, be influenced by the wearer's general state of health. The writer believes that a turquoise, like an egg, can never be restored to its original state.

---

[132] Burke, "The Medicine-men of the Apache," Ninth Annual Report of the Bureau of Ethnology, 1887–1888, Washington, 1892, p. 589.

[133] Fernie, "Precious Stones for Curative Use," Bristol, 1907, p. 269.

IV

On the Use of Engraved and Carved Gems as Talismans

THE virtue believed to be inherent in precious stones was thought to gain an added potency when the stone was engraved with some symbol or figure possessing a special sacredness, or denoting and typifying a special quality. This presupposes a considerable development of civilization, since the art of engraving on precious stones offers many mechanical difficulties and thus requires a high degree of artistic and mechanical skill. It is true that the earliest engraved stones, the Babylonian cylinders and the Egyptian scarabs, were both designed to serve an eminently practical purpose as well, namely, that of seals; but in a great number of instances these primitive seals were looked upon as endowed with talismanic power, and were worn on the person as talismans.

The scarab, so highly favored by the Egyptians as an ornamental form, is a representation of the *scarabæus sacer,* the typical genus of the family *Scarabæidæ.* They are usually black, but occasionally show a fine play of metallic colors. After gathering up a clump of dung for the reception of the eggs, the insect rolls this along, using the hind legs to propel it, until the material, at first soft and of irregular form, becomes hardened and almost perfectly round. A curious symbolism induced the Egyptians to find in this beetle an emblem of the world of fatherhood and of man. The round ball wherein the eggs were deposited typified the world, and, as the Egyptians thought that the scarabæi were all males, they especially signified the male principle in generation, becoming types of fatherhood and man. At the same time, as only full-grown beetles were observed, it was believed these creatures represented a regeneration or reincarnation, since it was not realized that the eggs or larval and pupa stages had anything to do with the generation of the beetle. Thus the scarab was used as a symbol of immortality.

While, however, this was the popular view, it seems unlikely that such close observers as were the more cultured Egyptians should have been entirely unfamiliar with the real genesis of the *Scarabæus sacer;* but, in this case also, there would have been no difficulty in finding it emblematic of immortality in the various stages through which it passed. The larval stage might well signify the mortal life; the pupa stage, the intermediate period represented by the mummy, with which the soul was conceived to be vaguely connected, in spite of its wanderings through the nether-world; and, lastly, the fully developed beetle could be regarded as a type of the rebirth into everlasting life, when the purified and perfected soul again animated the original and transfigured form in a mysterious resurrection.

Scarabs are frequently engraved with the hieroglyph  (*anch,* "life") and  (*ha,* "increase of power"). The emblem of stability  (*tet*) is also employed, as well as many others. In addition to these simple symbols, many scarabs bear legends supposed to render them exceptionally luck-bringing. The following are characteristic specimens.[1]

[1] From "The Sacred Beetle," by John Ward, London, 1902, Plate VIII, Nos. 46, 58, 89, 275, 276, 446.

 *maat ankh neb,* "Lord of Truth and Life."

 "abounding in graces" (very deeply cut as a seal).

"May thy name be established; mayst thou have a son."

 (within ornamental border), "good stability."

 *ikht neb nefer,* "All good things."

(Inlaid). "A good day" (a holiday).

"A mother is a truly good thing" or "Truth is a good Mother."

The scarab, for the Egyptians a type of the rising sun and hence of the renewal of life after death, was copied by the Phœnicians from the Egyptian types and modified in various ways to suit the religious fancies of the various lands to which they bore the products of their art. Much of the original significance of this symbol must have been lost; probably in many cases little was left but a vague idea that an amulet of this form would bring good luck to the wearer and guard from harm.

Funeral scarabs were often made of jasper, amethyst, lapis-lazuli, ruby, or carnelian, with the names of gods, kings, priests, officials, or private persons engraved on the base; occasionally monograms or floral devices were engraved. Sometimes the base of the scarab was heart-shaped and at others the scarab was combined with the "utat," or eye of Horus, and also with the frog, typifying revivification. Set in rings they were placed on the fingers of the dead, or else, wrapped in linen bandages, they rested on the heart of the deceased, a type of the sun which rose each day to renewed life. They were symbols of the resurrection of the body.[2]

[2] Budge, "The Mummy," Cambridge, 1894, pp. 234–235.

Some of the Egyptian scarabs were evidently used as talismanic gifts from one friend to another. Two such scarabs are in the collection of the Metropolitan Museum of Art in New York. One bears the inscription "May Ra grant you a happy New Year," the text of the other reading as follows: "May your name be established, may you have a son," and "May your house flourish every day." It is a curious fact that the modern greeting "Happy New Year" was current in Egypt probably three thousand years ago.[3]

On the Egyptian inscribed scarabs used as signets were engraved many of the symbols to which a talismanic virtue was attributed. The uræus serpent, signifying death, is sometimes associated with the knot, the so-called *ankh* symbol, denoting life. Often the hieroglyph for *nub*, gold, appears; this symbol is a necklace with pendant beads, showing that gold beads must have been known in Egypt in the early days when the hieroglyph for gold was first used. All these symbolic figures, of which a great number occur, served to impart to the signet a sacred and auspicious quality which communicated itself to the wearer, and even to the impression made by the seal, this in its turn acquiring a certain magic force. Few of us would be willing to confess to a belief in the innate power of any symbol, but the suggestive power of a symbol is as real to-day as it ever was. Any object that evokes a high thought or serves to emphasize a profound conviction really possesses a kind of magical quality, since it is capable of causing an effect out of all proportion to its intrinsic worth or its material quality.

Many scarabs and signets exist made of the artificial

[3] The Metropolitan Museum of Art; the Murch Collection of Egyptian antiquities; supplement to the Bulletin of the Met. Mus. of Art, January, 1910.

cyanus, which was an imitation lapis-lazuli made in Egypt. This was an alkaline silicate, colored a deep blue with carbonate of copper. Often a wonderful translucent or opaque blue glass was used. The genuine lapis-lazuli was also used to a considerable extent for scarabs and cylinders, in Egypt and Assyria, and gems were also cut from it in imperial Roman times.[4] A notable instance of the use of lapis-lazuli in ancient Egypt was as the material for the image of Truth (*Ma*), which the Egyptian chief-justice wore on his neck, suspended from a golden chain.[5]

In Roman times some of the legionaries are said to have worn rings set with scarabs, for the reason that this figure was believed to impart great courage and vigor to the wearer.[6]

The Egyptian amulets of the earliest period, up to the XII dynasty (circa 2000 B.C.), differ considerably from those made and worn after the beginning of the XVIII dynasty (1580 B.C.). Those of the earlier period are not numerous and present but a small number of types, animal forms or the heads of animals constituting the most favored models. The precious stone materials are principally carnelian, beryl, and amethyst. After the close of the so-called Hyksos period, the age during which foreign kings ruled over Egypt, came the brilliant revival and development of Egyptian civilization that characterized the XVIII dynasty. Some of the old forms were entirely cast aside while others were greatly modified in form and significance, the animal forms losing much of their fetich-

---

[4] Middleton, "Engraved Gems of Ancient Times," Cambridge, 1891, p. 151.

[5] Diodori Siculi, "Bibliothecæ historicales," ed. Dindorf, Parisiis, 1842, vol. i, p. 65; lib. i, cap. 75.

[6] Æliani, "De animalibus," lib. x, cap. 15.

istic quality and coming to be more and more regarded as images of the multifarious divinities worshipped in this later period. In many cases the animal type was entirely or partially discarded and the amulets figured the conventional types given to the various divinities. However, while some of these images were wholly human, many of them show a human body with an animal head. Various symbolic designs were also favored, one believed to signify the blood of Isis having the form of a knot or tie. A frog fashioned out of lapis-lazuli and having eyes of gold is one of these amulets of the XVIII dynasty or later.

An interesting Egyptian talisman in the Louvre is engraved with a design representing Thothmes II seizing a lion by the tail and raising the animal aloft; at the same time he brandishes in the other hand a club, with which he is about to dash out the lion's brains. The Egyptian word *quen,* "strength," is engraved beneath the design and indicates that the virtue of the talisman was to increase the strength and courage of the wearer, the inscription being a kind of perpetual invocation to the higher powers whose aid was sought.[7]

The children of Israel, when in the desert, were said to have engraved figures on carnelian, "just as seals are engraved."[8] This statement, repeated by many early writers, may perhaps have arisen from an identification of carnelian with the first stone of the breastplate, the *odem,* unquestionably a red stone, and very possibly carnelian. There can be no doubt that this was one of the

---

[7] Hoernes, "Urgeschichte der bildenden Kunst," Wien, 1898, pp. 155, 156.

[8] Konrad v. Megenberg, "Buch der Natur," ed. Pfeiffer, Stuttgart, 1861, p. 448; see also Johannis de Cuba, "Hortus Sanitatis" [Strassburg, 1483], tractatus de lapidibus, cap. xliii.

first stones used for ornamental purposes and for engraving, as a number of specimens have been preserved from early Egyptian times. Because of the cooling and calming effect exercised by carnelian upon the blood, if worn on the neck or on the finger, it was believed to still all angry passions.[9]

A class of amulets even older than the Egyptian scarabs is represented by the engraved Assyrio-Babylonian cylinders. There has been much discussion among scholars as to the original purpose for which these cylinders were made, some holding that they were exclusively employed as seals or signets, while others incline to the belief that many of them were intended only for use as amulets or talismans.

These cylinders are perforated and were worn suspended from the neck or wrist, as is most frequently the case with talismans, and the engraved designs often represent religious or mythological subjects, the accompanying inscription merely consisting of the names of the gods. Cylinders of this type could not have been used as personal signets, and it is quite possible that Dr. Wiedemann is right in supposing that their imprint on a document was considered to impart a certain mystic sanction to the agreement, and render the divinities or spirits accountable for the fulfilment of the contract.[10]

The oldest known form of seal is the cylinder. Babylonian and Assyrian cylinder-seals are known of a date as early as 4000 B.C. From the earliest period until 2500 B.C. they were made of black or green serpentine, conglomerate, diorite, and frequently of the central core of

---

[9] Marbodei, "De lapidibus," Friburgi, 1531, fol. 19.

[10] Fischer and Wiedemann, "Ueber Babylonische 'Talismane' aus dem hist. Mus. im steierisch-landschaftl. Joanneum zu Graz," Stuttgart, 1881, p. 9.

a large conch shell from the Persian Gulf. From 2500 B.C. to 500 B.C the cylindrical form was prevalent, and the materials include a brick-red ferruginous quartz, red hematite (an iron ore), and chalcedony, a beautiful variety of the last-named stone known as sapphirine being sometimes used. On the cylinders produced from 4000 B.C. to 2500 B.C. the designs most frequently represent animal forms; on those dating from 2500 B.C. to 500 B.C. are generally inscribed five or six rows of cuneiform characters. Up to the last-named date the work was all done by the sapphire point, and not by the wheel, and it is not until the fifth century B.C. that wheel work is apparent in any Babylonian or Assyrian stone-engraving. In the course of the sixth century B.C. the cylindrical seals became less frequent, and the tall cone-like seals came into use.[11]

A new type makes its appearance about the fifth or sixth century B.C., namely, the scaraboid seal introduced from Egypt. From the third century B.C. until the second or third century A.D., the seals became lower and flatter, and the perforation larger, until they sometimes assumed the form of rings; later the ring form becomes general. They are usually hollowed a little in the middle, which gives them the shape and size of the lower short joints of a reed; indeed, it has been suggested that the original seal was rudely patterned after a reed joint. The materials used for these cylinders include lapis-lazuli, very freely used and probably from the Persian mines, jasper, rock-crystals, chalcedony, carnelian, agate, jade, etc.; a hard, black variety of serpentine is perhaps the most common of all the materials used for this purpose.[12]

---

[11] See Ward, "The Seal Cylinders of Western Asia," Carnegie Institution Pub., Washington, D. C., 1910, pp. 1–5.

[12] Ward, l. c., p. 5 and pp. 5–8.

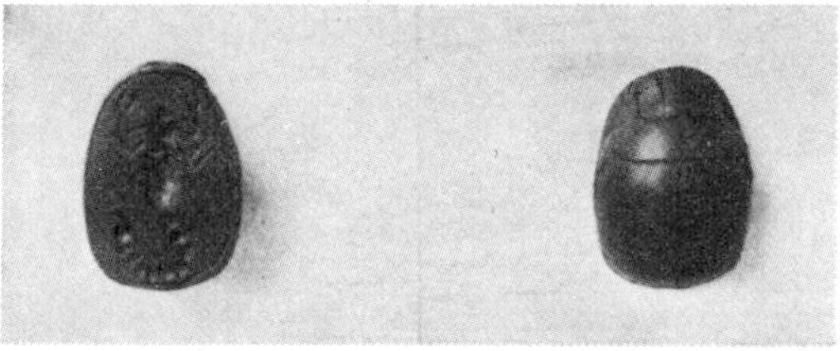

PHOENICIAN SCARAB, WITH ENGRAVED SCORPION. (See page 115.)

ANCIENT BABYLONIAN CYLINDER IMPRESSION, BEARING FIGURES OF THE GOD NEBO AND A WORSHIPPER, AND SYMBOLS OF SUN AND MOON.

From Fischer and Wiedemann "Ueber Babylonische Talismane," Stuttgart, 1881, Pl. 1, fig. 3

A SMALL JADE CELT ENGRAVED WITH GNOSTIC INSCRIPTIONS IN THE FOURTH CENTURY.

On one side are seven lines of characters, principally consisting of the seven Greek vowels used to denote the Ineffable Name. On the reverse is cut a laurel branch with 18 leaves, enclosed within each of which are characters expressing the name of one of the personifications of Gnostic theosophy. Brought from Egypt and deposited by its possessor, General Lefroy, in the Rotunda at Woolwich. Now in the Egyptian Department of the British Museum. (See page 129.)

A good example of these talismanic cylinders shows the figure of the god Nebo, seated on a throne and holding a ring in his left hand. Before him are two altars, over which appear, respectively, a star and the crescent moon; in front of the god is the figure of a man in an attitude of adoration. Borsippa, where the cylinder was found, was the special seat of the worship of Nebo, whose name appears in those of the kings Nebuchadnezzar, Nebopalasser, and Nabonaid. Regarded as the inventor of writing and as the god of learning, Nebo was the lord of the planet Mercury, and this shows a close connection between Babylonian and Græco-Roman ideas in reference to the god associated with that planet. Nebo was also believed to be the orderer of times and seasons, and this character is indicated by the star and the crescent.[13]

The Cretan peasants of to-day set a high value upon certain very ancient seals—dating perhaps from as early as 2500 B.C.—which they find buried in the soil. These seals are inscribed with symbols supposed to represent the prehistoric Cretan form of writing. Of course these inscriptions, which have not yet been deciphered by archæologists, are utterly incomprehensible for the peasants, but they undoubtedly serve to render the stones objects of mystery. The peasants call them *galopetræ,* or "milk-stones," and they are supposed to promote the secretion of milk, as was the case with the galactite.[14] The careful preservation of these so-called *galopetræ* by Cretan women has served the purpose of archæological research, as otherwise so large a supply of these very interesting seals would not now be available.

---

[13] Fischer and Wiedemann, "Ueber Babylonische Talismane," Stuttgart, 1881, p. 11. See Pl. I, fig. 3.

[14] A. Evans, in "Journal of Hellenic Studies," vol. xiv (1893), p. 270.

Many engraved stones of the Roman imperial period bore the figures of Serapis and of Isis, the former signifying Time and the latter Earth. On other stones the symbols of the zodiacal signs appear, referring to the natal constellation of the wearer. The astrologers, who derived their lore from the Orient, were consulted by all classes of the Roman people, and it is therefore very

1. ENGRAVED HELIOTROPE.

Head of Serapis surrounded by the twelve Zodiacal symbols. From Gori's "Thesaurus Gemmarum Antiquarum Astriferarum," Florence, 1750. Vol. i, Pl. XVII.

2. ENGRAVED RED JASPER.

Head of Medusa, Museum Cl. Passerii.

natural that the signet, or the ring worn as an amulet, should frequently have been engraved with astrological symbols. These designs were usually engraved on onyxes, carnelians, and similar stones, in Greek and Roman times; but occasionally the emerald was used in this way, and more rarely the ruby or the sapphire. Here the costliness of the material was probably thought to enhance the value of the amulet. The emerald ring of Polycrates must have possessed some other than a purely artistic value in his eyes, when it could be regarded by him as the most precious of his possessions.

In Roman times the image of Alexander the Great was looked upon as possessing magic virtues, and it is related that when Cornelius Macer gave a splendid banquet in the temple of Hercules, the chief ornament of the table was an amber cup, in the midst of which was a portrait of Alexander, and around this his whole history figured in small, finely engraved representations. From this cup Macer drank to the health of the pontifex and then ordered that it should be passed around among the guests, so that each one might gaze upon the image of the great man. Pollio, relating this, states that it was a common belief that everything happened fortunately for those who bore with them Alexander's portrait executed in gold or silver.[15] Indeed, even among Christians coins of Alexander were in great favor as amulets, and the stern John Chrysostom sharply rebukes those who wore bronze coins of this monarch attached to their heads and their feet.[16]

Nowhere in the world was the use of amulets so common as in Alexandria, especially in the first centuries of our era, and the types produced here were scattered far and wide throughout the Roman world. Amulets made from various colored stones had been used for religious purposes in Egypt from the very earliest period of its history, so that the custom was deeply rooted in that land. When, therefore, Alexandria was founded in

---

[15] Trebelii Pollionis, De XXX tyrannis, Lipsiæ, p. 295.

[16] Ad illum. catech., Hom. II, 5.

the fourth century B.C., and became a great commercial centre, attracting men of all races and all religions, it is not surprising that the population eagerly adopted the various amulets used by the adherents of the different religions. The result was a combining and confusion of many different types. With the rapid rise and growth of the Christian religion, a new element was introduced. Unquestionably the leading Christian teachers were strongly opposed to such superstitious practices, but the rank and file of the faithful clung to their old fancies.

In the second century the Gnostic heresy gave a new impulse to the fabrication of amulets. This strange eclecticism, resulting from an interweaving of pagan and Christian ideas, with its complicated symbolism, much of which is almost incomprehensible, found expression in the creation of the most bizarre types of amulets, and the magic virtues of the curious designs was enhanced by inscriptions purposely obscure. The incomprehensible always seems to have a mysterious charm for those devoted to the magic arts, and the adepts willingly catered to this taste, so that we can often only guess at the signification of the words and names engraved upon the Gnostic or Basilidian gems. So widespread was their use throughout the Roman Empire, that there were factories entirely devoted to the production of these objects.[17]

Regarding the sacred name Abrasax, which was inscribed on so many Gnostic gems, we read in St. Augustine's treatise De hæres., vi, "Basilides asserted that there were 365 heavens; it was for this reason that he regarded the name Abrasax as sacred and venerable."

[17] Krause, "Pyrgoteles," Halle, 1856, pp. 197–8.

1

2

3

4

5

1. Gnostic gem, heliotrope, with Abraxas god. Gorlaeus Collection. From the "Abraxas seu Apistopistus" of Macarius (L'Heureux) Antwerp, 1657, Pl. II.

2. Another type; with seven stars.

3. Gnostic gem. Type of Abraxas god and mystic letters I A W. From Gori's "Thesaurus Gemmarum Antiquarum Astriferarum," Florence, 1750, vol. i, Pl. CLXXXIX.

4. Abraxas gem, jasper, mystic letters I A W. From Gorlaeus, "Cabinet de Pierres Gravées," Paris, 1778.

5. Jasper engraved with the symbol of the Agathodaemon Serpent. The type of amulet noted by Galen as that used by the Egyptian king "Nechepsus" (Necho 610–594 B.C.). Original at one time in the collection of Johann Schinkel. From the "Abraxas seu Apistopistus" of Macarius (L'Heureux) Antwerp, 1657, Pl. XVII. See page 385.

According to the Greek notation the letters comprising this name give that number:

$$
\begin{aligned}
\alpha &= 1\\
\beta &= 2\\
\rho &= 100\\
\alpha &= 1\\
\sigma &= 200\\
\alpha &= 1\\
\xi &= 60\\
\hline
&\ 365
\end{aligned}
$$

It is, however, not unlikely that the 365 days in the solar year are signified; and this enigmatical name might thus be brought into connection with Mithra, the solar divinity, who was worshipped throughout the Persian and Roman empires in the first and second centuries of our era.

A very recondite but ingenious explanation of the Gnostic name Abrasax is given by Harduin in his notes to Pliny's "Natural History."[18] He sees in the first three letters the initials of the three Hebrew words signifying father, son, and spirit (*ab, ben, ruah*), the Triune God; the last four letters are the initials of the Greek words ἀνθρώπους σώζει ἁγίῳ ξύλῳ or "he saves men by the sacred wood" (the cross). This seems rather far-fetched, it must be confessed, and yet to any one familiar with the vagaries of Alexandrine eclecticism, and with the tendency of the time and place to make strange and uncouth combinations of Greek and Hebrew forms, there is nothing inherently improbable in the explanation. Indeed, the Hebrew and Greek words in this composite sen-

---

[18] Caii Plinii Secundi, Naturalis Historia, ed. Harduin, Parisiis, 1741, vol. ii, p. 489.

tence might have been regarded as typifying the union of the Old and New Testaments, and such an acrostic would certainly have been looked upon as possessing a mystic and supernatural power.

Many explanations have been offered as to the origin and significance of the characteristic figure of the Abrasax god engraved on a number of Gnostic amulets. There seems to be no doubt that this figure was invented by Basilides, chief of the Gnostic sect bearing his name, and who flourished in the early part of the second cen-

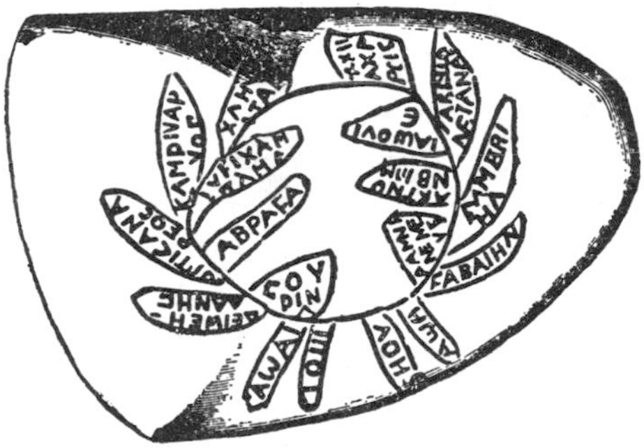

ANTIQUE JADE CELT CONVERTED INTO A GNOSTIC TALISMAN

Enclosed within the outlines of the 18 leaves are as many names of the personifications of Gnostic Theosophy.

tury A.D. While the details of the type as perfected were undoubtedly borrowed from the eclectic symbolism of the Egyptian and western Asiatic world it is almost impossible to conjecture the reasons determining the selection of this particular form.

A jasper engraved with the famous Gnostic symbol was set in the ring worn by Seffrid, Bishop of Chichester (A.D. 1159). This ring was found on the skeleton of the bishop and is now preserved in the treasury of the Cathedral of Chichester. Undoubtedly the curious symbolic figure was given a perfectly orthodox meaning, and, indeed, it was not really a pagan symbol, as the Gnostics were "indifferent Christians," although their system was a fanciful elaboration of the doctrines of the late Alexandrian school of Greek Philosophy and an adaptation of this to the teachings of Christian tradition. In many cases, however, gems with purely pagan designs were worn by Christians, designs such as Isis with the child Horus, which was taken to be the Virgin Mary with the infant Jesus.

A curious amulet, apparently belonging to the Gnostic variety, and intended to bring success to the owner of a racehorse, is now in the collection of the Metropolitan Museum of Art, in New York. The material is green jasper with red spots. On the obverse the horse is figured with the victor's palm and the name Tiberis; on the reverse appears the vulture-headed figure of the Abraxas god and the characters, "ZACTA IAW BAPIA," which have been translated, "Iao the Destroyer and Creator."[19] Possibly this amulet may have been attached to the horse during his races to insure victory, as we know that amulets of this kind were used in this way.

As illustrating the eclectic character of some of the amulets used in the early Christian centuries, we may note one in the Cabinet de Médailles, in Paris. This has upon the obverse the head of Alexander the Great; on the reverse is a she-ass with her foal, and below this a scorpion and the name Jesus Christ. Another amulet of this class, figured by Vettori,[20] also has the head of Alex-

---

[19] King, Catalogue of Engraved Gems, Metropolitan Museum of Art, p. 81, No. 302, 1885.

[20] Dissert. apol. de quibusdam Alexandri Severi numismat., p. 59. Cited in Dictionnaire de l'arch. chrét., vol. i, Pt. II, Paris, 1907, cols. 1789, 1790, where the amulet is figured.

ander on the obverse, while the reverse bears the Greek monogram of the name Christos.

After the third or fourth century of our era the art of gem-engraving seems to have been lost, or at least to have been very seldom practised, and it is noteworthy in the matter that after this period writers who treat of the virtues of engraved gems as talismans rarely, if ever, use the words "if you engrave" such or such a figure on a stone, but write "if you find" such a figure.

The figures engraved on precious stones were supposed to have a greater or lesser degree of efficacy in themselves independent of the virtues peculiar to the stone on which they were engraved, and this efficacy depended largely upon the hour, day, or month during which the work was executed. For the influence of the planet, star, or constellation which was in the ascendant was thought to infuse a subtle essence into the stone while the appropriate image was being engraved. However, to exert the maximum power, the virtue of the image must be of the same character as the virtue inherent in the material, and the gem became less potent when this was not the case. Certain images, those symbolizing the zodiacal signs for instance, were looked upon as possessing such power that their peculiar nature impressed itself even upon stones inherently of different quality; others again were only efficacious when engraved on stones the quality of which was in sympathy with them.[21]

Naturally, many of the ancient gems which had been preserved from Greek and Roman times were recognized as being purely products of art, but in medieval and later times the idea of the magic quality of all engraved gems had become so deeply rooted that in many cases a magical

[21] Camilli Leonardi, Speculum Lapidum, Venetia, 1502.

character was ascribed to them entirely foreign to the intention of the engraver. Great ingenuity was often displayed in seeking and finding some analogy between the supposed significance of the design and the fancied power of the stone itself. Taking the agate as an illustration, Camillo Leonardo says that its many different varieties had as many different virtues, and he finds in this an explanation of the multiplicity of images engraved on the various kinds of agate, without realizing that the true reason was that this material lent itself more readily to artistic treatment than did many others.

The idea that some special design should be engraved upon a given stone became quite general in the early centuries of our era. The emerald, for instance, according to Damigeron, was to be engraved with a scarab, beneath which was to be a standing figure of Isis. The gem, when completed, was to be pierced longitudinally and worn in a brooch. The fortunate owner of this talisman was then to adorn himself and the members of his family, and, a consecration having been pronounced, he was assured that he would see "the glory of the stone granted it by God."[22] Possibly this may have meant that the stone would become luminous.

A list of these symbolic designs is said to have been given in the "Book of Wings," by Ragiel, one of the curious treatises composed about the thirteenth century under the influence of Hebrew and Greco-Roman tradition. Although it owes its origin to the Hebrew "Book of Raziel," it bears little if any likeness to that work. As will be seen in the following items, the fact that the design is on its appropriate stone is always insisted on:

---

[22] Pitra, "Specilegium Solesmense," Parisiis, 1885, vol. iii, pp. 326, 327.

MOSS AGATE MOCHA STONES, HINDOOSTAN.

The beautiful and terrible figure of a dragon. If this is found on a ruby or any other stone of similar nature and virtue, it has the power to augment the goods of this world and makes the wearer joyous and healthy.

The figure of a falcon, if on a topaz, helps to acquire the good-will of kings, princes, and magnates. The image of an astrolabe, if on a sapphire, has power to increase wealth and enables the wearer to predict the future.

The well-formed image of a lion, if engraved on a garnet, will protect and preserve honors and health, cures the wearer of all diseases, brings him honors, and guards him from all perils in travelling.

An ass, if represented on a chrysolite, will give power to prognosticate and predict the future.

The figure of a ram or of a bearded man, on a sapphire, has the power to cure and preserve from many infirmities as well as to free from poison and from all demons. This is a royal image; it confers dignities and honors and exalts the wearer.

A frog, engraved on a beryl, will have the power to reconcile enemies and produce friendship where there was discord.

A camel's head or two goats among myrtles, if on an onyx, has the power to convoke, assemble, and constrain demons; if any one wears it, he will see terrible visions in sleep.

A vulture, if on a chrysolite, has the power to constrain demons and the winds. It controls demons and prevents them from coming together in the place where the gem may be; it also guards against their importunities. The demons obey the wearer.

A bat, represented on a heliotrope or bloodstone, gives the wearer power over demons and helps incantations.

A griffin, imaged on a crystal, produces abundance of milk.

A man richly dressed and with a beautiful object in his hand, engraved on a carnelian, checks the flow of blood and confers honors.

A lion or an archer, on a jasper, gives help against poison and cures from fever.

A man in armor, with bow and arrow, on an iris stone, protects from evil both the wearer and the place where it may be.

A man with a sword in his hand, on a carnelian, preserves the place where it may be from lightning and tempest, and guards the wearer from vices and enchantments.

A bull engraved on a prase is said to give aid against evil spells and to procure the favor of magistrates.

A hoopoo with a tarragon herb before it, represented on a beryl, confers the power to invoke water-spirits and to converse with them, as well as to call up the mighty dead and to obtain answers to questions addressed to them.

A swallow, on a celonite, establishes and preserves peace and concord among men.

A man with his right hand raised aloft, if engraved on a chalcedony, gives success in lawsuits, renders the wearer healthy, gives him safety in his travels and preserves him from all evil chances.

The names of God, on a *ceraunia* stone, have the power to preserve the place where the stone may be from tempests; they also give to the wearer victory over his enemies.

A bear, if engraved on an amethyst, has the virtue of putting demons to flight and defends and preserves the wearer from drunkenness.

A man in armor, graven on a magnet, or loadstone, has the power to aid in incantations and makes the wearer victorious in war.[23]

An Italian manuscript, dating from the fourteenth century, gives the following talismanic gems:

If thou findest a stone on which is graven or figured a man with a goat's head, whoever wears this stone, with God's help, will have great riches and the love of all men and animals.

If a stone be found on which is graven or figured an armed man or the draped figure of a virgin, bound with laurel and having a laurel branch in her hand, this stone is sacred and frees the wearer from all chañges and haps of fortune.

When thou findest a stone on which is graven the figure of a man holding a scythe in his hand, a stone like this imparts strength and power to the wearer. Every day adds to his strength, courage and boldness.

Hold dear that stone on which thou shalt find figured or cut the moon or the sun, or both together, for it makes the wearer chaste and guards him from lust.

A jewel to be prized is that stone on which is graven or figured a man with wings having beneath his feet a serpent whose head he

[23] Camilli Leonardi, "Speculum Lapidum," Venetia, 1502, ff. lvi–lvii.

holds in his hand. A stone of this kind gives the wearer, by God's help, abundant wealth of knowledge, as well as good health and favor.

Shouldst thou find a stone on which is the figure of a man holding in his right hand a palm branch, this stone, with God's help, renders the wearer victorious in disputes and in battles, and brings him the favor of the great.

Finding the stone called jasper, bearing graven or figured a huntsman, a dog, or a stag, the wearer, with God's help, will have the power to heal one possessed of a devil, or who is insane.

A good stone is that one on which thou shalt find graven or figured a serpent with a raven on its tail. Whoever wears this stone will enjoy high station and be much honored; it also protects from the ill-effects of the heat.[24]

The original meaning of the swastika emblem has been variously explained as a symbol of fire, of the four cardinal points, of water, of the lightning, etc. Still another explanation is given by Hoernes, who inclines to the belief that it is simply a conventionalized representation of the human form, the lower shaft being the two legs joined together, the two horizontal shafts the outstretched arms, and the upper shaft the trunk of the body; the four projections would stand for the feet, the two hands and the head.[25]

The Egyptian crux ansata, the hieroglyphic symbol for "life," and the Phœnician Tau symbol, the "mark" that was to be stamped upon the foreheads of the faithful in Jerusalem (Ezek. ix, 4), and which in Early Christian art was frequently substituted for the usual cross, are both explained by Hoernes in a similar way, and he notes the fact that the swastika symbol does not appear in

[24] From an anonymous Italian treatise in a fourteenth century MS. in the author's collection; fol. 40 verso, 41 recto.

[25] Hoernes, "Urgeschichte der bildenden Kunst," Vienna, 1898, p. 338.

Egyptian or Phœnician art, drawing the inference that all three symbols originated in the same form or figure.[26] To all these symbols were attributed talismanic virtues and they were frequently engraved on precious stones.

The so-called "Monogrammatic Cross" was very freely used in work of the fifth century. This is simply a modification of the monogram formed of the first two

MONOGRAM OF THE NAME OF CHRIST ENGRAVED ON AN ONYX GEM.
From the "Cabinet de Pierres Antiques Gravées," of Gorlaeus, Paris, 1778, Pl. XCV.

letters of the name Christ as written in Greek, a device which first appeared after the time of Constantine the Great (d. 337 A.D.). This monogram usually assumed the following form: ☧, and the "Monogrammatic Cross" was made by changing the position of the Greek X (chi), and making one of its arms serve as the straight stroke of the P (r), thus giving the following form: ⳨.

[26] Hoernes, Urgeschichte der bildenden Kunst," Vienna, 1898, p. 338.

A curious amulet to avert the spell of the Evil Eye is an engraved sard showing an eye in the centre, around which are grouped the attributes of the divinities presiding over the days of the week. Sunday, the dies Solis, is represented by a lion; Monday, the dies Lunæ, by a stag; Tuesday, the dies Martis, by a scorpion; Wednesday, the dies Mercurii, by a dog; Thursday, the dies Jovis, by a thunderbolt; Friday, the dies Veneris, by a snake; and Saturday, the dies Saturni, by an owl.[27] In this way the wearer was protected at all times from the evil influence.

Because of its peculiar markings, some of which suggest the form of an eye, malachite was worn in some parts of Italy (*e.g.*, in Bettona) as an amulet to protect the wearer from the spell of the Evil Eye. Such stones were called "peacock-stones," from their resemblance in color and marking to the peacock's tail. The form of these malachite amulets is usually triangular, and they were mounted in silver. It is curious to note, as a proof of the persistence of superstitions, that in an Etruscan tomb at Chiusi there was found a triangular, perforated piece of glass, each angle terminating in an eye formed of glass of various colors.[28]

On many of the amulets fabricated in Italy for protection against the dreaded jettatura, or spell of the Evil Eye, the cock is figured. His image was supposed in ancient times to assure the protection of the sun-god, and his crowing was regarded as an inarticulate hymn of praise to this deity. He was also a type of dauntless courage. All this contributed to make him a defender of

---

[27] King, "The Gnostics and their Remains," London, 1864, p. 238, figure opp. p. 115.

[28] Catalogue de l'Exposition de la Société d'Anthropologie (Exposition de 1900), p. 286.

the weak, especially of women and children, against the wiles of the spirits of darkness.[29] Rostand, in his "Chantecler," has enlarged this conception, and endows the cock with the proud conviction that it is to his matu-

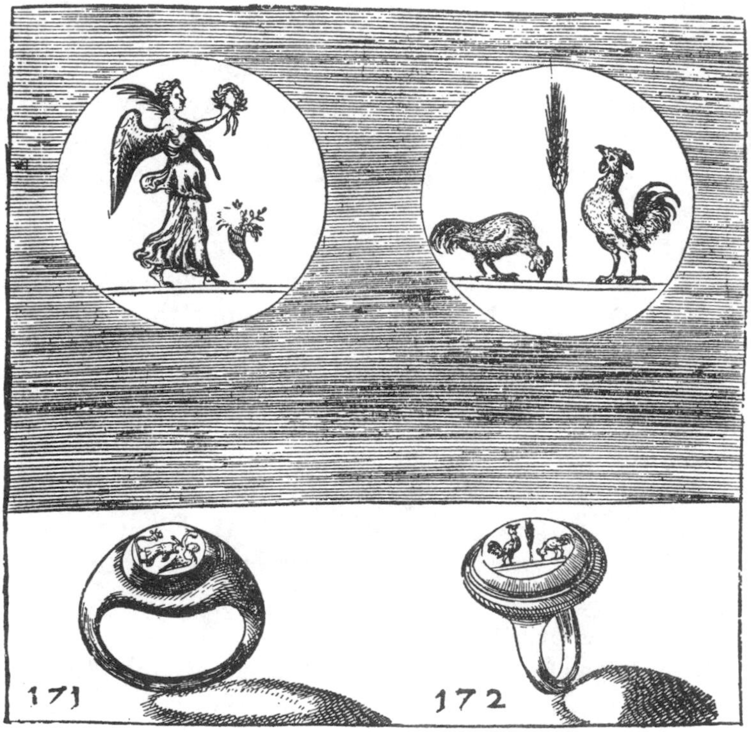

TWO GOLD RINGS SET WITH ENGRAVED ONYX GEMS.

On the right, a Victory; on the left, game-cocks. From the Dactyliotheca, of Gorlaeus, Delft, 1601, Figs. 171, 172.

tinal chant alone that the world owes the daily recurrent phenomenon of the sunrise.

In Palestine the Evil Eye is supposed to be the baleful gift of men who have light-blue eyes, more especially if

---

[29] Elworthy, "The Evil Eye," London, 1895, pp. 353, 354.

they are beardless. Possibly this is the power in which some of our blond and beardless "mashers" repose their trust. As an antidote to the awful influence of these blue-eyed monsters, the Syrian women decorated themselves with blue beads, on the principle *similia similibus curantur.* A maiden with beautiful hair will tie a blue ribbon about it, or wear a blue bead in it, so as to ward off any evil spell cast by the blue eye that might rob her of her fair dower.[30]

It is a well-known fact that many amulets were made in forms suggesting objects offensive to our sense of propriety. These were thought to protect the wearers by denoting the contempt they felt for the evil spirits leagued against them. Some such fancy may have induced the peculiar designs of certain of the jewels alleged to have been pawned in Paris by the ex-Sultan Abdul Hamid for the sum of 1,200,000 francs ($240,000). According to rumor, these pledges must be sold, as the sultan has failed to redeem them, but the designs are so *risqué* that they cannot be offered at public sale; therefore the stones and pearls are to be removed and the gold settings are to be melted and sold as metal.

It is not exclusively characteristic of our commercial and industrial age that the price paid for a work of art should influence the popular estimation of the merits of the work, as appears in an anecdote related by Pliny. An emerald (smaragd), upon which was engraved a figure of Amymone (one of the Danaidæ), having been offered for sale in the Isle of Cyprus, at the price of six golden denarii, Ismenias, a flute-player, gave orders to

[30] Stern, "Medizin, Aborglaube und Geschelechtsleben in der Turkei," Berlin, 1903, vol. i, p. 235.

purchase it. The dealer, however, reduced the price and returned two denarii; upon which Ismenias remarked, "By Hercules! he has done me but a bad turn in this, for the merit of the stone has been greatly impaired by this reduction in price."[31]

A variant of the design directed by Damigeron to be placed on the emerald is recommended in a thirteenth century manuscript, where we read that to fit this stone for use as a talisman, it should be engraved with the form of a scarab, beneath which there should appear a crested paroquet.[32] According to the same manuscript, a jasper should bear the figure of Mars fully armed, or else that of a virgin wearing a flowing robe and bearing a laurel branch. It should then be "consecrated with perpetual consecration." The mythical author Cethel asserts that the owner of a jasper engraved with the sacred symbol of the cross would be preserved from drowning.[33]

A curious quid pro quo appears in a fifteenth century treatise on gems written in French. Here, in a list of engraved gems suitable for use as amulets, we read, "If you find a dromedary engraved on a stone with hair flowing over its shoulders, this stone will bring peace and concord between man and wife." The original Latin text read, "If you find Andromeda on a stone with hair flowing over her shoulders, etc.[34] The translator's art which could turn Andromeda into a dromedary almost equalled that of the enchantress Circe.

---

[31] Plini, "Historia naturalis," lib xxxvi, cap. 3.

[32] Archæologia, vol. xxx, p. 541, London, 1844; MS. Harl. No. 80, folio 105, recto.

[33] Pitra, "Specilegium Solesmense," Parisiis, 1855, vol. iii, p. 336.

[34] De Mély, in La Grande Encyclopédie, vol. xxv, p. 885, art. Pierres précieuses.

A few even of the early writers were disposed to be sceptical as to the virtues ascribed to these engraved gems, and did not hesitate to assert that the Greek and Roman engravers executed their designs for ornamental purposes rather than to fit the gems for use as talismans. This was undoubtedly true in a large number of cases but nevertheless, as we have seen, many engraved talismans were really cut in the early centuries. As the art of gem engraving was not practised in the Middle Ages, some medieval writers suppose that the engraved talismanic gems current in their time were not works of art, but of nature, and Konrad von Megenberg accepting this view, gave it as his opinion that "God granted these stones their beauty and virtue for the help and comfort of the human race," adding that when he hoped to receive help from them he in no wise denied the grace of God.[35]

Damigeron writes of the sard that, if worn by a woman, it is a good and fortunate stone. It should be engraved with a design showing a grape-vine and ivy intertwined.[36]

A celebrated topaz was that noted by George Agricola as being in the possession of a Neapolitan, Hadrianus Gulielmus.[37] It bore, in ancient Roman characters, the terse and pregnant inscription:

Natura deficit,
Fortuna mutatur.
Deus omnia cernit.

---

[35] Konrad von Megenberg, "Buch der Natur," Stuttgart, 1861, p. 469.

[36] Pitra, "Specilegium Solesmense," Parisiis, 1855, vol. iii, p. 335.

[37] Agricola, "De natura fossilum," lib. vi, Basileæ, 1546, p. 291.

This was very freely rendered by Thomas Nicols as follows:[38]

> Nature by frailty doth dayly waste away.
> Fortune is turn'd and changed every day.
> In all, there is an eye know's no decay.
> Jah sees for aye.

There is in the Imperial Academy at Moscow a turquoise two inches in diameter, inscribed with a text from the Koran in letters of gold. This turquoise was formerly worn by the Shah of Persia as an amulet, and it was valued at 5000 rubles by the jeweller from whose hands it came.[39]

It is well known that Napoleon III was inclined to be superstitious, and there is not, therefore, anything inherently improbable in the report that he left the seal he wore on his watch-chain to his son, the unfortunate Prince Imperial, as a talisman. This seal is said to have borne an inscription in Arabic characters, signifying "The slave Abraham relying on the Merciful One (God)."[40] The talisman lost its virtue on that unlucky day when, in far-off Zululand, the heir to so many hopes was cut off in the first flush of early manhood (see page 64).

---

[38] Nicols, "Faithful Lapidary," London, 1659, p. 107.

[39] Kluge, "Edelsteinkunde," Leipsic, 1860, p. 366.

[40] Fernie, "Precious Stones for Curative Wear," Bristol, 1907, p. 109.

V

On Ominous and Luminous Stones

THE OPAL

Mother. Come, let me place a charm upon thy brow,
And may good spirits grant, that never care
Approach, to trace a single furrow there!
Daughter. Thy love, my mother, better far than charm,
Shall shield thy child—and yet this wondrous gem[1]
Looks as though some strange influence it had won
From the bright skies—for every rainbow hue
Shoots quivering through its depths in changeful gleams,
Like the mild lightnings of a summer eve.
Mother. Even so doth love pervade a mother's heart;
Thus, ever active, looks through her fond eyes.[2]

THERE can be little doubt that much of the modern superstition regarding the supposed unlucky quality of the opal owes its origin to a careless reading of Sir Walter Scott's novel, "Anne of Geierstein."[3] The wonderful tale therein related of the Lady Hermione, a sort of enchanted princess, who came no one knew whence and always wore a dazzling opal in her hair, contains nothing to indicate that Scott really meant to represent the opal as unlucky. Lady Hermione's gem was an enchanted stone just as its owner was a product of

[1] The opal is said to preserve its wearer from disease; and hence, in the East, is much used in the form of amulets.

[2] From "Gems of Beauty," by the Countess of Blessington, London, 1836.

[3] Sir Walter Scott, "Novels," The Janson Society, New York, 1907, vol. xxiii, pp. 126–138.

enchantment, and its peculiarities depended entirely upon its mysterious character, which might equally well have been attributed to a diamond, a ruby, or a sapphire. The life of the stone was bound up with the life of Hermione; it sparkled when she was gay, it shot out red gleams when she was angry; and when a few drops of holy water were sprinkled over it, they quenched its radiance. Hermione fell into a swoon, was carried to her chamber, and the next day nothing but a small heap of ashes remained on the bed whereon she had been laid. The spell was broken and the enchantment dissolved. All that can have determined the selection of the opal rather than any other precious stone is the fact of its wonderful play of color and its sensitiveness to moisture. Hence we are perfectly justified in returning to the older belief of the manifold virtues of the opal, only remembering that this gem is a little more fragile than many others and should be more carefully handled and guarded.

The opal, October's gem, recalls in its wonderful and varied play of color the glories of a bright October day in the country, when earth and sky vie with each other in brilliancy and the eye is fairly dazzled with the bewildering variety of color.

It rarely happens that Pliny gives any information as to particular jewels, almost all his notices of precious stones being confined to descriptions of their form and color, and data regarding what was popularly believed as to their talismanic or therapeutic power. In the case of the *opalus*, however, he writes as follows: "There exists to-day a gem of this kind, on account of which the senator Nonius was proscribed by Antony. Seeking safety in flight, he took with him of all his possessions this ring alone, which it is certain, was valued at

2,000,000 sesterces ($80,000)."[4] The stone was "as large as a hazel-nut."

This "opal of Nonius" would be the great historic opal if we had any assurance that it was really the stone to which we now give this name. As, however, the principal European source of supply in Hungary does not appear to have been available in classic times to the Romans, and as opals are not found in the places whence, according to Pliny, the *opalus* was derived, we are almost forced to the conclusion that he had some other stone in mind when he gave his eloquent description of the *opalus*. And yet, in spite of all this, Pliny's words so well describe the beauties of a fine opal that it is difficult to determine what other stone he could have meant. For it can well be said of opals that "There is in them a softer fire than in the carbuncle, there is the brilliant purple of the amethyst; there is the sea-green of the emerald—all shining together in incredible union. Some by their refulgent splendor rival the colors of the painters, others the flame of burning sulphur or of fire quickened by oil."[5] Possibly some brilliant varieties of iridescent quartz—"iris" quartz, possessing an internal fracture, displays with great brilliancy all the colors of the rainbow, sparkling with wonderful clearness in its field of transparent mineral—might excite the admiration of one who had never seen an opal. Referring again to these quartz crystals, they are often cut so as to form a dome of quartz and are even used as distinct jewels. The fact that Pliny could praise the Indian imitations of the *opalus* in glass, and could state that this stone was more successfully imitated than any other, is an almost de-

---

[4] Plinii, "Naturalis historia," lib. xxxvii, cap. 6.
[5] Plinii, l. c.

cisive argument against identifying the *opalus* with an opal, for it is well known that no stone is more difficult to imitate.

About the middle of the eighteenth century, a peasant found a brilliant precious stone in some old ruins at Alexandria, Egypt. This stone was set in a ring. It was as large as a hazel-nut and is said to have been an opal cut *en cabochon.* According to the report, it was eventually taken to Constantinople, where it was estimated to be worth "several thousand ducats."[6] The description given of this gem, its apparent antiquity, and the high value set upon it have contributed to induce many to conjecture that it was the celebrated "opal of Nonius." Of course this was nothing but a romantic fancy. It is also quite certain that an opal would scarcely hold its play of color or compactness for twenty centuries, for most opals lose their water—slowly perhaps, but surely—within a lesser space of time. Even the finest Hungarian opals show some loss of life and color within a century or even less, and some transparent Mexican opals lose their color and are filled with flaws within a few years' time.

The Edda tells of a sacred stone called the yarkastein, which the clever smith Volöndr (the Scandinavian Vulcan) formed from the eyes of children. Grimm conjectures that this name designates a round, milk-white opal. Certainly the opal was often called *ophthalmios,* or eye-stone, in the Middle Ages, and it was a common idea that the image of a boy or girl could be seen in the pupil of the eye.

Albertus Magnus describes under the name *orphanus*

---

[6] Hesselquist, "Voyages and Travels in the Levant," English trans., London, 1766, pp. 273, 274.

a stone which was set in the imperial crown of the Holy Roman Empire. This gem is believed to have been a splendid opal, and Albertus describes it as follows:

The orphanus is a stone which is in the crown of the Roman Emperor, and none like it has ever been seen; for this very reason it is called orphanus. It is of a subtle vinous tinge, and its hue is as

THE "ORPHANUS JEWEL" IN THE GERMAN IMPERIAL CROWN.
From the "Hortus Sanitatis" of Johannis de Cuba [Strassburg, Jean Pryss, ca. 1483]; De lapidibus, cap. xcii. Author's library.

though pure white snow flashed and sparkled with the color of bright, ruddy wine, and was overcome by this radiance. It is a translucent stone, and there is a tradition that formerly it shone in the night-time; but now, in our age, it does not sparkle in the dark. It is said to guard the regal honor.[7]

[7] Alberti Magni, Opera Omnia, ed. Borgnet, Parisiis, 1890, vol. v, p. 42.

Evidently this imperial gem was regarded as *sui generis,* for Albertus has just described the *ophthalmus lapis,* a name frequently bestowed upon the opal in medieval times, reciting the virtues usually ascribed to the opal for the cure of diseases of the eye, and the magic power of the stone to render its wearer invisible, wherefore it was denominated *patronus furum,* or "patron of thieves."

In the Middle Ages the opal mines of Cernowitz, in Hungary, were very actively exploited, and at the opening of the fifteenth century more than three hundred men are said to have been employed here in the search for opals. At that time, and for many centuries after, no breath of suspicion ever tarnished the fame of the opal as not only a thing of rare beauty, but also a talisman of the first rank. We are told that blond maidens valued nothing more highly than necklaces of opals, for while they wore these ornaments their hair was sure to guard its beautiful color. The latter superstitions probably arose from the frangibility of the stone and its occasional loss of fire.

From the earliest times the baleful influence of the Evil Eye has struck terror into the souls of the ignorant and superstitious. It is believed by some that the name "opal"—written "ophal" in the time of Queen Elizabeth—was derived from *ophthalmos,* the eye, or *ophthalmius,* pertaining to the eye, and that hence the foolish superstition regarding the ill luck of the opal had some connection with the belief in the Evil Eye. However, this is altogether incorrect, since the stone called *ophthalmius* by early writers, and which seems to have been the opalus of the ancients and our opal, was believed to have a wonderfully beneficial effect upon the sight, and if it

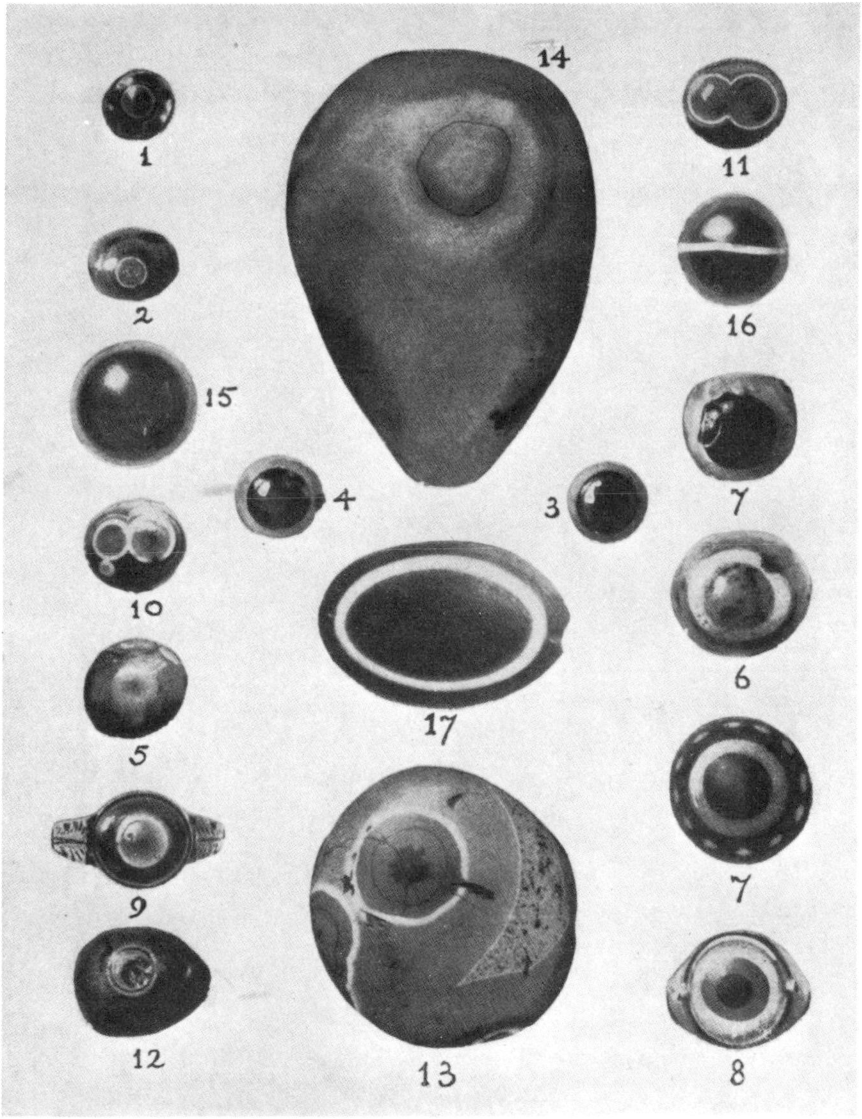

1, 2, 3, 4, 5. Eye agates, Aleppo stones, Arabia.
6 and 7. Antique eye agates, with double zone.
8 and 9. Aleppo stones set in rings.
10 and 11. Double eye agates, Aleppo stones, Arabia.
12. Natural pebble, showing eye from Isle Royal, Lake Superior.
13 and 14. Natural agates with eye-like effect, East Indian. Had been used as votive charms.
15. Eye agate, Brazil.
16. Agate called Oriental agate, eye effect, from Brazil.
17. Ancient eye of idol, agate variety sardonyx. Had been pierced lengthwise and worn as a charm on the arm. East Indian.

was thought to render the wearer invisible, this was only an added virtue of the stone.

The eye-agates were sometimes used to form the eyes of idols. At a later period some of these "agate-eyes" were removed from the statues and cut with a glyptic subject on the lower side. Some of the most interesting antique gems are of this kind. In Aleppo (and elsewhere in the East) there is a certain type of sore known as the "Aleppo button" or "Aleppo boil." The boil frequently does not appear for a long period after infection has taken place. It often appears as a swelling surrounded by a white ring, and there is a belief among the natives that there are "Aleppo stones," these being the so-called "eye-agates" frequently produced by cutting a three-layer, naturally pale yellow or pale gray agate, with intervening white zones in such a way that it looks like an eye or a double-eye, and such stones are used in alleviation of the Aleppo sore. What beneficial influence they may have is due to the fact that the agate is cold and furnishes a little relief for the time.

This "Aleppo boil" or "Oriental sore" so prevalent in many parts of western Asia, is produced, according to the best authorities, by a pathogenic organism *Leishmania tropica* (Wright) 1903. As to the means by which this organism is introduced into the human subject nothing very definite is known, but mosquitoes or *Phlebotomus* have been suggested as possible transmitting agencies.[8]

The eye of some invisible monster, the eye of the dragon, the eye of the serpent, were all regarded as possessed of malign power. It is well known that in the East Indies a peacock's feather is thought to bring ill-luck,

[8] Communication of Dr. Frederick Knab, citing Castellani and Chalmers, "Manual of Tropical Medicine," 1910.

the eye in the feather being the baleful point. Even in our own time, and among those for whom this primitive superstition has no terrors, the humorous use of the idea—as shown, for instance, in the "Dick Dead-Eye" of Gilbert and Sullivan's "Pinafore"—proves that the Evil Eye is familiar to our thoughts. For this reason, stones such as those which have been named the cat's-eye, the tiger's-eye, or the oculus Beli, always possess a certain strange interest.

One of the earliest descriptions of the opal in English is that written in the reign of Queen Elizabeth by Dr. Stephen Batman (d. 1584). While the passage is essentially a translation from the "De proprietatibus rerum," of Bartolomæus Anglicus, the English version is interesting in itself as showing what was accepted by English readers of the time regarding the virtues of the opal. There is, of course, no trace of the foolish modern superstition touching the ominous quality of this beautiful gem. Batman writes:[9]

> Optallio is called Oppalus also, and is a stone distinguished with colors of divers precious stones, as *Isid.* saith. . . . This stone breedeth onely in *Inde* and is deemed to have as many virtues, as hiewes and colours. Of this *Optallius* it is said in Lapidario, that this *Optallius* keepeth and saveth his eyen that beareth it, cleere and sharp and without griefe, and dimmeth other men's eyen that be about, with a maner clowde, and smiteth them with a maner blindnesse, that is called *Amentia,* so that they may not see neither take heede what is done before their eyen. Therefore it is said that it is the most sure patron of theeves.

The opal seems to have appealed to Shakespeare as a fit emblem of inconstancy, for in "Twelfth Night" he makes the clown say to the Duke:[10]

---

[9] Batman, "Uppon Bartholome," London, 1582, p. 264, lib. xvi, cap. 73.

[10] Shakespeare, "Twelfth Night," Act ii, Sc. 4.

Now the melancholy God protect thee, and the Tailor make thy garment of changeable taffeta, for thy mind is very opal.

That the beauty of the opal was fully appreciated in the sixteenth century is shown by the words of Cardano, who states that he once bought one of these stones for fifteen gold crowns and found as much pleasure in its possession as he did in that of a diamond that had cost him five hundred crowns.[11] Although superstitious beliefs were rather the rule than the exception in Cardano's time, none of the silly fancies regarding the ominous quality of the opal were then current. It was reserved for the nineteenth century to develop these altogether unreasonable—and indeed almost inexplicable—superstitions. The ownership of so fair an object as a fine opal must certainly be a source of pleasure, and hence add to the good fortune of the owner.

Although opal has been considered by some a stone of misfortune, black opal is regarded as an exceptionally lucky stone. Formerly black opals were artificially made by dipping the light-colored stone into ink, or by allowing burnt oil to enter cracks in the stone produced by heating. About the year 1900, however, a number of deposits of natural black opals were found in the White Cliff region of New South Wales, whence exceedingly beautiful gems have been secured, with wonderful flames of green, red, and blue in a black field. Some of these have sold for $1000 and even for a higher price, the smaller ones bringing from a few dollars upward each. It has been claimed that $2,000,000 worth have been sold from New South Wales. A remarkable example is figured on the frontispiece of this volume.* The late F. Marion Crawford was a great admirer of this strangely beautiful variety of opal.

That ill-luck and good-luck are relative terms is shown

---

[11] Cardani, "De subtilitate," Basileæ, 1560, p. 445.

*Reference is to first color plate, following page 40.

as published of an opal by Paris newspapers. A shopgirl, plainly clad, in crossing the Place de l'Opéra, when the street traffic was at its greatest, stopped at one of the "refuges" halfway across the street. To the girl's great surprise, an elegantly attired lady standing there slipped an opal ring from her finger and gave it to the girl, who took it to a jeweller's shop to sell it. Here she was arrested on suspicion of having stolen it. The magistrate before whom she appeared was inclined to believe her story and ordered a "personal" in a widely read journal asking the lady to clear the girl of the charge. A titled lady presented herself, substantiating the girl's statement. She feared ill-luck would befall her if she wore or kept the ring, which was returned to the shopgirl.

A possible explanation of the superstitious dread the opal used to excite some time ago may be found in the fact that lapidaries and gem-setters to whom opals were entrusted were sometimes so unfortunate as to fracture them in the process of cutting or setting. This was frequently due to no fault on the part of the cutters or setters, but was owing to the natural brittleness of the opal. As such workmen are responsible to the owners for any injury to the gems, they would soon acquire a prejudice against opals, and would come to regard them as unlucky stones. Very widespread superstitions have no better foundation than this, for the original cause, sometimes a quite rational one, is soon lost sight of and popular fantasy suggests something entirely different and better calculated to appeal to the imagination.

The belief that the diamond fractured the teeth if it were put in the mouth, and ruptured the intestines if it were swallowed, already appears in pseudo-Aristotle,[12]

[12] Rose, "Aristoteles De lapidibus und Arnoldus Saxo," in Zeitschr. für D. Alt., New Series, vol. vi, p. 391. See also Avicenna, "Liber canonis," Basileæ, 1556, p. 182, lib. ii, Tract. ii, cap. 20.

and can therefore be dated back to the ninth and perhaps to the seventh century. This fancy evidently owes its origin to the fact that the diamond, because of its hardness, was used to cut all other stones, and the idea of its destructive quality was strengthened by the old legends regarding the venomous serpents which guarded the place where it was found. Hence the firm conviction that it would bring death to any one who swallowed it.

According to Garcias ab Orta (1563), the diamond was not used for medicinal purposes in the India of his time, except when injected into the bladder to break up vesical calculi. He notes, however, the prevalent belief that diamonds, or diamond dust, when taken internally, worked as a poison. As a proof of the falsity of this belief, Garcias adduces the fact that the slaves who worked in the diamond mines often swallowed diamonds to conceal them, and never experienced any ill effects, the stones being recovered in a natural way. The same author notes the case of a man who suffered from chronic dysentery and whose wife had for a long time administered to him doses of diamond dust. If this did not help him, neither did it injure him; finally, by the advice of the doctors, this strange treatment was abandoned. The man eventually died of his disease, but many days after the doses of diamond dust had been discontinued.[13]

The Hindus believed that a flawed diamond, or one containing specks or spots, was so unlucky that it could even deprive Indra of his highest heaven. The original shape of the stone was also considered of great importance, more especially in early times, when but few, if any, diamonds, were cut. A triangular stone was said to cause

[13] Garcias ab Orta, "Aromatum historia" (Lat. version by Clusius). Antverpiæ, 1579, p. 172. The Portuguese original was published in Goa, in 1563.

quarrels, a square diamond inspired the wearer with vague terrors; a five-cornered stone had the worst effect of all, for it brought death; only the six-cornered diamond was productive of good.[14]

The Turkish sultan Bejazet II (1447–1512) is said to have been done to death by a dose of pulverized diamond administered to him by his son Selim, who mixed the diamond dust with the sultan's food.[15] It is also related that the disciples of Paracelsus (1493–1541) spread the report that he died from the effects of a dose of diamond dust. Ambrosius[16] conjectures that this was only an excuse to explain the demise of the master in the prime of life—he was but forty-eight years old at the time of his death—although he had promised long life to all who made use of his medicaments.

While Benvenuto Cellini (1500–1571), the unrivalled goldsmith, was imprisoned in Rome, in 1538, he strongly suspected that his enemies were seeking to poison him by tampering with his food. Cellini shared the belief of his contemporaries that there was no more deadly poison than diamond dust. One day, while eating his noonday meal, he felt something grate between his teeth. He paid no particular attention to this, but when he had finished eating his eye was caught by some bright particles on the plate. Picking up one of these and examining it carefully, he was terrified to find what he supposed to be a diamond splinter, and he straightway gave himself up for lost, thinking that he had swallowed a quantity of diamond dust. He prayed to God for an hour and finally

---

[14] Surindro Mohun Tagore, "Mani Málá," Pt. I, Calcutta, 1879, pp. 122, 125.

[15] Justi Lepsii, "De fraude et vi," cap. v, §8; cited in Pindar, "De adamante," Berolini, 1829, p. 58.

[16] Aldrovandi, "Museum metallicum," Bononiæ, 1648, p. 949.

became reconciled to the thought of dying, but suddenly it occurred to him that he had not tested the hardness of the fragment he had found in his food. He immediately took the splinter and tried to crush it between his knife and the stone window-sill; to his joy the attempt succeeded, and he became convinced that what he had swallowed was not diamond dust. Later, after his release, Cellini learned that an enemy had given a diamond to a certain Lione Aretino, a gem-cutter, instructing him to grind it up so that the dust could be placed in Cellini's food. The gem-cutter was very poor and the diamond was worth a hundred scudi, so the man yielded to temptation and substituted a citrine for the diamond. To this circumstance alone did Cellini attribute his escape from death.[17]

In England, more than seventy years after Cellini's experience, diamond dust was selected as a poison to do away with a luckless prisoner. Sir Thomas Overbury had incurred the bitter animosity of the Countess of Essex, because he opposed her marriage with the favorite of James I, Robert Carr, Viscount Somerset, whom he had befriended and whose career he had furthered. The marriage took place, however, and, in 1613, Overbury was imprisoned in the Tower, through the machinations of the countess. She then sought the aid of one James Franklin, an apothecary, directing him to concoct a slow and deadly poison, which should be mixed with Overbury's food. In the minutes of Franklin's confession, he is said to have stated that the countess asked him what he thought of white arsenic. His reply was that this poison would prove too violent. "What say you (quoth she) to powder of diamonds?" He answered, "I know

[17] Vita di Benvenuto Cellini, ed. Carpani, Milano, 1806, p. 445.

not the nature of that." She said that he was a fool, and gave him pieces of gold, and bade him buy some of that powder for her. It appears, however, from the testimony, that a number of ingredients were employed, quite probably small doses of mercury, cantharides, etc., as well as the baleful diamond dust. Poor Overbury lingered on for more than three months, but was finally put out of his misery by a clyster of corrosive sublimate.[18]

As a proof of the deadly effects caused by the diamond, the Portuguese Zacutus relates the case of a merchant's servant who surreptitiously swallowed three rough diamonds belonging to his master. On the following day this man was seized with violent abdominal pains, all the remedies administered to him were without effect, and he soon died from the extensive internal ulceration produced by the sharp edges of the diamonds.[19]

This old fancy that diamonds or diamond dust had deadly effects when swallowed is pretty well exploded by this time, little or no confirmation being afforded by the instances cited in the matter. However, quite recently it has been shown that swallowing a diamond can prove fatal to a fowl. While a prize-winning cockerel was being fondled by his proud owner, it spied a flashing diamond set in a ring on his hand, and immediately pecked out the stone and swallowed it. Not long after, the fowl died—not, however, because it was poisoned by the diamond, but because it was chloroformed to insure the speedy recovery of the stone.

An old English ballad, treating of the loves of Hind Horn and Maid Rimnild, recounts that when Hind Horn,

---

[18] Amos, "The Great Oyer of Poisoning," London, 1846, pp. 336 sqq.

[19] Aldrovandi, "Museum metallicum," Bononiæ, 1648, p. 949.

who loved and was beloved by the king's daughter, went to sea to escape the wrath of the king, the princess gave him a ring set with seven diamonds. We are told that when far from home:

> One day he looked his ring upon
> He saw the diamond pale and wan.

Hereupon, he hastened back, for the paleness of the stone was a sign the loved one was unfaithful to him. On his return, he succeeded in preventing her marriage to another, and everything ended happily.[20]

In a fourteenth century MS. of the Old English romance upon which the ballad is founded, the stone in the ring is not named; in giving it Rimnild says:[21]

> Loke thou forsake it for no thing;
> The ston it is well trewe.
> When the ston wexeth wan
> Than chaungeth the thought of thi leman,
>
> Take than a newe.
> When the ston wexeth rede,
> Than have Y lorn mi maidenhed,
> Oghaines[22] the untrewe.

In this older form of the tale, the stone either grows pale or red as a sign of misfortune. It is interesting to note that Epiphanius, writing a thousand years earlier, states that the *adamas* of the high-priest grew red as a presage of bloodshed and defeat for the Jews.

Regarding the old fancy that a serpent could not look

---

[20] Child, "The English and Scottish Popular Ballads," Boston, 1882–96, vol. i, pp. 187 sqq.

[21] Child, l. c.

[22] Against thee.

upon an emerald without losing its sight, the Arabian gem dealer, Ahmed Teifashi, in 1242 writes as follows: [23]

After having read in learned books of this peculiarity of the emerald, I tested it by my own experiment and found the statements exact. It chanced that I had in my possession a fine emerald of the *zabâbi* variety, and with this I decided to make the experiment on the eyes of a viper. Therefore, having made a bargain with a snake-charmer to procure me some vipers, as soon as I received them I selected one and placed it in a vessel. This being done, I took a stick of wood, attached to the end a piece of wax, and embedded my emerald in this. I then brought the emerald near to the viper's eyes. The reptile was strong and vigorous, and even raised its head out of the vessel, but as soon as I approached the emerald to its eyes, I heard a slight crepitation and saw that the eyes were protruding and dissolving into a humor. After this the viper was dazed and confused; I had expected that it would spring from the vessel, but it moved uneasily hither and thither, without knowing which way to turn; all its agility was lost, and its restless movements soon ceased.

Wolfgang Gabelchover, in his commentary on the sixth book of the treatise "De Gemmis," by Andrea Baccio, gives the following account of a strange and tragic experience in regard to a ruby: [24]

It is worthy of note that the true Oriental ruby, by frequent changes of color and by growing obscurity, announces to the wearer some impending misfortune or calamity; and the obscurity and opacity is greater or less according to the extent of the coming ill-fortune. Alas! that what I had often heard proclaimed by learned men, I should myself experience; for as, on the fifth of December, 1600, I was travelling from Stuttgart to Calw with my beloved wife Catherine Adelmann of pious memory, I plainly observed in the course of the journey that a very beautiful ruby which she had given me, and which I wore on my hand, set in a gold ring, once and again lost

[23] Ravii, "Specimen Arabicum," Trajecti ad Rhenum, 1784, pp. 97, 98.

[24] Andreæ Baccii, "De gemmis et lapidibus pretiosis," Latin trans. by Wolfgang Gabelchover, Francofurti, 1603, pp. 63, 64.

its splendid coloring and became obscure, changing its brightness for a dark hue. This dark hue continued not for one or two days only, but so long that I was greatly terrified, and, removing the ring from my finger, concealed it in a case. Wherefore, I repeatedly warned my wife that some great calamity was impending either for her or for myself, the which I inferred from the change and variation of the ruby. Nor was I deceived, for within a few days she was seized with a dangerous illness, which resulted in her death.

A story explaining one at least of these supposedly ominous changes of color in precious stones, is given by Johann Jacob Spener, who states that it was told him by a trustworthy informant: [25]

There was a jeweller, expert, prudent, and rich, three essential qualities in a jeweller. One day, after having washed his hands, this man sat at a table, when, glancing at a ruby ring he wore on his finger, he remarked that the stone, which usually delighted the eye with its splendor, had lost its brilliancy and become dull. Since he believed what others had related to him, he was firmly persuaded that some misfortune threatened him, and, having removed the ring from his finger, he placed it in its case. A fortnight later, one of this man's sons died of varioloid. Reminded by this event of the phenomenon observed in the ruby, the jeweller took it from the case and found, on examination, that it had regained its pristine brilliancy. This fact confirmed him in his belief in the ominous quality of the stone. Once more, shortly after washing his hands, he remarked anew that the splendor of the ruby was dimmed, and he again fell a prey to anxiety, lest some fresh misfortune was impending. Since, however, his apprehensions proved vain and no untoward event happened, he investigated the matter carefully, and discovered that the obscuration of the color was due to a drop of water which had penetrated between the ruby and the foil, as the jewellers call it, and that the former brilliancy returned when the water had evaporated.

The ominous character of the onyx is especially noted in Arabic tradition, as is shown by the Arabic name for the stone, *el jaza*, "sadness." The following passage

[25] "De gemmis errores vulgares," Lipsiæ, 1688, sect. ii, §12.

from pseudo-Aristotle offers an illustration of the strength of this prejudice against the onyx, which was said to come from China and the Magreb: [26]

> Those who are in the land of China fear this stone so much that they dread to go into the mines where it occurs; hence none but slaves and menials, who have no other means of gaining a livelihood, take the stone from the mines. When it has been extracted, it is carried out of the country and sold in other lands. Those men of the Magreb also who are gifted with any wisdom will not wear an onyx or place it in their treasuries. Indeed, no one is willing to wear it, unless he be bereft of his senses; for whosoever wears it, either set in a ring or in any other way, will have fearful dreams and be tormented by a multitude of doubts and apprehensions; he will also have many disputes and lawsuits. Lastly, whoever keeps an onyx in his house, or places it in a vessel, or puts it in food or drink, will suffer loss of energy and capacity.

An ominous character was attributed to the red coral, especially the more highly colored varieties. If worn so that the substance came in direct contact with the skin, it was asserted that the color would pale, the coral also losing its brightness if the wearer became ill, or even if he were only threatened with severe illness. The same effect was said to be induced if some deadly poison had been taken. Cardano writes that he more than once observed this phenomenon, and he thinks that in these cases, where the wearer was not yet attacked by disease, its threatening "vapor," though not strong enough to provoke decided symptoms in the human body, was sufficiently powerful to offset the more delicate and subtle essence of the mineral substance. Of course, for us the mineral would be much less sensitive than flesh and blood, but the sixteenth century writers, and to a still greater

[26] Rose, Aristoteles De lapidibus and Arnoldus Saxo, Zeitschr. für D. Alt., New Series, vol. vi, 1875, pp. 360, 361.

degree those of an earlier time, attributed to stones not only life in a general way, but old age, disease, and death, in a very positive sense.[27]

Rabbinical tradition tells of a wonderful luminous stone placed by Noah in the Ark. This stone shone more brilliantly by day than by night, and served to distinguish the day from the night when, during the flood, neither sun nor moon could be seen.[28] According to another Jewish legend, Abraham is said to have built a city for the six sons Hagar bore to him. The wall with which this city was surrounded was so lofty that the light of the sun was cut off, and to offset this Abraham gave to his sons enormous precious stones and pearls. These exceeded the sun in brightness, and will be used in the time of the Messiah.[29]

Ælian relates the following tale of a luminous stone. A woman of Tarentum, named Heracleis, who was a pattern of the domestic virtues, lost her husband and mourned sincerely for him. Her grief made her compassionate, for when a young stork just learning to fly lost its strength and fell to the ground before her, Heracleis picked up the helpless bird and tended it carefully until its strength returned and it was able to fly away. A year later, when the woman was outside the house enjoying the bright warm sunshine, she saw a stork flying toward her. As the bird passed over her head, it let fall a precious stone into her lap. Heracleis took the

---

[27] Cardani, "De subtilitate," Basileæ, 1554, lib. vii, pp. 191, 205.

[28] Ginsburg, "Legends of the Jews," Eng. trans., Phila., 1909, vol. i, p. 162. See also Levy, "Dictionary of the Targumim," etc., New York and London, 1903, vol. ii, p. 836, s. v. מַרְגָּלִית. Pirke d'R. El., ch. xxiii.

[29] Ginsburg, l. c., p. 298.

stone with her into the house, feeling by an infallible instinct that the stork which had dropped it was the one she had cared for in the previous year. During the night she woke up, and was astonished to see that the room was lighted up as though by many torches, the radiance proceeding from the stone bestowed by the stork as a proof of its gratitude.[30]

In German, the stone called *Donnerkeil* (thunderbolt) has several synonyms; among these is *Storchstein* ("stork-stone"). It is evident that the stone of Heracleis was identical with the precious and brilliant variety of *ceraunię* mentioned by Pliny, "which drew to themselves the radiance of the stars." The flashing and ruddy light of the ruby suggested an igneous origin, and induced the belief that rubies were generated by a fire from heaven,—in other words, by the lightning flash.[31]

The analogy between the flame of a lamp or the glow of a burning coal and the radiance of a ruby, suggested some of the names given to this stone, or those resembling it in color, as, for instance, the Greek *anthrax* and the Latin *carbunculus* and *lychnis*. Probably the fancy that such stones were luminous in the dark was nothing more than the logical result of the quasi-identification of them with fire in some of its manifestations. Still, it is a well-known fact that some stones possess a high degree of phosphorescence. This circumstance must have been observed by chance, and may have had something to do with the legends of luminous stones, although this peculiarity is not characteristic of the ruby.

According to Pliny, the lychnis, perhaps a spinel, was

[30] Claudii Æliani, "De animalium natura," lib. viii, cap. 22, ed. Gesner, Tiguri, 1568, pp. 182, 183.

[31] Grimm, "Wörterbuch," vol. ii, col. 1244.

so called *a lucernarum accensu* (from the lighting, or the light, of lamps). The author of the poem "Lithica" says that the diamond (*adamas*), like the crystal, when placed on an altar, sent forth a flame without the aid of fire.[32] If this did not refer to the use of rock-crystal as a burning-glass, we might see in the passage an indication that the phosphorescence of the diamond had already been noted before the second or third century of our era.

From the Lydian river Tmolus a marvellous stone was taken which was said to change color four times a day. This surpasses the properties of the "saphire merveilleux" which changed its hue at night. Only innocent young girls could find the Lydian stone, and while they wore it they were defended from outrage.[33] Is it possible that the ancient writer intended to hint at the proverbial fickleness of woman, when stating that this changeable stone could only be discovered by one of the fair sex?

The temple of the Syrian goddess Astarte contained an image of this divinity crowned with a diadem in which was set a luminous stone. Such was the splendor of the light emitted by this gem that the whole sanctuary was lighted up as though with a myriad of lamps. Indeed, the stone itself bore the name *lychnos* ("lamp"). In the daytime this light was fainter, but was still very noticeable, as a fiery glow.[34]

Two fabulous stones are noted by pseudo-Aristotle, and one of these, the "sleeping-stone," must have possessed marvellous soporific power. It was a luminous stone of a bright ruddy hue, and shone in the darkness with a bright light. If a small quantity of this stone were

---

[32] "Lithica," line 270.

[33] De Mely, "La traité des fleuves de Plutarche," in Revue des Études Grecques, vol. v (1892), p. 331.

[34] Luciani, "De Syria dea," cap. 32.

hung about a person's neck, he would sleep uninterruptedly for three days and nights, and, when awakened on the fourth day, he would still be almost overcome by sleep. The other stone, of a greenish hue, had the opposite quality and induced prolonged wakefulness; so long as it was worn, sleep was banished. Our author gravely states that "some men who must watch at night suffer greatly from lack of sleep." If, however, they wore the "waking-stone," they suffered no inconvenience from their enforced vigils.[35] Evidently this stone would be a precious possession for night-watchmen, and a more satisfactory guarantee for their employers than "time-clocks" or other tests of wakefulness.

In his commentary on Marbodus, Alardus of Amsterdam relates the history of a wonderful luminous stone, a "chrysolampis," which, with many other precious stones, was set in a marvellous golden tablet dedicated to St. Adelbert, apostle of the Frisians and patron of the town of Egmund (d. 720–730), by Hildegard, wife of Theodoric, Count of Holland. The gift was made to the Abbey of Egmund, where the saint's body reposed. Alardus tells us that the "chrysolampis" shone so brightly that when the monks were called to the chapel in the night-time, they could read the Hours without any other light. This wonderful stone was stolen by one of the monks, whom Alardus terms "the most rapacious creature who ever went on two legs"; but, fearing to keep so valuable a gem with him, he cast it into the sea and it was never recovered.[36]

---

[35] Rose, "Aristoteles de lapidibus und Arnoldus Saxo," Zeitschr. für D. Alt., New Series, vol. vi, 1875, pp. 375, 376.

[36] The abbey to which Hildegard gave the tablet was probably that built by Theodoric II and destroyed by the Reformers in 1572. The first building was of wood and was erected by Theodoric I in 923 or 924; this was ravaged by the Frisians not many years later.

Strange tales were told of a luminous "carbuncle" on the shrine of St. Elizabeth (d. 1231) at Marburg. This stone was set above the statuette of the Virgin, and it was said to emit fiery rays at night. However, Creuzer informs us that it was only a very brilliant rock crystal of a yellowish-white hue. The shrine was an elaborate work of art in silver gilt, and was literally covered with precious stones to the number of 824, besides two large pearls and a great many smaller ones. All these gems were stripped from their settings when the shrine was taken from Marburg to Cassel in 1810.[37]

At the Dusseldorf Exhibition of 1891, the writer saw what was called "The Ring of St. Elizabeth," purporting to be set with her miraculously luminous ruby. The stone in the setting proved, however, to be a large almost flat carbuncle garnet of no great brilliancy, set in a narrow rim of gold.

After noting the reports of medieval travellers regarding the wonderful luminous rubies of the sovereigns of Pegu and repeating the tale that the night was illumined by their splendor, Cleandro Arnobio adds that it did not appear that any such rubies were to be found in his day. Nevertheless, he had heard from an ecclesiastic of a certain jewel that shone brightly at night. This stone, however, was not a ruby, but was of a pale citron hue, and hence Arnobio inclines to believe that it was either a topaz or a yellow diamond.[38] This probably refers to the Marburg "carbuncle."

The luminous "ruby" of the King of Ceylon is noted by Chau Ju-Kua,[39] a Chinese writer of about the middle

---

[37] Creuzer, "Antik geschnittene Steine vom Grabmahl der heiligen Elizabeth," Leipsic and Darmstadt, 1834, pp. 25, 26.

[38] Arnobio, "Il tesoro delle gioie," Venice, 1602, p. 34.

[39] See the English translation of his "Chu-fan-chï," by Friedrich Hirth and W. W. Rockhill, St. Petersburg, 1911, p. 72.

of the thirteenth century and hence a contemporary of the Arab Teifashi. He says: "The king holds in his hand a jewel five inches in diameter, which cannot be burned by fire, and which shines in the night like a torch." This gigantic luminous gem was also believed to possess the virtues of an elixir of youth, for we are told that the king rubbed his face with it daily and by this means would retain his youthful looks even should he live more than ninety years.

The glories of Emperor Manuel's (ca. 1120–1180) throne are celebrated by the Hebrew traveller Benjamin of Tudela, who visited Constantinople in 1161 A.D. This splendid throne was of gold studded with precious stones and, suspended from the canopy by gold chains, hung a magnificent golden crown set with jewels of incalculable value and so bright and sparkling that their glitter rendered needless any other illumination at night.[40]

When Henry II of France (1519–1559) made his solemn entry into the city of Boulogne, a stranger from India presented to the sovereign a luminous stone. It was rather soft, had a fiery brilliance, and could not be touched with impunity. According to De Thou, this story was vouched for by J. Pipin, who saw the stone himself and described it in a letter to Antoine Mizauld, a writer on occult themes, well known in his day.[41]

Although Garcias ab Orta did not believe in the tales current in his time regarding luminous rubies, he relates a story of such a stone told to him by a gem-dealer. This man stated that he had purchased a number of fine but

---

[40] "Die Reisebeschreibung des R. Benjamin von Tudela," ed. by L. Grünhut and Marcus N. Adler, Jerusalem, 1903, pt. ii, trans., p. 17.

[41] Beckmann, "History of Inventions," English trans., London, 1846, vol. ii, p. 433.

small rubies from Ceylon, and had spread them out over a table. When he gathered them up again, one of the stones remained hidden in a fold of the table-cloth. In the night he remarked something like a flame emanating from the table. Lighting a candle, he approached the table and found there the small ruby; when this was removed and the candle extinguished, the light was no longer visible. Garcias admits that the gem-dealers were fond of telling good stories, but he concludes with the dictum, "we must trust in them nevertheless."[42]

Not only the ruby, but the emerald also had the reputation of being a luminous stone, for, besides the shining "emerald" pillar in the temple of Melkart at Tyre, Pliny records the tale of a marble lion, with eyes of gleaming emeralds, which was set over the tomb of "a petty king called Hermias." This tomb was on the coast, and the flashing light from the emerald eyes frightened away the tunny-fish, to the great loss of the fishermen.[43] Whether the eyes of the magnificent chryselephantine statue of Athene by Phidias were supposed to be luminous we do not know, but they were incrusted with precious stones.[44]

The collection of works by the English alchemists, published by Elias Ashmole, contains the tale of a worthy parson who lived in a little town near London, and who wished to immortalize himself by building across the Thames a bridge which would always be lighted at night. After relating several expedients which suggested themselves to him, the poet continues:

[42] Garcias ab Orta, "Aromatum historia" (Lat. version by Clusius), Antverpiæ, 1579, lib. i, p. 174.

[43] Plinii, "Naturalis historia," lib. xxxvii, cap. 17.

[44] Platonis, "Hippias major," ed. Didot, vol. i, p. 745.

At the laste he thought to make the light,
For the Bridge to shine by nighte,
With *Carbuncle Stones,* to make men wonder,
With double reflexion above and under:
Then new thought troubled his Minde
Carbuncle Stones how he might finde;

And where to find wise men and trewe,
Which would for his interest pursue,
In seeking all the Worlde about,
Plenty of Carbuncles to find out;
For this he took so mickle thought,
That his satt flesh wasted nigh to naught.[45]

It is scarcely necessary to add that the poor parson never realized his dream, but the story shows how popular was the belief that carbuncles or rubies shone with their own light.

A luminous or phosphorescent stone, which has been named the Bologna stone, is the subject of a treatise published by the physician Mentzel in 1675.[46] The writer describes various experiments made to test the peculiar qualities of this mineral, which is partly a radiated or crystalline sulphate of barytes, and phosphoresces when calcined. It was sometimes called the "lunar stone" (*lapis lunaris*), because, like the moon, it gave out in the darkness the light it received from the sun. Mentzel also relates that the stone was first discovered, in 1604, by Vincenzio Casscioroli, an adept in alchemy, who believed that it would be a great aid in the transmutation of the baser metals into gold, on account of its solar quality. The place of its occurrence was Monte Paterno, near

[45] Norton's "Ordinall"; in Ashmole "Theatrum Chemicum Britannicum," London, 1652, p. 27.

[46] Christiani Mentzelli, "Lapis Bononensis," Bilefeldiæ, 1675.

Bologna, where it appeared in the fissures of the mountain, after torrential rains.

The various phenomena of fluorescence and phosphorescence undoubtedly explain some at least of the legends

SPECIMEN

DE

GEMMARVM

ORIGINE & VIRTVTIBVS.

IN QVO

Proponuntur & historicè illustrantur quædam Conjecturæ circa consistentiam materiæ *Lapidum Pretiosorum*, & subjecta, in quibus eorum præcipuæ virtutes consistunt.

Ab Honoratissimo

ROBERTO BOYLE

NOBILI ANGLO, è SOCIETATE REGIA.

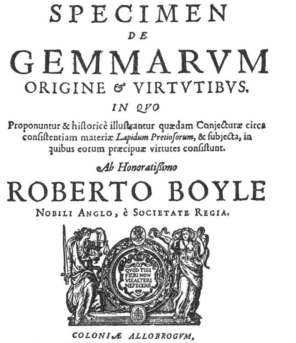

COLONIÆ ALLOBROGVM,

Apud SAMVELEM DE TOVRNES.

M. DC. LXXX.

TITLE PAGE OF ROBERT BOYLE'S WORK ON THE ORIGIN AND VIRTUES OF GEMS.

Printed in Cologne in 1680.

regarding luminous stones, superstition or fantasy having here as in most other cases a certain substratum of fact. This class of physical phenomena has been made the subject of special investigation by the author, as many as 13,000 specimens of various minerals having been subjected to the most searching tests in order to determine their qualities in this respect.[47] His interest in this field of research was greatly stimulated by a fortuitous happening. In 1891 his wife, while hanging up a gown in a closet one evening, saw that the diamond in a ring she was wearing gave off a faint streak of light which was very noticeable in the dark, and this fact led to a long series of experiments on the fluorescence, phosphorescence, and triboluminescence of the diamond.[48] More than two centuries before, Robert Boyle made a similar set of experiments at night with a diamond which must have been an Indian stone, and which he describes as table cut, about one-third of an inch long and somewhat less in width; he remarks that it was a dull stone of very bad water, having a blemish with a whitish cloud covering nearly a third of the stone.[49]

The "Journal des Sçavans" for 1739 gives certain tests of the luminous quality of diamonds made by Mons. Du Fay. In order successfully to observe this phenomenon, he prescribes that the experimenter shall remain in a darkened room for fifteen minutes, taking the additional precaution of closing one or both of his eyes. The diamond to be tested should be exposed to the sun's rays, or to strong daylight, for less than a minute, and when taken into darkness the luminosity, if observable, lasts twelve

---

[47] See Kunz, "The Phosphorescence of the Diamond," Trans. N. Y. Academy of Sciences, vol. x, p. 50, 1890–91; Kunz and Baskerville, "The Action of Radium, Actinium, Roentgen rays, and Ultra Violet Light in Minerals and Gems," Science, vol. xviii, No. 468, pp. 769–783, December 18, 1903.

[48] See page 172.

[49] Boyle, "Works," London, 1744, vol. ii, p. 85. The experiments were made October 27, 1663, and the results were communicated to the Royal Society the next day, the diamond which had been used being shown to the members at that time.

or thirteen minutes at longest. Not all diamonds show this quality, and nothing in their form or appearance serves to determine their possession of it. However, Mons. du Fay observed that the yellow diamonds, of which he tried a considerable number, were luminous. A single emerald, out of twenty that were tested, proved to be luminous.[50]

Boyle's experiments led to the discovery that some diamonds, when rubbed against wood or other hard substances, and even against cloth or silk, will emit a ray of light which seems to follow them; this is what is called triboluminescence.

The power of absorbing sunlight or artificial light and then giving it off in the dark is only possessed by certain diamonds. These are Brazilian stones, slightly milky in tint, or blue-white as they are often termed, and it is an included substance and not the diamond itself that possesses the power of storing up light and then giving it out. Willemite, kunzite, sphalerite (sulphide of zinc) and some other minerals possess the same power. Their peculiar property may be due to the presence of a slight quantity of manganese or to that of some of the uranium salts. That it is only the ultra-violet rays that are thus absorbed by these diamonds is proved by the fact that the phenomenon is not observable when a thin plate of glass is interposed between the sunlight or artificial light and the diamond, as glass is not traversed by these rays. The still undetermined substance to whose presence in diamonds of this type the special class of phenomena must be due, was named by the author

---

[50] "Journal des Sçavans," 1739, pp. 438, 439, of Amsterdam edition, citing "Hist. de l'Acad. Roy. des Sciences," 1735 (vol. xxxviii).

tiffanyite, in honor of the late Charles L. Tiffany (1812–1902), founder of the firm of Tiffany & Company.[51]

On the other hand all diamonds phosphoresce when exposed to the rays of radium, polonium, or actinium, even when glass is interposed. Treating of some of the aspects of phosphorescence in diamonds, Sir William Crookes says:[52]

> In a vacuum, exposed to a high-tension current of electricity, diamonds phosphoresce of different colours, most South African diamonds shining with a bluish light. Diamonds from other localities emit bright blue, apricot, pale blue, red, yellowish-green, orange, and pale green light. The most phosphorescent diamonds are those which are fluorescent in the sun. One beautiful green diamond in my collection, when phosphorescing in a good vacuum, gives almost as much light as a candle, and you can easily read by its rays. But the time has hardly come when diamonds can be used as domestic illuminants!
>
> By permission of Mrs. Kunz, wife of the well-known New York mineralogist, I will show you perhaps the most remarkable of all phosphorescing diamonds. This prodigy diamond will phosphoresce in the dark for some minutes after being exposed to a small pocket electric light, and if rubbed on a piece of cloth a long streak of phosphorescence appears.

The luminescence produced by heat is wonderfully marked in the case of chlorophane, a variety of fluorite. A Siberian specimen of a pale violet color emitted a white light merely from the heat of the hand; boiling water caused it to give out a green light, which was so greatly intensified when the specimen rested on a live coal that the radiance could be discerned from a considerable distance. Similar phenomena were observable in the case

---

[51] See Transactions of the New York Academy of Sciences, vol. xiv, p. 260; 1895.

[52] "Diamonds," a lecture delivered before the British Association at Kimberley, Sept. 5, 1905; London, 1905, p. 37. See also the same author's "Diamonds," London and New York, 1909, pp. 96–101.

of chlorophane from Amelia Court House, Va., and the writer found that specimens from this source also exhibited strong triboluminescence, resulting either from contact with one another, or with any hard substance.[53]

As the terms fluorescence and phosphorescence are sometimes rather carelessly employed, it may be well to note here that while both terms are used to denote the luminescence of a non-luminous body resulting from the action of light rays, of the electric current, or of radiant energy of any kind, as well as from heat, fluorescence signifies a luminosity which only continues so long as the exciting cause is present, while phosphorescence means a luminosity persisting for a longer or shorter period after the exciting cause has ceased to operate directly. The latter term therefore denotes a luminous energy stored up in the formerly non-luminous body and emitted by it for a certain time, at the expiration of which it again becomes non-luminous. Other special designations of induced luminosity in minerals are triboluminescence, the emission of light as a result of friction and thermoluminescence, a term used to denote light-emission excited by moderate heating, even by the warmth of the hand.

An old treatise in Greek, said in its title to come from "the sanctuary of the temple," and containing material, partly of Egyptian origin, may help us to understand something of the processes employed by a temple priest to impress the common people by the sight of luminous gems. The writer of the treatise declares that for the production of "the carbuncle that shines in the night" use was made of certain parts (he says "the

---

[53] Kunz, "Gems and Precious Stones of North America," New York, 1890, pp. 183, 184.

bile'') of marine animals whose entrails, scales and bones exhibited the phenomenon of phosphorescence. If properly treated, precious stones (preferably carbuncles) would glow so brightly at night ''that anyone owning such a stone could read or write by its light as well as he could by daylight.[54]

In the *Annales de Chimie et Physique,* the great French chemist, M. Berthelot, discusses this matter and expresses the following opinion:[55]

> "The texts leave no room for doubt as to the employment by the ancients of precious stones rendered phosphorescent in the dark by the superficial application of tinctures composed of materials whose phosphorescent quality is known to us. Although this luminescence, due to an application of organic oxidizable materials, could not well be durable, still it might be made to last several hours, perhaps several days, and it could always be renewed by repeating the application."

The use of jewelled ornaments to heighten by their luminosity in obscurity or in darkness the effect produced by a sacred image, and to stimulate religious awe in the beholder, is testified to by the ultra-Protestant traveller, Fynes Moryson, Gent., who went to Italy in 1594. Of his visit to the Santa Casa in Loreto, he says that he himself and two Dutchmen, his companions, were permitted to enter the inner chapel of the sanctuary, ''where,'' he proceeds, ''we did see the Virgin's picture, adorned with pretious Jewels, and the place (to increase religious horror) being darke, yet the Jewels shined by the light of wax candles.'' Although there is no question here of naturally luminous gems, this might have

[54] "Collection des anciens alchemistes grecs," ed. by M. Berthelot, trans., p. 336–338; text pp. 351, 352, Paris, 1887, 1888.

[55] "Sur un procédé antique pour rendre les pierres précieuses et les vitrifications phosphorescentes," Annales de Chimie et Physique, 6th ser., vol. xiv, pp. 429–432.

been the impression produced upon a more sympathetic pilgrim.[56]

Writing of the traditions in regard to luminous stones, Sir Richard F. Burton says, "There may be a basis of fact to this fancy, the abnormal effect of precious stones upon mesmeric sensitives."[57] However, while some instances are recorded of psychic impression produced by precious stones on the minds of persons possessing a highly sensitive nervous system, it seems likely that some legends of luminous stones had their origin in the refractive powers of cut gems, by means of which a dim and distant light would be reflected from the surface of the stones and would seem to spring from them. Quite possibly, in other instances, there was a disposition to cater to the popular belief by placing a light so that the hidden beams traversed the stone and appeared to emanate from it.

---

[56] Moryson, "An Itinerary containing his Ten Yeeres Travell through the Twelve Dominions," etc., Glasgow, 1907–8, vol. i. p. 216.

[57] Burton, "Supplementary Nights," London, 1886, vol. iii, p. 354, note.

VI

On Crystal Balls and Crystal Gazing

WE have evidence of the use of crystal balls as means of divination in medieval times, and "scrying" in some of its many forms was by no means rare in the Greek and Roman periods. The essential requisite for the exercise of this species of divination is a polished surface of some sort upon which the scryer shall gaze intently; for this purpose mirrors, globules of lead or quicksilver, polished steel, the surface of water, and even pools of ink, have been employed and have been found to insure quite as satisfactory results as the crystal ball. The points of light reflected from the polished surface (*points de repère*) serve to attract the attention of the gazer and to fix the eye until, gradually, the optic nerve becomes so fatigued that it finally ceases to transmit to the sensorium the impression made from without and begins to respond to the reflex action proceeding from the brain of the gazer. In this way the impression received from within is apparently projected and seems to come from without. It is easy to understand that the results must vary according to the idiosyncrasy of the various scryers; for everything depends upon the sensitiveness of the optic nerve. In many cases the effect of prolonged gazing upon the brilliant surface will simply produce a loss of sight, the optic nerve will be temporarily paralyzed and will as little respond to stimulation from within as from without; in other cases, however, the nerve will be only deadened as regards external impressions, while retaining sufficient activity to react against a stimulus

ROCK-CRYSTAL BALL PENETRATED BY CRYSTALS OF RUTILE. MADAGASCAR.

from the brain centres. It is almost invariably stated that, prior to the appearance of the desired visions, the crystal seems to disappear and a mist rises before the gazer's eye.

The Achaians, as Pausanius relates, frequently used a mirror to divine diseases or to learn whether there was danger of sudden death. Of the Temple of Demeter, or Ceres, at Patras, he writes:[1]

> In front of the temple of Demeter there is a well. A stone wall separates this well from the temple, but steps lead down to it from the outside. Here there is an infallible oracle, although it does not answer all questions, but only those touching diseases. They attach a slender cord to a mirror and let it down into the well, balancing it carefully so that the water does not cover the face, but only touches the rim. Then, after making a prayer to the goddess and burning incense to her, they look into the mirror, and it shows whether the sick person will die or recover. Such is the power of truth in this water.

This sacred well with its oracle of the magic mirror must have been in Lucian's mind when, in his description of the palace of the Moon-King, he says:[2]

> Another wonderful thing I saw in the palace. Suspended over a rather shallow well there is a large mirror, and anyone who goes down into this well will hear every word that is spoken on earth, while, if he gazes on the mirror, he will see there every city and every nation just as clearly as though he were looking down upon them from a slight elevation. At the time I was there, I saw my native country and its inhabitants. Whether I myself was seen by them in turn, I am not sure.

Lucian adds, with a fine touch of irony, "Anyone who doubts this assertion needs only to go there himself and he will find out that I speak the truth." As no one has

---

[1] Pausaniæ, "Descriptio Græciæ," ed. Schubart, vol. ii, Lipsiæ, 1883, pp. 54, 55, lib. ii, cap, 21, 12.

[2] Luciani, "Vera Historia," lib. i, 26.

yet made a trip to the moon, the assertion is still uncontradicted.

In their religious legends the ancient Mexicans taught that their god Tezcatlipuco had a magic mirror in which he saw everything that happened in the world.[3] He was sometimes named Necocyautl, "sower of discord," because he often stirred up war and strife among men, but he was also lord of riches and prosperity, which he bestowed and took away again at his will. To the influence of this divinity were attributed many omens and certain strange visions, announced by repeated knockings.[4]

In the Orphic poem "Lithica," a magic sphere of stone is described. The substance is called "sideritis" or "ophitis," and is said to be black, round, and heavy; possibly some metal, rather than a stone, is designated by these names. Helenus, the Trojan soothsayer, is said to have used this sphere to foretell the downfall of his native city. He fasted for twenty-one days and then wrapped the sphere in soft garments, like an infant, and offered sacrifices to it until, by the magic of his prayers, "a living soul warmed the precious substance."

A strange variety of divination by means of mirrors placed on the heads of boys, who, with eyes blindfolded, were supposed to perceive forms or signs of some description in the mirrors, is noted by Spartianus in his life of the Emperor Didius Julianus (ca. 133–193). This ruler is said to have resorted to this form of divination, and the boy entrusted with the task is asserted to have

[3] Balz, "Die sogenannte magische Spiegel und ihr Gebrauch"; Archiv für Anthrop. N.S., vol. ii, p. 45, 1904.

[4] Sahagun, "Historia general de las cosas de Nueva España," Mexico, 1829, vol. i, pp. 2, 3; vol. ii, pp. 6, 12, 16, 17; lib. i, cap. 3; lib. v, cap. 3, 9, 11, 12.

announced the approaching accession of Septimius Severus (146–211) and the dethronement of Didius Julianus.[5]

An indication that the usage of divination by means of a silver cup existed among the primitive Hebrews has been found in the story of Joseph and his brethren. In Genesis xliv, 1–5, we read that Joseph concealed a silver cup in the sack of grain borne away by Benjamin, making of this a pretext for requiring the return of his brethren. He sent messengers to overtake them and directed them to demand the return of the cup, using these words: "Is not this it in which my lord drinketh, and whereby indeed he divineth?"

The Arabic author, Haly Abou Gefar, tells of a golden ball used by "the Magi, followers of Zoroaster," in their incantations. It was incrusted with celestial symbols and set with a sapphire, and one of these magicians, after attaching it to a strip of bullhide, swung it around, reciting at the same time various spells and incantations.[6] Probably the magician, by fixing his gaze upon the brilliant revolving sphere, gradually fell into a hypnotic trance, during which visions appeared to him. These he could afterward interpret to those who had sought his aid to read the future, or obtain information regarding things that were happening for away.

An important side-light on the beliefs of Western Europe, in the fifth century, regarding crystal-gazing, is afforded by one of the canons of the synod held about 450 A.D. by St. Patrick and the bishops Auxilius and Issernanus. Here it is decreed that any Christian who believes there is a Lamia (or witch) in the mirror is to be anathe-

[5] Spartiani, "Vita Didii Juliani," cap 7.

[6] Reichelti, "De amuletis," Argentorati, 1676, p. 36.

matized, and is not to be again received into the Church unless he shall have renounced this belief and shall have diligently performed the penance imposed upon him.[7] In this case, as in many others, the vision in the crystal or mirror did not represent some former or contemporaneous happening, but the figure of an evil spirit, who, either by signs or words, imparted to the scryer the information he was seeking.

The power to see images of evil spirits on the surface of water was claimed by those called *hydromantii* in the ninth century. This is attested in a work composed about 860 A.D. by Hincmar, Archbishop of Rheims, who characterizes the supposed appearances as "images or deceptions of the demons." These diviners asserted that they received audible communications from the spirits, and they therefore evidently believed that the appearances were realities.[8]

Although, as we have seen, many different materials were used for scrying, the preference was often given to polished spheres of beryl; in modern times, however, the rock-crystal is considered the best adapted for the purpose.

In his introduction to "Crystal Gazing," by N. W. Thomas,[9] Andrew Lang writes of what he terms hypnagogic illusions—images which appear when the eyes are closed and before sleep supervenes. When faces appeared to him in this way, they were always unfamiliar ones, with the single exception of having once seen his own face in profile. The same was almost invariably true

[7] "Synodum episcorporum Patricii, Auxilii et Issernani," in Migne, Patr. Lat., vol. liii, Parisiis, 1865, col. 825.

[8] Hincmari, "Opera Omnia," in Migne, Patr. Lat., vol. cxxv, col. 7; De devortio Lotharii et Tetbergæ.

[9] London, 1905, pp. xxiv, xxx.

of landscape and inanimate objects. These forms seemed to grow out of the bright points of light which frequently appear when the eyes are closed, and Lang suggests a similar origin for the visions of the "scryers"—namely, the development of the images from dark or light points in the glass.

In regard to this, we have an interesting passage in the works of Ibn Kaldoun, a Persian writer, born in 1332, who gives the following very acute analysis of the phenomena accompanying crystal-gazing.[10]

Some believe that the image perceived in this way takes form on the surface of the mirror, but they are mistaken. The diviner looks at this surface fixedly until it disappears, and a curtain, like a mist, is interposed between him and the mirror. Upon this curtain are designed the forms he wishes to see, and this permits him to give indications, either affirmative or negative, concerning the matter on which he is questioned. He then describes his perceptions as he has received them. The diviners, while in this state, do not see what is really to be seen (in the mirror); it is another kind of perception, which is born in them and which is realized not by sight but by the soul.

As to the character and quality of the crystal to be used, Abbot Tritheim, the master of the famous Cornelius Agrippa, says:[11]

Procure of a lapidary a good, clear, pellucid crystal of the bigness of a small orange,—*i.e.*, about one inch and a half in diameter; let it be globular, or round each way alike; then you have got this crystal fair and clear, without any clouds or specks. Get a small plate of pure gold to encompass the crystal round one-half; let this be fitted on an ivory or ebony pedestal. Let there be engraved a circle round the crystal; afterwards the name: Tetragrammaton. On the other side of the plate let there be engraved, Michael, Gabriel, Uriel, Raphael, which are the four principal angels ruling over the Sun, Moon, Venus, and Mercury.

---

[10] Ibn Kaldoun, in Notices et Ext. de MSS. de la Bib. Imp., vol. xix, p. 221.

[11] See Barrett, "The Magus," London, 1801, p. 135.

The four letters constituting the Tetragrammaton are the Hebrew characters *yôdh, hê, wâw* and *hê,* יהוה. As this divine name was regarded in later Judaism as too sacred to be pronounced, the word lord, *adonai,* was substituted for it in the reading of the Scriptures. For this reason, when the vowel signs were added to the text to indicate the traditional pronunciation, the consonants Yhwh were provided with the vowels of *adonai* and the name was therefore read Jehovah by Christian scholars.

The Persian poet Jâmi writes thus of a magic mirror in the poem "Salamân and Absal": [12]

> Then from his secret Art the Sage Vizyr
> A Magic Mirror made; a Mirror like
> The bosom of All-wise Intelligence,
> Reflecting in its mystic compass all
> Within the sev'nfold volume of the World
> Invol'd; and looking in that Mirror's face
> The Shah beheld the face of his Desire.

Roger Bacon (1214–1292) was probably the most gifted man of the thirteenth century, and his writings testify to an extraordinarily clear perception of the essential principles of scientific research. However, his true greatness was not generally appreciated in his own age, and popular fancy wove about his name a fabric of legend in which he appeared as an arch-necromancer and magician. The curious old work entitled "The Famous Historie of Fryar Bacon" gives a number of the strange recitals which became current in England in regard to Bacon's wonderful powers.

One of these treats of a marvellous "glass" made by the friar, in which events happening at far-distant places

---

[12] Jâmi's "Salamân and Absal," trans. by Edward Fitzgerald, Boston, 1899, p. 84.

GLASS BALL, PERFORATED AND MOUNTED IN METAL, SO THAT IT CAN BE SUSPENDED AND USED FOR OCCULT AND CURATIVE PURPOSES.

Period of about tenth or twelfth century. Collection of Sir Charles Hercules Read.

BALL OF JET, PERFORATED, MOUNTED IN METAL, SO THAT IT CAN BE SUSPENDED AND USED FOR OCCULT AND CURATIVE PURPOSES.

Period of about tenth or twelfth century. Collection of Sir Charles Hercules Read.

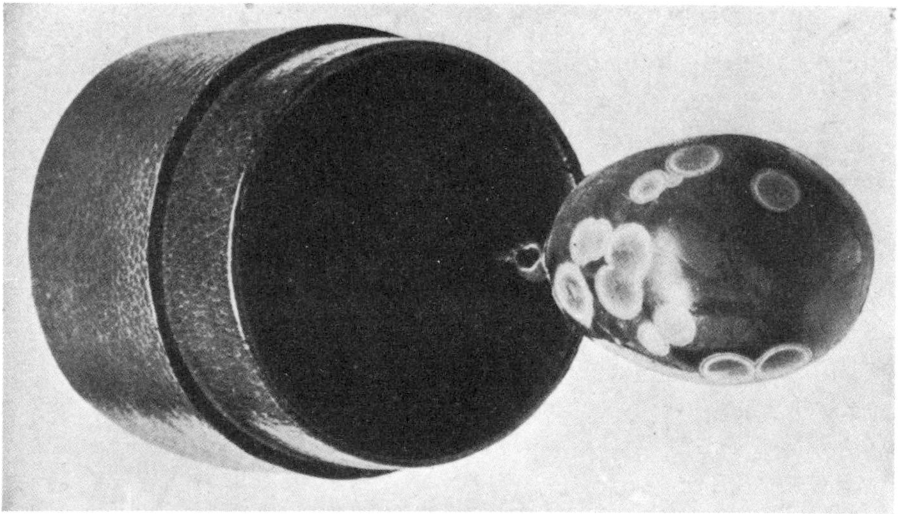

EYE AGATE, SHOWING A NUMBER OF CIRCULAR MARKINGS.

Mounted in metal and kept in a box, as a votive or curative stone. About fourteenth century. British Museum. (See page 149.)

were mirrored. On one occasion two young men, between whom the friendliest feelings existed, came to Bacon and requested him to let them see in the mirror what their fathers were doing at the time. The friar consented, but the experiment, while successful, was the cause of a terrible misfortune. The story is as follows:

> The Fathers of these two Gentlemen (in their Sonnes absence) were become great foes: this hatred betweene them was growne to that height, that wheresoever they met, they had not onely wordes, but blowes. Just at that time, as it should seeme, that their Sonnes were looking to see how they were in health, they were met, and had drawne, and were together by the eares. Their Sonnes seeing this, and having been alwayes great friends, knew not what to say to one another, but beheld each other with angry lookes. At last one of their Fathers, as they might perceive in the Glasse, had a fall, and the other, taking advantage, stood over him ready to strike him. The Sonne of him that was downe could then containe himselfe no longer, but told the other young man, that his Father had received wrong. He answered againe, that it was faire. At last there grew such foule words betweene them, and their bloods were so heated, that they presently stabbed the one the other with their Daggers, and so fell downe dead.

The sceptre of the Scottish regalia is surmounted by a crystal globe, two inches and a quarter in diameter, and the mace by a large crystal beryl. In former times these stones were regarded as amulets and their use was traced back to the Druids. Sir Walter Scott tells us that in his time they were still known among the Scottish Highlanders as "Stones of Power."[13]

The testimony of John of Salisbury (1120?–1180) shows that in the twelfth century, in England, divination by means of the arts of the *specularii* was often practised. The prelate writes that when a boy, he himself and a companion a few years older received instruction from

---

[13] Description of the Regalia of Scotland, by Sir Walter Scott, Bart., Edinburg, n. d., p. 13.

a priest who was addicted to the use of these magic arts. This priest used to polish the finger-nails of the boys with a consecrated oil or ointment, and then direct them to look upon the polished surface until some figure or form should appear. Sometimes the smooth, polished surface of a basin was used. John of Salisbury regarded it as a mark of divine favor that he himself saw nothing upon the smooth and lustrous surface, but he states that his companion observed certain vague and shadowy forms. Certain names pronounced by the priest on these occasions terrified the boy, for he believed them to be the names of evil spirits; indeed, such was his reluctance to participate in the unholy rites that his presence was believed to interfere with the production of the phenomena.[14]

In another part of his "Policraticus," John of Salisbury states that the specularii claimed that their gift of seeing visions on polished surfaces was never used to injure any one, but was often useful in the detection of theft and in counteracting magic spells.[15]

Under the comprehensive chapter heading: "How to conjure the crystal so that all things may be seen in it," Paracelsus (1493–1541) declares that "to conjure" means nothing more than "to observe anything rightly, to learn and to understand what it is." The crystal was of the nature of the air, and hence all things movable and immovable that could be seen in the air could also be seen in the crystal or *speculum*.[16]

---

[14] Johannis Saresberensis, "Policraticus," Lyon, 1513, fols. lxxvii, verso, lxxviii, recto, lib. ii, cap. 28.

[15] Johannis Saresberensis, l. c., fol. lxxvi, recto, lib. ii, cap. 28.

[16] "The Hermetic and Alchemical writings of Aureolus Philippus Theophrastus Bombast of Hohenheim, called Paracelsus the Great," trans. by Arthur Edward Waite, London, 1894, vol. i, p. 224.

Paracelsus showed keen insight, and his conclusions are excellent. One might add, however, that it is a fact that these are images condensed in the double convex lens, forming as it were, an internal crystal sphere. These images are reversed, distorted and twisted, and when they become visible to one who is expecting strange things, they form mental impressions which it is often very difficult to erase. Many crystal gazers are frequently very highly wrought, nervous and susceptible, and other influences uniting with the impressions produced, may give the brain for a time the power to evolve kaleidoscopic effects.

Directions for the use of an Erdenspiegel, or "earth-mirror," are given in an old German manuscript written in 1658 by a Capuchin priest.[17] The mirror is to be set about two inches above a board, and the questions to be answered are to be placed beneath it. The scryer is recommended to place three grains of salt upon his tongue, whereupon he is to repeat a prayer and cross himself. He now takes the mirror in his hand and breathes upon it three times, repeating the words, "In the name of the Father, of the Son, and of the Holy Spirit. Amen."

These preliminaries having been accomplished, the following prayer, or rather invocation, is repeated:

> O thou holy Archangel N. N., I pray to thee most fervently through the great and unsearchable name of the Lord of all Lords and King of all Kings, Jod, He, Vau, He, Tetragrammaton, Adonay, Schaday, receive my greeting and give ear to the humble petition which I offer in the name of the great and highest God, Elohim, Zebaoth, that thou shalt appear to me in the world-mirror, and give me knowledge and instruction in answer to my questions.

---

[17] "Unterricht vom Gebrauch des Erdspiegels, 1658" (Aus dem Kapuziner-Kloster in Immenstat. Eine Handschrift des Kapuziner-Paters Franziscus Seraph. Heider daselbst); in "Handschriftlichen Schätze aus Kloster Bibliotheken," Köln am Rhein, 1734–1810 (reprint).

The strong religious tone of these directions for the use of the mirror and the fact that it is a priest who gives them, shows that there was a disposition to tolerate the employment of such "white magic."

In medieval times it was believed that the vision in the crystal was produced through the agency of an indwelling spirit, and, therefore, it was necessary to use some very potent spell to force this spirit to enter the stone. Many of these ancient spells have been preserved, and they contain a strange and incongruous mixture of religious and magical formulas. In one of these, dating from the end of the fifteenth century, after a recitation of a long and rambling conjuration, we read, "And y^{en} ask ye chylde yf h^{e} seethe any thyng, and yf no, let the m^{r} begin his conjuratyō agayn." As usual the scrying was done by a child, the conjuration being spoken by the minister. An important part of the conjuration consisted in the repetition of a number of divine names, most of them originally Hebrew, but so much corrupted by reciters who did not know their meaning that it is now exceedingly difficult to interpret them correctly.

A proof that this form of magic was often regarded as quite compatible with religion is offered us in a passage from a sixteenth century manuscript,[18] where we read that the crystal should be laid on the altar "on the Side that the gospell is read on. And let the priest say a mass on the same Side." If the conjuration is successful, the same manuscript tells us that "these angells being once appeared will not depart the glasse or stone untill the Sonne be sett except you licence them." It also seems that "scrying" was looked upon as a special gift, only granted to a favored few as a peculiar privilege, and we

[18] Sloane MS. 3851, f. 50b.

read that "Prayer and a good beleefe prevailed much. For faith is the cay to this and all other works, and without it nothing can be effected." The child scryer, either maid or boy, should not be more than twelve years old.

That a certain religious spirit, however mistaken, often animated the crystal-gazers of the sixteenth century, is shown in the case of the "speculator" of John a Windor, who confessed that when he led an impure life the "dæmons" would not appear to him in his glass. He would then proceed to fumigate the apartment, as though believing that the very air was contaminated by the sins of the operator. We may hope that the seer was not content with this, but also tried to reform his evil ways. Another scryer, a woman named Sarah Skelhorn, declared that the spirits that appeared to her in the glass would often follow her about the house from room to room, so that she at last became weary of their presence.[19] Both of these scryers had regular employment, for it was quite customary for a gentleman to have a household seer, just as he would have a body-physician, if he could afford it.

A sixteenth century work on magic, the "Höllenzwang" of Dr. Faustus, whose name has been immortalized for all ages by Goethe, gives very particular and detailed directions for the preparation and consecration of a crystal, whether glass or quartz. Faust asks his "Mephistophelis" whether such crystals can be made, and the spirit replies: "Yes, indeed, my Faust," and directs Faust to go, on a Tuesday, to a glass-maker, and get the latter to form a glass. It was requisite that this

---

[19] Jonson, "The Alchemist," ed. Hathaway, New York, 1903, pp. 101, 145, note.

work should be done in the hour of Mars, that is, in the first, eighth, fifteenth or twenty-second hour of Tuesday. The crystal when completed must not be accepted as a gift, but a price must be paid for it. When the object had been secured, Mephistopheles directs that it be buried in a grave, where it must be left for the space of three weeks; it was then to be unearthed; if a woman purchased it, she must bury it in a woman's grave. However, these preliminaries only served to prepare the crystal for the final consecration, as the mere material mass was regarded as inert and possessing no virtue until certain spirits were summoned to dwell within it. Mephistopheles confesses that he alone would not be powerful enough, and he directs Faust to call upon the spirits Azeruel and Adadiel also. Faust is assured that the three spirits will show him in the crystal whatever he may wish to know. If anything has been stolen, the thief will appear; if any one is suffering from disease, the character of his malady will be revealed, etc.[20]

Another way of preparing a crystal glass or mirror is given in the same work. After the glass has been bought it is to be immersed in baptismal water in which a first-born male child has been baptized, and therein it is to remain for three weeks. The water is then to be poured out over a grave and the sixth chapter of the Revelation of St. John is to be read. Hereupon the following conjuration should be pronounced:

O crystal, thou art a pure and tender virgin, thou standest at one off the gates of heaven, that nothing may be hidden from thee; thou standest under a cloud of heaven that nothing may be hidden from

[20] Keisewetter, "Faust in der Geschichte und Tradition," Leipzig, 1893, p. 472.

thee, whether in fields or meadows, whether master or servant, whether wife or maid. Let this be said to thee in the name of God, as a plea for thy help.[21]

The visions seen in crystal gazing were often supposed to be the work of evil spirits, seeking to seduce the souls of men by offering the promise of riches or by according them an unlawful glimpse into the future. Here, as in other magical operations, there was both white and black magic, recourse being had in some cases to good, and in others to evil spirits. As an illustration of the latter practice, a sixteenth century writer relates that in the city of Nuremberg, some time during the year 1530, a "demon" showed to a priest, in a crystal, the vision of a buried treasure. Believing in the truth of this vision, the priest went to the spot indicated, where he found an excavation in the form of a cavern, in the depths of which he could see a chest and a black dog lying alongside it. Eagerly the priest entered the cavern, hoping to possess himself of the treasure, but the top of the excavation caved in and he was crushed to death.[22]

The famous charlatan, Dr. Dee, who was for a time a prominent figure at the court of Emperor Rudolph II, was highly favored by Queen Elizabeth. The queen visited him several times, and even appears to have consulted him on political matters. In his diary the doctor relates that the queen called at his house shortly after his wife's death, which took place March 16, 1575. Of this visit he gives the following details:

The Queen's Majestie, with her most honorable Privy Council, and other the Lords and Nobility, came purposely to have visited my library: but finding that my wife was within four hours before buried

[21] Keisewetter, "Faust in der Geschichte und Tradition, p. 473.

[22] Wieri, "De prestigiis demonum," Basileæ, 1563, p. 121.

out of the house, her Majestie refused to come in; but willed to fetch my glass so famous, and to show unto her some of the properties of it, which I did. Her Majestie being taken down from her horse by the Earle of Liecester, Master of the Horse, at the church wall of Mortlake, did see some of the properties of that glass, to her Majestie's great contentment and delight.[23]

It was at Mortlake, on December 22, 1581, that Dr. Dee made his first essay with his crystal ball. The proceedings were conducted with a certain religious ceremonial, and began with a pious invocation to the angel of the stone. This celestial being soon graciously deigned to manifest himself in the stone and—presumably by the voice of the scryer—answered the questions put by those present.

There can be little doubt that Dee used more than one crystal in the course of his experiments; that now in the British Museum is of cairngorm, or "smoky-quartz." This variety of quartz may have been chosen because of the Scotch superstitions regarding its virtues; for, as a rule, charlatans seek to avail themselves of already existing superstitions in order to make their innovations more acceptable.

To give assurance to those who consulted such crystals that no diabolical agency was involved in the production of the phenomena, it was customary that a child should be the crystal-gazer. In Dr. Dee's experiments, however, it was usually the notorious Kelley, his *âme damnée,* who undertook this task of interpreting the crystal visions. The description given by Dee of a little girl who frequently acted as the intermediary of the higher powers suggests one of the fanciful

[23] "The Private Diary of Dr. John Dee," ed. by Halliwell, London, 1842 (Camden Soc. Pub.), p. 9, note ("Compendious Memorial," p. 516).

DR. DEE'S SHEW STONE.

Natural size. British Museum. This sphere of smoky-quartz came to the British Museum in 1700 with the Cottonian Library, donated at that time by the grandson of the original collector, Sir Robert Bruce Cotton (1571-1631).

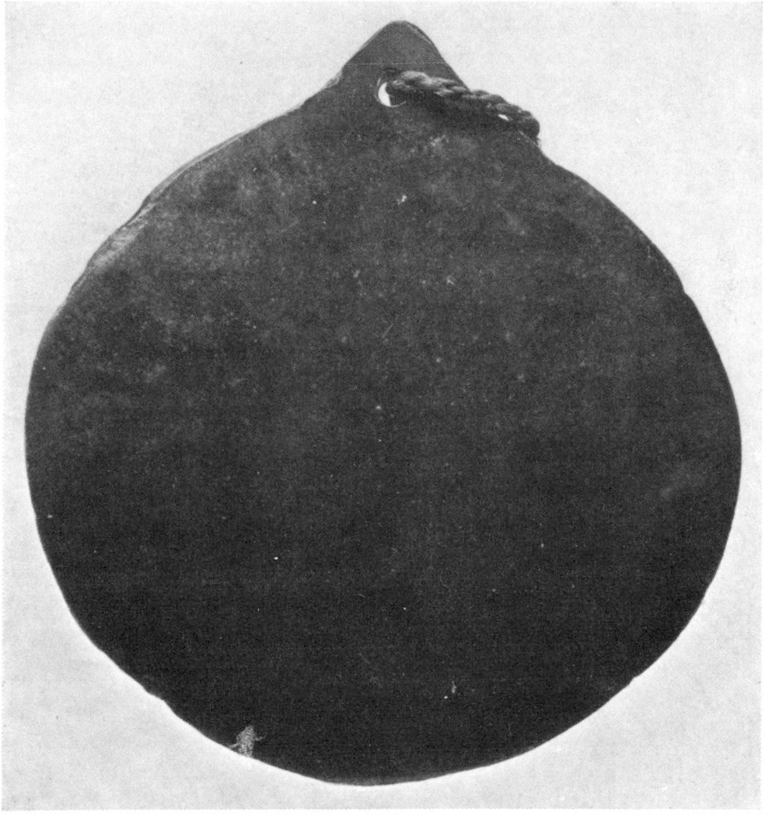

OBSIDIAN MIRROR, WITH NATIVE TEXTILE STRING.

Used by Aztecs and ancient Mexicans for various purposes. British Museum. Identical in shape and size with that known as "Dr. Dee's Mirror," now in the possession of Prince Alexis Soltykoff, of Russia. This was enclosed in a leather-covered case.

creations of our great novelist Hawthorne. Her mystic name was Madimi, and she is depicted as a pretty girl about eight years old, and with long flowing hair. To make her appearance more conspicuous, she was attired in a silk dress with chatoyant effects in red and green. At times, during the séances, this gay little figure could be seen flitting about the study, rendered even more whimsical and strange from its contrast with the piles of dusty old books, the curiosities, and the magical instruments collected there.[24]

This visionary maiden Madimi, of whom Dee relates so much in his diary, was apparently a child of fancy, a creation of Kelley's fertile brain. The diary is somewhat obscure in this particular and easily misunderstood; but there can be little doubt that where Madimi is represented as speaking, it is Kelley's voice that transmits to Dee her revelations. One passage, often overlooked, gives evidence of this. Madimi has appeared and is addressing her remarks to Kelley and to Dee by turns; finally, Dee says, "I know you see me often and I see you only by faith and imagination." To this Madimi quickly retorts, "pointing to E. K." (Kelley), "That sight is perfecter than his." Evidently we must understand this to signify something that Kelley has told Dee, for the latter's words show that he did not himself see the little fairy pointing to his friend. In many respects little Madimi may recall another "spiritual" maiden of whom we heard much a few years ago, the sprightly little Indian spirit "Bright Eyes," whose love for candy and jewelry was so very earthly.

---

[24] A true and faithful Relation of what passed for Many Yeeres between Dr. John Dee and Some Spirits. With preface by Meric. Casaubon, London, 1659, p. 1.

Not only the quality of the crystal had to be considered, but also its support and surroundings. Of this we have an interesting instance in the case of Dr. Dee's crystal. In one of his manuscripts is recorded the fact that on the 10th of March, 1582, Kelley saw in the crystal a representation of the form and arrangement of the table on which it should be set; particular instructions on the matter were also directly imparted to the scryer by the angel Uriel. The table was to be square, measuring two cubits each way and two cubits in height; and it was to have four feet. The material was to be "swete wood" and upon it was to be placed the Sigillum Dei (Seal of God) impressed upon the purest, colorless wax, the disk being 1⅛ inches thick and 9 inches in diameter. It bore a cross and the magic letters A. G. L. A., a transliteration into Roman characters of the initials of the Hebrew words signifying "Thou are great forever, O Lord." Four other and smaller seals were to be provided, one to be placed under each leg of the table; each of these seals being impressd with geometrical figures within or upon which were the seven sacred names of God and the names of the seven angels ruling the seven planetary heavens; Zabothiel, Zedekiel, Madiniel, Semeliel [Semeshiel], Nogabiel, Corabiel [Cocabiel] and Levaniel, the angels, respectively, of Saturn, Jupiter, Mars, the Sun, Venus, Mercury and the Moon. There then appeared to the scryer the figure of the table with the crystal resting upon it. Of this it is said:[25]

"Under the table did seeme to be layd red sylk to lye four square somewhat broader than the table, hanging down with four knops or tassells at the four corners

---

[25] See B. M. Dalton's notes in the Proceedings of the Society of Antiquaries, 2d ser., vol. xxi, 380–383. Sloane MS. A. 3188.

thereof. Uppon the uppermost red silk did seme to be set the stone with the frame, right over and uppon the principal seal, saving that the sayd sylk was betwene the one and the other."

It therefore seems that the prejudice in favor of a black or at least a dark background for the crystal did not appeal to Dr. Dee, and indeed the effect of color may perhaps better serve to neutralize troublesome reflections than does black.

The personages Kelley pretended to see in or around the magic crystal were described by him to Dr. Dee in the greatest detail, and this undoubtedly served to lend more reality and authority to their communications. As an illustration of Kelley's inventiveness in this matter, we may take his description of "Nalvage," a spirit that first appeared while the doctor and his famulus were in Cracow, April 10, 1584, and was subsequently a frequent visitor. The seer introduces his new "control" as follows:[26]

> He hath a Gown of white silk, with a Cape with three pendants with tassels on the end of them all green; it is fur, white, and seemeth to shine, with a wavering glittering. On his head is nothing, he hath no berd. His phisiognomy is like the pictures of King Edward the Sixth; his hair hangeth down a quarter of the length of the Cap, somewhat curling, yellow. He hath a rod or wand in his hand, almost as big as my little finger; it is of Gold, and divided into three equal parts, with a brighter Gold than the rest. He standeth upon his round table of Christal, or rather Mother of Pearl.

When reading the words spoken by Kelley and so carefully preserved by Dr. Dee, we are reminded, aside from the archaic turn of speech, of the minute descriptions so glibly given by modern mediums. It is true that

---

[26] Casaubon's "Relation," p. 73.

lately, in America, the spirits of the former owners of the land, of the blameless aborigines, seem to have acquired a quasi monopoly of the intercourse with the other world.

Most of the early records of crystal-gazing show conclusively enough that the images revealed in the stone were produced by the expectations, the hopes, or the fears of the gazer. In many cases, indeed, the vision is only prophetic because it determines the future conduct of the person who consults the stone. Fully persuaded that what has been seen must come to pass, he, or she, proceeds more or less consciously to make it happen, to fulfil the prediction.

As an instance of this we may take from an old German book [27] the tale of a lovelorn maiden who seeks the aid of an enchantress to learn whether she will marry her lover, upon whom her parents look with disfavor. The mystic crystal is brought out wrapped in a yellow handkerchief, and is placed in a green bowl beneath which is spread a blue cloth, the reflections from these different colors being probably calculated to stimulate the optic nerve and favor the appearance of some picture upon the polished surface of the crystal. The young girl, in rapt attention, looks long and earnestly; at last she cries out that she sees her own form and that of her lover. Both look pale and sad, and they appear to be about to set forth upon a long and perilous journey, for the lover wears riding-boots and carries a brace of pistols. The girl is so terrified at the sight that she faints away. The sequel of this vision is a runaway match, and we can easily understand that when the lover proposed

---

[27] Rist, "Die Aller-Edelste Zeit-Verkürtung der ganzen Welt," Franckfurt on dem Mayn, 1668, p. 255.

this adventure, the girl believed that it was written in the book of fate and willingly agreed to undertake it.

The great humorous poem "Hudibras," wherein all the foibles of the seventeenth century are castigated, does not fail to make mention of Dee and Kelley and their crystal. Of the sorcerer whose aid Hudibras seeks we are told:[28]

> He'd read Dee's prefaces before,
> The Dev'l and Euclid o'er and o'er;
> And all th' intrigues 'twixt him and Kelley,
> Lascus and th' Emperor, would tell ye.
>
> Kelley did all his feats upon
> The devil's looking-glass, a stone
> Where, playing with him at bo-peep
> He solved all problems ne'er so deep.

In his experiments in crystal-gazing, Dr. Dee evidently used more than one crystal, and did not indeed confine the operations of his scryer or scryers to brilliant spheres. In the collection of Horace Walpole, at Strawberry Hill, was a polished slab of black stone, obsidian, from Mexico. This came into the possession of Mr. Smythe Piggott and later (1853) into that of Lord Londesborough; it is now in the collection of Prince Alexis Soltykoff. Horace Walpole wrote a label for the stone, in which he says that it had long been owned by the Mordaunts, Earls of Petersborough, and was described in the catalogue of their collection as the black stone into which Dr. Dee used to call his spirits. Later it was owned by John Campbell, Duke of Argyle, who gave it to Horace Wal-

[28] Butler, "Hudibras," Part II, Canto III, 11, 235–8, and 631–4. This second part was issued in 1663, four years after Casaubon's publication of Dee's journal.

pole.[29] Undoubtedly any polished surface, whether flat or convex, might serve the purpose of the scryer almost equally well; the possible advantage of a convex or a spherical form consists in the multiplying of the reflections and light points so that the sight is induced to wander from point to point, and that forms and even motions are suggested by the superposition and combination of the various reflections. Often, too, a light point visible to one eye will not be so to the other, this sometimes provoking the phenomenon of binocular vision, which asserts itself for a moment or two, when the diverse images coalesce again, though imperfectly, giving an impression of movement. For one gifted with imagination and the natural quality of visualizing brain-pictures, these shifting light-points and the more or less definite and repeated reflections of surrounding objects offer abundant material out of which to construct life-like pictures apparently seen in the crystal. That the brain-pictures thus thrown out, so to speak, upon the crystal, may or may not have a peculiar psychic value, other than their value as mere phenomena, depends upon the significance we are inclined to attribute to the processes of the subconscious intelligence; of its existence, indeed, there can be no doubt, and many of our best thinkers incline to the belief that through it the narrow limits of our personality are occasionally transcended.

The following history and description of a crystal ball is given by John Aubrey (1626–1697):

I have here set down the figure of a consecrated Beryl—now in the possession of Sir Edward Harley, Knight of the Bath, which

---

[29] Miscellanea graphica: Representations of Ancient Medieval and Renaissance remains in the Possession of Lord Londesborough; introd. by Thomas Wright, London, 1857, p. 81.

1

2

3

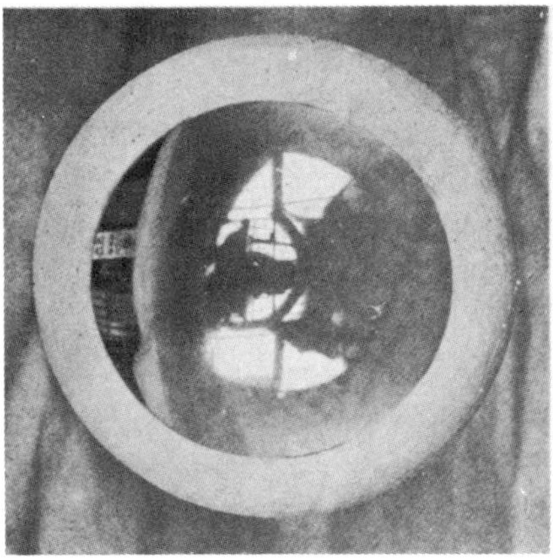

4

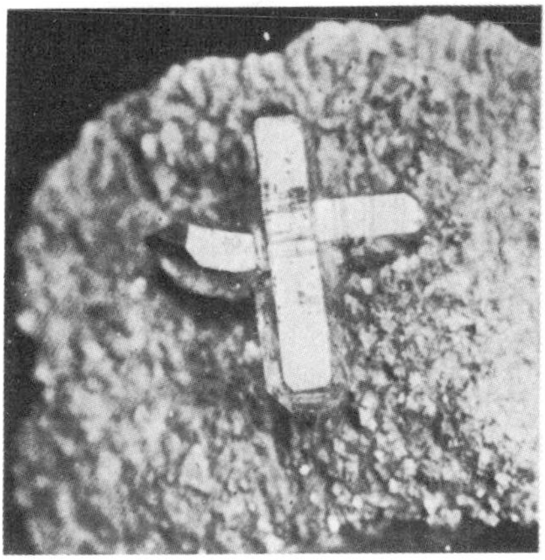

1, 2, 3. Rock-crystal spheres having portions of the surface ground so that they are rendered partially opaque.

4. Natural cross of rock-crystal. On dolomite, Ossining, New York.

he keeps in his closet at Brampton Bryan in Herefordshire amongst his Cimelia, which I saw there. It came first from Norfolk; a minister had it there, and a call was to be made with it. Afterwards a miller had it and he did work great cures with it (if curable), and in the Beryl they did see, either the receipt in writing, or else the herb. To this minister, the spirits or angels would appear openly, and because the miller (who was his familiar friend) one day happened to see them, he gave him the aforesaid Beryl and Call; by these angels the minister was forewarned of his death. This account I had from Mr. Ashmole. Afterwards this Beryl came into somebody's hand in London who did tell strange things by it; insomuch that at last he was questioned for it, and it was taken away by authority (it was about 1645). This Beryl is a perfect sphere, the diameter of it I guess to be something more than an inch; it is set in a ring, or circle, of silver, resembling the meredian of a globe; the stem of it is about ten inches high, all gilt. At the four quarters of it are the names of four angels, viz: Uriel, Raphael, Michael, Gabriel. On the top is a cross patee.[30]

In his "Sudducismus Triumphatus," Joseph Glanvil writes that "one Compton of Summersetshire, who practised Physick, and pretends to strange Matters," demonstrated his power to evoke the image of a distant person on the surface of a mirror. Glanvil relates that Compton offered to show to a Mr. Hill any one the latter wished to see. Hill "had no great confidence in his talk," but replied that he desired to see his wife who was many miles distant. "Upon this, Compton took up a Looking-glass that was in the Room, and setting it down again, bid my Friend look in it, which he did, and then, as he most solemnly and seriously professeth, he saw the exact Image of his Wife, in that Habit which she then wore and working at her Needle in such a part of the Room (then represented also) in which and about which time she really was, as he found upon enquiry when he came

[30] Aubrey, "Miscellanies," London, 1890, pp. 156, 157. (There is a figure on p. 156.)

home. The Gentleman himself averred this to me, and he is a very sober, intelligent, and credible Person. Compton had no knowledge of him before, and was an utter stranger to the Person of his Wife. He was by all accounts a very odd Person." [31]

A contemporary record recites that when a certain Sir Marmaduke Langdale (of the seventeenth century) was in Italy, he went to a sorcerer and was shown in a glass his own figure kneeling before a crucifix. Though a Protestant at this time, he shortly after became a Catholic.[32] If we exclude all idea of trickery, it is likely enough that the idea of becoming a Catholic was already present to the scryer's mind and called up this picture before him.

The celebrated Cagliostro, a Sicilian whose real name was Giuseppe Balsamo, among his other arts to excite curiosity and play upon the superstition of his contemporaries, had recourse to a species of crystal-gazing. In the only authentic biography of this extraordinary impostor occurs the following passage, which we give in Carlyle's version: [33]

Cagliostro brought a little Boy into the Lodge, son of a nobleman there. He placed him on his knees before a table, whereon stood a Bottle of pure water, and behind this some lighted candles: he made an exorcism round the boy, put his hand on his head and both, in this attitude, addressed their prayers to God for the happy accomplishment of the work. Having bid the child look into the Bottle, directly the child cried that he saw a garden. Knowing hereby that Heaven assisted him, Cagliostro took courage, and bade the child ask of God the grace to see the angel Michael. At first the child said: "I see some-

---

[31] Glanvil, "Sadducismus Triumphatus," London, 1726, p. 281.

[32] Aubrey, "Miscellanies," London, 1890, p. 155.

[33] Carlyle, "Works," Ashburton ed., vol. xvi, p. 509; from Vie de Joseph Balsamo, traduite d'après l'original Italien, ch. ii, 111 (Paris, 1791).

thing white; I know not what it is." Then he began jumping, stamping like a possessed creature, and cried: "There now! I see a child like myself, that seems to have something angelical." All the assembly, and Cagliostro himself, remained speechless with emotion. . . . The child being anew exorcised with the hand of the Venerable on his head, and the customary prayer addressed to Heaven, he looked into the Bottle, and said he saw his sister at that moment coming down stairs, and embracing one of her brothers. That appeared impossible, the brother in question being then hundreds of miles off; however, Cagliostro felt not disconcerted; said they might send to the country-house where the sister was, and see.

Taken all in all this experiment does not seem very satisfactory; but we have in it all the essential phases of crystal-gazing. Excitement and expectation produced their usual effect upon an impressionable child, and suggestion did the rest; the final vision may have been corroborated in some way, or, if not, it would be explained so as to convince those present at the experiment that the child had really seen a representation of some actual happening.

During the Terror, among those upon whom fell the suspicions of the Jacobins was General Marlière. He knew that a trial and quite probably a condemnation awaited him. A few days before the date fixed for his appearance before his judges, he met a colonel in the French army, who had served in the American Revolutionary War, and who was a firm believer in the truth of the visions seen in crystal balls. In the course of the conversation this subject was alluded to, and the general immediately declared that he was eager to put the matter to the test, and learn, if possible, what fate was in store for him. The colonel was at first very unwilling to undertake the experiment, probably he thought that General Marlière's doom was sealed, and, believing as he did in the revelations of the crystal, he dreaded the results; however, the general insisted and the experiment took place. As usual, the medium was an "innocent child." In the crystal appeared a man wearing a private's uniform of the National Guard struggling with one wearing a general's uniform. The child was much excited and terrified by the sight, exclaiming that the general's assailant had thrown him down and was beheading him. That the vision portended the general's execution was clear enough, but the peculiar dress of the executioner was a mystery to those present at the test, for the official garb bore no resemblance whatever to a soldier's uniform. The prediction was, however, fulfilled to the letter. General Marlière was tried, found guilty, and guillotined. This in itself did not mean much in view of the innumerable executions in the time of the Terror; but, on the day of this execution, Samson, the official executioner, desiring to gratify his personal vanity and to attract the gaze of the spectators, dressed himself in the uniform of a national guardsman.[34] That this altogether unusual circumstance, which could scarcely have been known to any of those who assisted at the crystal-gazing, should have been revealed in the crystal, is certainly very mysterious. If we had positive assurance that the events narrated happened exactly in the way they are said to have happened, this would be one of the few instances in which the vision seen in the crystal reproduced something entirely unknown to the scryer.

Many extraordinary visions are said to have been seen in crystal balls by a French scryer whose grandmother had clairvoyant powers and was sometimes consulted by Napoleon I. It is claimed that the grandson

---

[34] Kiesewetter, "Faust in der Geschichte und Tradition," Leipzig, 1893, p. 476.

has enjoyed the patronage of many royal personages, and had predicted, in a more or less definite way, the assassination of King Humbert of Italy, and the attempted assassination of Alfonso XIII and of his young bride, when they were returning to the palace after the conclusion of the marriage ceremony. This French scryer has stated that he is powerfully affected when he is consulted by any one destined to die a violent death; on such occasions he feels, in his own organism, a modified form of the particular kind of suffering they are fated to experience. This exceptional sensitiveness to occult influences was also shown when the crystal-gazer went to the Boulaq Museum in Cairo, and gazed upon the rows of mummies exhibited there; he immediately felt, as intensely as though it were a personal experience, the mingled sorrow and rage of the disembodied spirits at seeing their embalmed bodies exposed to the view of the idle crowd, when they should have been permitted to rest in their tombs until the hour of the Resurrection.

In England all those who attempted, with a greater or less degree of success, to reveal the hidden secrets of the future, were expressly designated as rogues and vagabonds according to the terms of an act passed June 21, 1824.[35] Such offenders, on being duly convicted before the Justice of the Peace, could be committed to the House of Correction, "there to be kept at hard Labour for any time not exceeding Three Calendar Months." This class of undesirable citizens comprised all using "any subtle Craft, Means, or Device, by Palmistry or otherwise" for the deception of his Majesty's subjects.

The *h'men,* or diviner, of Yucatan, places great re-

---

[35] George IV, cap. lxxxiii.

liance upon his *zaztun,* or "clear stone." This may be a quartz crystal, or else some other translucent stone; but in order to serve for divining purposes it must be sanctified according to special rites, gum-copal being burned before it, and certain magic formulas recited, which have been transmitted from generation to generation in an archaic dialect. When thus rendered fit for use, the diviner claims to be able to see in the depths of the crystal the whereabouts of lost articles, and also what absent persons are doing at the time he makes his observation. Not only this, but the future is also laid bare before his eyes. As these stones are supposed to possess such miraculous powers we need not be surprised that one of them should be found in almost every village in Yucatan.[36]

The Apache medicine-men are also fully persuaded that crystals possess the virtue of inducing visions, and they have used them for the purpose of finding lost property. To aid in the recovery of stolen ponies is one of the most important tasks of the Apache medicine-man, and to this end his crystal offers great assistance. Capt. John G. Burke relates that he made a great friend of a medicine-man named Na-a-che by giving him a large crystal of denticulated spar, much superior to the crystal he had been in the habit of using for his visions. That this was thoroughly satisfactory to the medicine-man at least, is shown by his statement to Capt. Burke that by looking into his crystal he could see everything he wanted to see. Of the way this came about he did not attempt any explanation.[37]

---

[36] Brinton, "Essays of an Americanist," Philadelphia, 1890, p. 165.

[37] Burke, "The Medicine-men of the Apache," Ninth Annual Report of the Bureau of Ethnology, 1887–1888, Washington, 1892, p. 461.

The magic power supposed to dwell within rock-crystal has been recognized in a peculiar way by some natives of New South Wales. They have the barbarous custom of knocking out one or more of the front teeth of their boys at the obligatory initiation ceremonies, and on one occasion Dr. Howitt was entrusted with the care of a number of these teeth, which are believed to preserve a certain undefined connection with the health and fortunes of their former possessors, and on this account great fear was expressed lest the custodian should place the precious teeth in the same bag with some rock-crystals, for the natives thought that the magic power of these crystals would injuriously affect the teeth, and through them the boys, from whose jaws they had been broken.[38]

In a paper entitled "The Origin of Jewelry," read before the British Association, Professor W. Ridgeley says:

> Australians and tribes of New Guinea use crystals for rain-making, although they cannot bore them, and this stone is a powerful amulet in Uganda when fastened into leather. Sorcerers in Africa carry a small bag of pebbles as an important part of their equipment. So it was in Greece. The crystal was used to light the sacrificial fire and was so employed in the church down to the fifteenth century. Egyptians used it largely under the XII Dynasty, piercing it along its axis after rubbing off the pyramid points of the crystal, sometimes leaving the natural six sides, or else grinding it into a complete cylinder. From this bead came the artificial cylindrical glass beads made later by the Egyptians.

Professor Ridgeley believes that the primary use of all these objects was because of their supposed magic powers. He holds the same view in regard to cylinders and rings, considering that the use of these as signets

---

[38] Fraser, "The Golden Bough," pt. i, "The Magic Art," vol. i, London, 1911, p. 176.

only became habitual at a later time, and he finds a proof of this theory in the fact that unengraved Babylonian cylinders and Mycenean gems have been discovered. This is, of course, perfectly true, but does not in the least prove that such ornaments may not have been originally worn simply for purposes of adornment; unquestionably, the custom of engraving them so as to render them signets must have arisen at a much later date.

Flacourt stated that the natives of Madagascar used crystals to aid them in divining. These stones, which were said to have fallen from heaven, were attached to the corners of the boards whereon the sorcerers produced their geomantic figures.[39] Here, however, the crystals were not directly used, but were only supposed to attract influences propitious to the diviner's efforts.

In the notes to the 1888 edition of the Chinese criminal code, some curious details are given of a practice called Yuan-kuang-fuchou (the magic of the round glittering). While this designation certainly seems to indicate the use of a polished sphere of some description, the details given refer to a different practice. We are told that when anything was stolen appeal was sometimes made to a certain Sun-Yuan Sheng, who would then hang up a piece of white paper and utter a spell, while a boy gazed upon the paper until he saw the figure of the thief. This magician was punished for carrying on an unlawful practice.[40]

The Mexicans made images of their god Tezcatlipoca of obsidian, and the name of this divinity is interpreted as signifying "shining mirror." This is supposed to refer to, or to have been expressed by, the brilliant effect

[39] Lang, "The Making of Religion," London, 1898, pp. 91–92.

[40] Thomas, "Crystal Gazing," London, 1905, p. 48.

BABYLONIAN CYLINDERS AND PERSIAN BEADS.

hematite, rock-crystal, lapis-lazuli, chalcedony, banded agate, and other stones. From 3000 B.C. to the Christian era. (See page 121.)

of the polished surface of the obsidian. Mirrors of this material are said to have been used for divination in ancient Mexico and the neighboring countries.[41] One of these Mexican mirrors seems to have been employed by Dr. Dee in his experiments in crystal vision.

A remarkable series of tests in the art of scrying, given in the presence of Lane, the great Arabic scholar, and translator of the Arabian Nights, illustrates the fallibility of most of the evidence adduced in such matters, for, at first, Lane was strongly impressed by the exhibition. Although no crystal was used, the process of scrying was precisely the same as in crystal-gazing,—that is to say, the vision called for by the visitors was seen by the scryer on a polished surface. The master of ceremonies was an Arab magician, though, of course, he did not do the scrying himself, but employed a boy for this purpose, for it is generally thought that half-grown boys or girls are more receptive. Although Lane himself was perfectly familiar with Arabic, an interpreter was always present in the interest of the other Europeans who assisted at the experiments.

After invoking many mysterious geniuses and burning incense and scraps of paper inscribed with magic formulas, the magician drew a magic square on a large sheet of paper and dropped a quantity of ink in the centre. On this the boy was directed to fix his gaze, and after he had shown that he was thoroughly under the magician's influence, by describing the images suggested to him, the visitors were permitted to ask him questions. The answers were successful in most cases; a single instance will suffice. When the boy was asked to describe Admiral

[41] Nuttall, "The Fundamental Principles of Old and New World Civilization," Cambridge, Mass., 1901, p. 80.

Nelson, he replied: "I see a man clothed in a dark garb; there is something strange about him, he has but one arm." Then, quickly correcting himself, he added: "No, I was mistaken, he has one of his arms across his breast." This correction impressed those present more than the first statement, for it was well known that Nelson usually had the empty sleeve of his coat pinned to his breast. It also seemed as though there could be no collusion, for both the magician and the boy were ignorant of everything English and evidently knew nothing of Nelson. Unfortunately, however, for those who would fain believe that there is something supernatural in scrying, it was later discovered that the interpreter was a renegade Scotchman, masquerading as an Arab, and there can be little doubt that he managed to suggest the boy's answer. The fact that no satisfactory results were obtained when this interpreter was absent, makes this explanation almost certainly the correct one.

The Armenians sometimes practised divination by watching the images that appeared, or were supposed to appear, on the smooth surface of the waters of a well, and the person who saw such images was called *hornaiogh*, "he who looks into a well." An Arab woman living in the neighborhood of Constantinople enjoyed a great reputation for her power in this respect, and was frequently consulted by Armenians and by other dwellers in the Turkish capital. Whoever wished to question this woman regarding the cause of an illness, the whereabouts of stolen objects, etc., usually took along a child of the household, and the actual scrying was generally performed by this child, who would describe or identify the forms it saw on the water's surface. If, however, for one reason or another, no child was brought, the witch herself did the scrying. In regard to illness, a distinction was made between "natural" maladies and those directly caused by some spirit. Should the spirit (*peri*) supposed to cause the dire malady known as *drsévé,* a kind of consumption, be seen to glide over the surface of the water, the sorceress would find it necessary to invoke the whole race of *peris* to come to the aid of the patient, who was expected to pay more than the usual fee for this very special service.[42]

The *peris* of Armenian legend were sometimes good and sometimes evil spirits; in the former case these were supposed to perform the functions of guardian angels, and every one was said to have a peri especially delegated to watch over him. This found expression in the fact that when one Armenian felt at first sight an instinctive sympathy for another, he would say, "My peri loves you dearly (*peris chad siretz kezi*)." In the contrary case, the feeling of antipathy was also attributed to the attitude assumed by the guardian spirit toward the new acquaintance.[43] These spirits were therefore supposed to encourage or discourage greater intimacy with newcomers in accord with the true interests of those over whom they watched.

The power to see images in a crystal does not appear to depend to any great extent upon a morbid nervous condition of the seer, for many of the most successful experimenters have been of good and even of exceptionally vigorous physique. Indeed, illness seems to diminish or destroy this power, at least in the case of those who are habitually healthy.[44] This does not imply that some highly nervous and even hysterical individuals have not

[42] Tcheraz, "Notes sur la mythologie Arménéenne," in Trans. of the Ninth Cong. of Orient. (1892), London, 1893, vol. ii, p. 832.

[43] Tcheraz, l. c., p. 835.

[44] Proc. Soc. of Psych. Research, vol. viii, p. 470.

been favored with "crystal visions." Very probably the rule here is the same as in ordinary hypnotism. Those persons who have a strong will and sound nerves are able to hypnotize themselves, while those whose nerves are disordered are subject to the hypnotic influence of others.

A well-known lady in New York City, in conversation with the writer, a few years ago, on the subject of crystal balls, was advised by him to try a ball herself and see what results she obtained. At the end of two years she found that by concentration she had been able to better her understanding of herself; and this effect is not only obtainable now by means of a crystal ball, but by fixing her gaze upon any bright object. This visual fixation has centred her whole being in such a way that her health has notably improved.

What are the laws that govern the production of these phenomena? That the "visions" are real enough has been proven time and again, but it seems almost certain that they do not offer anything but the ideas or impressions existing in the minds or optic nerves of the gazers. One of the most painstaking students of the subject, Miss Goodrich-Freer, gives many instances in proof of this, which show how easy it would be for a less critical observer to suppose that the crystal revealed something unknown to the gazer. On one occasion this lady was at a loss to remember the correct address of a friend whose letter, received a few days before, she had torn up. She resorted to her crystal, and after a few minutes saw in it, in gray letters on a white ground, the address she had forgotten. She mailed her answer to this address, and the reply came duly to hand, with the address stamped in gray upon the white paper of the note, which was

ROCK-CRYSTAL SPHERES. JAPANESE. (See page 217.)

identical with that she had first received.[45] The visual impression had been stirred up and "externalized" itself when she gazed upon the crystal. We believe that this explains the larger number of such visions, and that the rest are only inexplicable because the scryer has forgotten the source of the impression that is projected on the surface of the crystal.

It is true that both Miss Goodrich-Freer and many other crystal-gazers note instances in which the vision appears to represent something the scryer does not and cannot know. However, even in these cases, when carefully examined, there is little difficulty in finding an explanation. Coincidence accounts for much, and imagination for more, since it is not the vision itself, but the memory of the vision, that is later brought into comparison with actual facts. We all know how exceedingly hard it is to repeat, after a short lapse of time, all the circumstances and details of any occurrence. There is a natural growth and modification of mental impressions, due to association of ideas, and where there exists the least wish to make the prophecy accord with the event, or the vision with the coincident happening, this growth and modification will be in the direction of agreement. This takes place quite unconsciously, and the informant will be fully persuaded that all the circumstances are related exactly as they occurred.

The attempt to identify either persons or scenes observed by the scryer with real persons and real scenes unknown to him, must always be open to the objection that the one who makes the identification has no photographic impression upon which to base his judgment, but merely the words of the scryer. When we remember

---

[45] Proc. of the Soc. for Psych. Research, vol. v, p. 507.

what mistakes have been made in identifying individuals from photographs, we can easily appreciate the great chances of error entailed by the use of a verbal description of a visionary experience, even when the person giving the description is both willing and able to make it as exact and adequate as possible.

A very impartial witness, Andrew Lang, states that, in the course of a series of experiments he made in crystal-gazing, he saw nothing himself, but found that a surprisingly large proportion of those who tried were successful in seeing pictures of some sort on the polished surface. Almost invariably, when the gazer fixed his eyes upon the sphere, it appeared to grow milky-hued and then became black; upon this dark background the pictures showed themselves. One of the scryers, a lady, said that as a child she had seen pictures in ink that she had spilled for the purpose.[46] This method has been much favored by Orientals. While Lang does not quite venture to assert that all the "visions" reported to him were genuine ones, he inclines to the belief that this was the case with many of them. Experience has shown, however, that not all of those who see pictures in, or on, a glass or crystal sphere, can also see them in ink.[47] Nevertheless, in view of the fact that the crystal sphere is said to appear black to the eye before the pictures are seen, it would seem that some naturally black surface would be particularly adapted for the purpose.

An interesting point regarding the phenomena of crystal-gazing is the effect produced by magnification upon the images seen in, or on, the crystal ball. As to

[46] Thomas, "Crystal Gazing," London, 1908, Lang's preface, pp. xi, xii.

[47] Thomas, l. c., p. xxi.

this matter there is considerable difference of opinion, for, while some experimenters assert that the interposition of a magnifying-glass enlarges the image, others have not remarked any difference in its size under these conditions. Indeed, one of the most critical witnesses, Mrs. A. W. Verrall, declares that her vision entirely disappeared when she held a magnifying-glass before her eyes. On the other hand, we have the case of a subject who had been told, while in the hypnotic state, that he would see a play-bill on the crystal. When he was awakened and the crystal ball was placed before him, he said that he could see only detached letters, but when he looked through a magnifying-glass he saw all the letters distinctly and read the name of the play, in perfect accord with the suggestion.[48]

This image may have been reflected from some part of the room where the gazer had not noticed it, and may have been either before or behind the operator. The magnifying-glass would naturally make the small, condensed letters legible, as a play-bill would be many times larger than a crystal ball, and its minute image naturally too small to read, being reduced by the circular surface.

Usually, however, the image is not on the surface of the crystal, but in the beholder's eye; therefore when this image appears more clearly under magnification, the result is due to the expectation of the gazer based upon his experience of an invariable rule. This acts as a stimulus upon the visual function, which must be in an exceedingly sensitive state to produce visions at all. When, however, no result or a negative result follows the use of the glass, then we can safely assume that

[48] Proc. of the Soc. for Psych. Research, vol. viii, p. 473.

the gazer was naturally of a critical turn of mind, and was disposed to distrust sensual impressions; hence the glass became a disturbing influence, interfering with or even completely obliterating the eye-picture.

Many attempts have been made to establish distinctions between the different materials used for crystals, proceeding on the theory that subtle emanations from them affected the gazer and played an important part in producing the desired vision. That the beryl produced a greater number of these visions than any other mineral was the old belief which is still upheld in some quarters to-day; one scryer, indeed, asserts that his clearest and most satisfactory visions were seen in a cube of blue beryl, the beautiful color appearing to dispose the soul to a harmonious unfolding of its latent aptitudes.[49]

Among the instructions given to a would-be crystal gazer, the question of a proper and wholesome diet is not overlooked, as anything which tends to disturb the serenity of the organism will also interfere with the due exercise of the special clairvoyant faculty that expresses itself in crystal visions. A curious special recommendation made by one of the exponents of the art is that good results can be had by drinking an infusion of mugwort (*Artemisia vulgaris*), or of chicory (*Cichorium intybus*), because of their tonic and antibilious qualities. Moreover, we are told that these herbs are under the influence of the zodiacal sign Libra, the sign controlling the virtues of the beryl.[50] Above all the portion of the lunar month when the moon is on the increase is said to be far the best season for scrying, as the old astrologers recognized an affinity between the moon and rock-crystal.

---

[49] Shepharial, "The Crystal and the Seer," London [1900?], p. 14.
[50] John Melville, "Crystal Gazing," London, 1910, pp. 20, 21.

The claim is made that the adept at crystal-gazing can determine by the apparent difference in proximity of the visions whether they refer to the present or to a more or less remote past or future, that is to say, are nearer or farther removed in time from the period when the vision appears. The distinction between past and future is admitted to offer greater difficulty and a decision as to this point must depend upon a kind of intuitive and undefined impression on the part of the scryer.

Those who have made a sympathetic study of crystal-gazing recognize that the "visions" seen in or on the crystal differ according to the mental and psychic temperament of the scryer. Two broad distinctions are sometimes established, the one class comprising those whose mental attitude is a "positive" one while the second class includes the "passive" subjects. In the former case the crystal visions are more apt to be symbols denoting some past or future event than a clear picture of the event itself, the mentality of the "positive" subject being, perhaps, too strong merely to mirror the image cast upon it. Instead of so doing it transforms the impression received from this image into some symbolic form. This process is not, however, consciously done, but the scryer of this type is supposed nevertheless to have an instinctive appreciation of the fact that what he sees is purely and simply a symbol, and he proceeds to interpret this in accord with certain generally received rules, or in accord with his own personal experience.

The passive subject on the other hand is more apt to see a clear and definite picture of the persons or events revealed to him. Sometimes that picture is distinctly perceptible on or about the surface of the crystal, while at other times the visual perception will be rather indefinite and clouded, although accompanied by a strong mental impression in itself equivalent to that which would have been induced by an actual and objective vision.[51]

The proper use of the crystal is the prime factor in the art of scrying and great attention is paid to this point by all those who treat seriously of the subject. Among other things they recognize that freedom from pain, or even from a sense of physical discomfort, is quite essential, for the mind must assume a purely passive and receptive attitude, and not be forced to take cognizance of bodily discomfort. Moreover the nervous system must be in repose, for which reason a reasonable time should be allowed to lapse after taking a meal, before trying for crystal visions.[52]

An author on "psychomancy" affirms that fixing the gaze upon a crystal ball is one of the very best means of bringing out the latent faculty of astral vision, and he finds a reason for this in the atomic structure, the molecular arrangement of the material. He does not, however, impart any definite information as to what special structural characteristics render glass or rock-crystal particularly efficient in this direction.[53] The help that may be derived from crystal-gazing by those who are striving to pierce the veil that separates the "real life" about us from that spiritual life which is so much more real for those who believe in it, is also admitted by many.[54]

We cannot refrain from citing here the words spoken by Sir Oliver Lodge at Birmingham, Sept. 10, 1913, before the British Association for the Advancement of Science,

---

[51] Shepharial, "The Crystal and the Seer," London [1900?] pp. 11–13.

[52] Melville, "Crystal Gazing," London, 1910, p. 47.

[53] Atkinson, "Practical Psychomancy and Crystal Gazing," Chicago [1908], p. 46.

[54] See Leadbeater, "The Astral Plane," London, 1910, p. 14.

affirming his conviction, as a result of scientific investigation of occult phenomena, "that memory and affection are not limited to that association with matter by which alone they can manifest themselves here and now, and that personality persists beyond bodily death."

One of the latest types of glass balls for crystal-gazing has a small, circular, flat surface on the sphere. This may possibly be of service in furnishing a better field for the expected vision, and may also lessen the troublesome and baffling reflections which interfere so seriously with the projection of the mental picture.

A method that has been recommended to crystal-gazers is to place the crystal on a table, protect it from the reflections of surrounding objects by means of a velvet screen, and set seven candlesticks with wax tapers in front of the screen. The tapers are then to be lighted, the room being otherwise in perfect darkness, and the would-be scryer is to seat himself comfortably before the table, laying his hands flat upon it, and to gaze fixedly upon the crystal for half an hour or longer. The light from the tapers will certainly ensure a multitude of light points in the crystal. That the molecules forming the sphere may always remain *en rapport* with the gazer, he is advised to put it beneath his pillow when retiring to rest.[55]

The crystal gazer is strongly advised by some to limit the duration of his experiment at first to five minutes, during which he is to avoid thinking of anything in particular while keeping his eyes fixed intently upon the ball, but without any undue straining of attention. Should the eyes "water" after the test is concluded, this is to be regarded as an indication that the gazer has persisted

---

[55] Verner, "How to Know Your Future," London [1910?], p. 16.

too long; for brain-fag is to be strictly avoided, as such a state depresses instead of arousing the hidden and higher psychic faculties. Even after considerable practice, the scrying should not be carried on for more than a few minutes at a time. The faculty of visualization plays a most important part in crystal-gazing. The image thought to be seen on, before, or behind the surface of the crystal, is in its essence a fancied projection of a purely mental image conceived in the brain; such an image as is present to the consciousness of many when they call to mind a scene of some vivid past experience, or the face of someone they have known, and *see* it as an element of consciousness. When it is possible to externalize this interior vision, then we have at least a beginning of successful scrying. That it may go far beyond this, that it may reveal to the gazer events happening in some distant place, or even events yet to transpire in the dim future, is often claimed. An acceptance of this claim must depend largely upon our attitude toward premonitions and prophecies in general. Here, as in the simple picture evolved by an image of the past, the crystal is merely the background upon which are cast the mind-pictures or soul-pictures arising within our being.[56]

A use of crystal gazing to aid literary composition has been reported in the case of an English authoress of note, who, if she lost the thread of the story she was writing, would resort to her crystal, and would see mirrored therein the scenes and personages of her tale, the latter carrying on the plot in dramatic action. Aided by this suggestion she was able to resume her composition and successfully terminate her story.

---

[56] See Hereward Carrington's Correspondence Course of Instruction in Psychic Development, Lesson 24, New York, 1912.

CRYSTAL BALL, SUPPORTED BY BRONZE DRAGON. JAPANESE.

In Japan the smaller rock-crystals were believed to be the congealed breath of the White Dragon, while the larger and more brilliant ones were said to be the saliva of the Violet Dragon. As the dragon was emblematic of the highest powers of creation, this indicates the esteem in which the substance was held by the Japanese, who probably derived their appreciation of it from the Chinese. The name *suisho,* used both in China and Japan to designate rock-crystal, reflects the idea current in ancient times, and repeated even by seventeenth century writers, that rock-crystal was ice which had been so long congealed that it could not be liquefied.

For the Japanese, rock-crystal is the "perfect jewel," *tama;* it is at once a symbol of purity and of the infinity of space, and also of patience and perseverance. This latter significance probably originating from an observation of the patience and skill shown by the accurate and painstaking Japanese cutters and polishers of rock-crystal.

A crystal ball, one of the largest perfect spheres ever produced, has been made from rock-crystal of Madagascar. It is a very perfect sphere and of faultless material. The diameter is 6⅛ inches and the ball was held at about $20,000.

Many fine crystal balls are made in Japan, the materials being found in large, clear masses in the mountains on the islands of Nippon and Fusiyama and also in the granitic rocks of Central Japan. It is stated, however, that much of the Japanese material really comes from China. The Japanese methods of working rock-crystals are extremely simple and depend more upon the skill and patience of the workers than upon the tools at their command. Our illustration, taken from a sketch made by an Oriental traveller, shows the process of manufacturing crystal balls. The rough mass of crystal is gradually rounded by careful chipping with a small steel hammer. With the aid of this tool alone a perfect sphere is formed. The Japanese workmen thoroughly understand the fracture of the mineral, and know just when to apply chipping and when hammering. The crystal, having been reduced to a spherical form, is handed to a grinder, whose tools consist of cylindrical pieces of cast iron, about a foot in length, and full of perforations. These cylinders are of different curvatures, according to the size of the crystal to be ground. Powdered emery and garnet are used for the first polishing. Plenty of water is supplied during the process, and the balls are kept constantly turning, in order to secure a true spherical surface. Sometimes they are fixed on the end of a hollow tube and kept dexterously turning in the hand until smooth. The final polishing is effected with crocus or rouge (finely divided hematite), giving a splendid lustrous surface. As hand labor is exclusively used, the manufacture of crystal objects according to the Japanese methods is extremely laborious and slow.[57]

In Germany and France and in the United States, the fabrication of rock-crystal is accomplished almost entirely by machinery. The crystal to be shaped into a ball is placed against a semicircular groove worn in huge grindstones. This is illustrated in the case of the method practised in Oberstein, Germany. The workman has his feet firmly braced against a support, and, resting upon his chest, presses the crystal against the revolving grindstone. It is unnecessary to add that the practice is

---

[57] Kunz, "The Occurrence and Manipulation of Rock Crystal," Scientific American, vol. lv, pp. 103, 104 (Aug. 14, 1886). Trans. N. Y. Acad. Sciences, May 30, 1886.

By permission of the "Scientific American."

METHOD OF GRINDING CRYSTAL BALLS AND OTHER HARD STONE OBJECTS IN GERMANY AND FRANCE.

By permission of the "Scientific American."

JAPANESE METHOD OF CHIPPING, GRINDING AND POLISHING ROCK-CRYSTAL BALLS.

extremely unwholesome and develops early consumption among the workers. A constant stream of water is kept flowing over the stone so that the crystal shall always be moist, as the friction would otherwise hurt it, and the subsequent addition of water would be liable to cause a fracture. The final polishing is done on a wooden wheel with tripoli, or by means of a leather buffer with tripoli or rouge.[58]

There are three fine crystal balls in the collection of the American Museum of Natural History. One, apparently perfect, measures 5½ inches in diameter and was cut from a crystal found in Mokolumne, Calaveras Co., California; the second is 6½ inches in diameter and is from the same locality, but not entirely perfect. These were shown in the department of the Tiffany Collection prepared by the author, and were exhibited at the Paris Exposition of 1900 as part of the J. Pierpont Morgan gift to the American Museum of Natural History. Another fine crystal ball is now to be seen in the American Museum of Natural History, New York; this was donated to the institution. It measures $4^{11}/_{16}$ inches in diameter, is of wonderful purity, and the cutting has been executed with such a high degree of precision that an ideally perfect sphere has been produced.[59]

Crystal balls have been found occasionally in tombs or in funerary urns, and their presence in sepulchres may perhaps be considered to have been due to a belief that they possessed certain magic properties. In the tomb of Childeric (ca. 436–481 A.D.), the father of Clovis, a rock-crystal sphere was found which was for a time preserved in the Bibliothèque Royale, Paris, and later in

[58] Kunz, "The Occurrence and Manipulation of Rock Crystal."

[59] Gratacap, "The Mystic Crystal Sphere," in the American Museum Journal, January, 1913, p, 24; plate on p. 22.

the Louvre Museum; it measures 1½ inches in diameter.[60] The chance discovery of a number of crystal balls is related by Montfaucon. Towards the end of the sixteenth century, the canons of San Giovanni in Laterano, Rome, wished to have some repairs made to a house they owned, just outside of the city walls, and sent thither some workmen with the order to break up or remove two large, superimposed stones, which were much in the way. The workmen proceeded to break the upper stone, but were much astonished to find embedded within it an alabaster funerary urn with its cover. This had been hidden between the two stones, a space for its reception having been hollowed out in the upper and lower stones, so that it fitted within them. Opening the urn there were found inside, mingled with the ashes, twenty crystal balls, a gold ring with a stone setting, a needle, an ivory comb, and some bits of gold wire. The presence of the needle was taken to indicate conclusively that the ashes were those of a woman.[61]

The discovery of the tomb of Childeric was made, May 27, 1653, by a deaf-mute mason, named Adrien Quinquin, while he was excavating for the restoration of one of the dependencies of the church of Saint Brice de Tournai. One of the most interesting objects found in the tomb was the golden signet of Childeric bearing his head and the legend *Childerici regis*. The earliest description is given in a work by Chiflet entitled "Anastasis Childerici," "Resurrection of Childeric," published by Plantin of Antwerp in 1655. The various ornaments were sent by the Spanish Governor-General of the Netherlands to the Austrian treasury in Vienna, and were not long afterward, in 1664, graciously donated by

---

[60] Montfaucon, Les monumens de la monarchie Française, Paris, 1729, p. 15.

[61] Montfaucon, l. c.

ROCK-CRYSTAL SPHERE.

Japan, five inches diameter. Morgan collection, American Museum of Natural History, New York.

Emperor Leopold I to King Louis XIV, at the instance of Johann Philip of Schonborn, Archbishop of Mainz, who was under great obligation to the French sovereign.

In Paris the various ornaments were preserved in the Bibliothèque Royale until the night of November 5–6, 1831, when many of them, with other valuables, were stolen by an ex-convict. Closely pursued by the police, the thief threw his booty into the Seine; much of the plunder was subsequently recovered, but the signet of Childeric was lost for ever. The crystal ball had not seemed of sufficient value to tempt the thief and was left undisturbed; it was later, in 1852, deposited in the Louvre Museum.[62]

In a personal communication to Abbé Cochet made in 1858 by Mr. Thomas Wright, the latter stated that he had seen at Downing in Flintshire with Lord Fielding five crystal balls, bearing labels declaring that they came from the sepulchres of the kings of France violated at the time of the French Revolution. They had been purchased about 1810 at the sale of the Duchess of Portland's effects.[63]

Among the crystal balls found in French sepulchres may be noted one discovered by Rigollot in 1853 at Arras, and preserved in the Museum of that city; this still has the original gold mounting serving to attach it to the necklace from which it had been worn suspended. Another found at or near Levas was in the possession of M. Dancoise, a notary of Hénin-Liétard, dept. Pas de Calais.[64] In the Bibliothèque at Dieppe there is a crystal ball, 32 mm. in diameter, found at Douvrend, dept. Seine-Inferieure, in 1838, in a Merovingian tomb; this is pierced

[62] Cochet, "Le tombeau de Childeric Ier roi des Francs," Paris, 1859, pp. 16 sqq.

[63] Cochet, op. cit., p. 305.

[64] Cochet, op. cit., p. 302; figure.

through.[65] The department of Moselle supplied three discoveries of this kind, crystal balls having been found in a tomb at St. Preux-la-Montagne, Sablon and Moineville near Briey, the latter measuring 36 mm. in diameter.[66]

The Saxon tombs of England have also furnished a contingent of crystal balls, for example at Chatham, at Chassel Down on the Isle of Wight, where four were discovered, at Breach Down, Barham, near Canterbury, at Fairford, Gloucestershire, and also in Kent.[67]

We should also note a crystal ball found in a funerary urn at Hinsbury Hill, Northamptonshire;[68] this as well as the one found at Fairford was facetted.[69] From St. Nicholas, Worcestershire, is reported a crystal ball 1½ inches in diameter.[70]

In his "Hydrotaphia, or Urn Burial," published in 1658, Sir Thomas Browne (1605–1682), author of the "Religio Medici," relates that there was at that time in the possession of Cardinal Farnese, an urn in which, besides a number of antique engraved gems, an ape of agate, and an elephant of amber, there had been found a crystal ball and six "nuts" of crystal.[71]

---

[65] Cochet, op. cit., p. 303, No. 1.

[66] Simon, "Observations sur les sépulchres antiques découverts dans plusieures contrées des Gaules," p. 5; pl. ii, fig. 14.

[67] See Wylie's Fairford Graves," pl. iv, fig. 1, pl. v, fig 2; Akerman's "Remains of Pagan Saxondom," Roach Smith's "Collectanea antiqua"; Douglas' "Nenia Brittanica," and Hillier's "Antiquities of the Isle of Wight."

[68] Akerman, op. cit., p. 10.

[69] Journal of the Archæological Institute, vol. ix, p. 179.

[70] Akerman, op. cit., pp. 39, 40.

[71] Miscellanies upon various subjects, by John Aubrey, to which is added "Hydrotaphia, or Urn Burial," by Sir Thomas Browne, London, 1890, p. 244; chap. ii.

One of the largest and most perfect crystal balls is in the Dresden "Grüne Gewölbe" (Green Vaults). This weighs 15 German pounds and measures 6 2/3 inches in diameter; it was undoubtedly used for purposes of augury. Ten thousand dollars was the price paid for it in 1780.

A crystal ball known as the Currahmore Crystal, because it is kept at the seat of that name belonging to the Marquis of Waterford, has long enjoyed and still enjoys the repute of possessing magical powers. It is of rock-crystal, and the legend runs that one of the Le Poers brought it from the Holy Land, where it had been given him by the great crusader Godefroy de Bouillon (1058–1100). The ball is a trifle larger than an orange and a silver ring encircles it at the middle. The chief and much-prized virtue of this crystal is its power to cure cattle of any one of the many distempers to which they are subject. Its application for this purpose is rather peculiar, for the cattle are not touched with it, but driven up and down a stream in which it has been laid. Not only in the immediate neighborhood of Currahmore is resort had to this magic stone by the peasants, but requests for its loan are often made from far distant parts of Ireland. The privilege is almost always accorded and has never been abused, the crystal being in every case conscientiously returned to its rightful owner.[72]

The names "ghost-crystals," "phantom-crystals," "spectre-crystals," "shadow-crystals," etc., are applied to a form of quartz in which the crystallization was interrupted from time to time, so that in the transparent successive layers there is an occasional opaque layer,

[72] Lady Wilde, "Ancient Legends, Mystic Charms, and Superstitions of Ireland," Boston, 1888, p. 209.

often no thicker than the finest possible dusting of a whiter material. Sometimes as many as fifteen or twenty of these successive growths are observable, one over the other. When these crystals are in the natural form, they show beautifully from the sides and ends. Sometimes such crystals are found after they have been rolled in the beds of mountain torrents until they have become entirely opaque, but when the surfaces are polished, the "phantom," "spectre," or "ghost," appears with wonderful beauty. Occasionally the entire crystal has been worn down to a small part of the original prism, in which case it is cut into a ball. The ball may seem to be absolutely pure, but when held in certain lights little tent-like markings can often be observed; sometimes only one marking is visible, but there may be as many as twenty. These are occasionally due to a layer of smoky material, and, though they add a charm to the ball, they detract from its value. Nevertheless, crystal-gazers may find an additional interest when the "ghostly" or "spectral" interior exists in a crystal ball. This growth is similar in kind to that seen at times in opaque quartz, forming what is known as cap-quartz; here the crystallizations can frequently be broken apart so that they fit one over the other in many successive layers. Occasionally the regular crystalline development will be interrupted, as it were, and in place of the original crystal continuing its growth harmoniously, a larger crystal will form on a smaller one, forming a sort of mushroom, or "cap," or "stilt" quartz, as it is termed.

"PHANTOM CRYSTAL" OF QUARTZ (ROCK-CRYSTAL) MADAGASCAR.
In possession of the author.

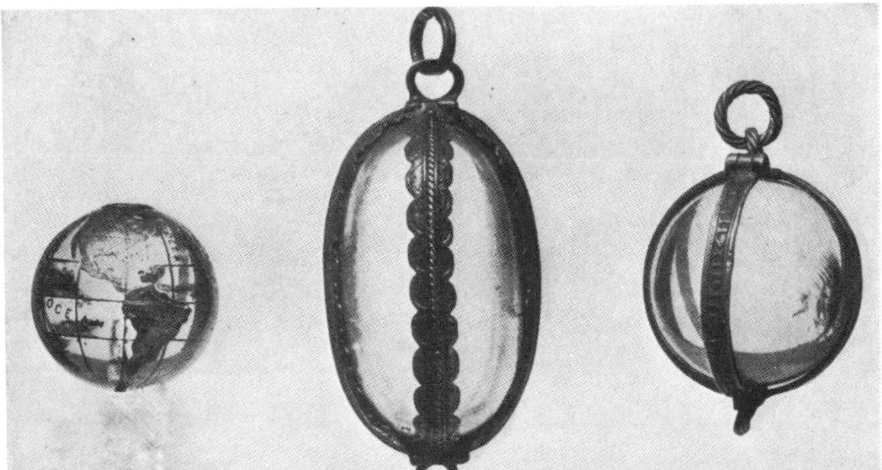

1. Rock-crystal, engraved with a map of the world. Russian work.
2, 3. Rock-crystal balls (one elipsoidal) mounted in silver. Probably twelfth or thirteenth century. Used for ornaments and possibly for scrying purposes. Collection of Sir Charles Hercules Read.

VII

Religious Uses of Precious Stones, Pagan, Hebrew, and Christian

THE use of stones for the decoration of images of the gods, and in religious ceremonies, more especially in those connected with the burial of the dead, can be traced back to a remote antiquity. Indeed, we may regard this religious use of precious or peculiar stones as the natural development of the original idea of their talismanic virtue. If a certain supernatural essence manifested itself in the stone, what more fit object could be imagined for the decoration of statues of the gods, or to bear engraved texts from the sacred writings, and to be placed with the bodies of the dead as "passports" to ensure the safe entry of the souls of the departed into the better land?

While this employment of mineral substances for religious purposes is practically universal, the earliest recorded instances come from Egypt, and concern the Egyptian custom of engraving texts from a very ancient ritual composition, called the Book of the Dead, upon certain semi-precious stones which had been cut into various symbolical forms. This "Book of the Dead," composed of a number of distinct chapters, each complete in itself, describes the passage of the soul of the deceased through the realm of the dead (Amenti). Here the soul addresses the gods and other beings who receive it, and the prayers and invocations recited in the chapters are supposed to procure a safe passage and protection from all evil influences or impediments.

One of the most usual of the engraved amulets is the buckle or tie (thet). This was generally of red jasper, carnelian, or red porphyry, or else of red glass or faience or of sycamore wood. The wood was symbolical of the blood of Isis, and the amulets were sometimes engraved with the 156th chapter of the Book of the Dead; they were placed on the mummy's neck. The formula engraved reads:

> Chapter of the buckle of carnelian which is put on the neck of the deceased.
>
> The blood of Isis, the virtue of Isis; the magic power of Isis, the magic power of the Eye are protecting this the Great one; they prevent any wrong being done to him.
>
> This chapter is said on a buckle of carnelian dipped into the juice of ankhama, inlaid into the substance of the sycamore-wood and put on the neck of the deceased.
>
> Whoever has this chapter read to him, the virtue of Isis protects him; Horus, the son of Isis, rejoices in seeing him, and no way is barred to him, unfailingly.[1]

Another amulet is the *tet*. The hieroglyph represents a mason's table and the word signifies "firmness, stability, preservation." These figures, made of faience, gold, carnelian, lapis-lazuli, and other materials, were placed on the neck of the mummy to afford protection.[2]

The "papyrus scepter," *uat*, is usually cut from matrix-emerald or made of faience of similar hue. *Uat* means "verdure, flourishing, greenness"; placed on the neck of the mummy it was regarded as emblematic of the

---

[1] Life Work of Sir Peter le Page Renouf, Paris, 1907, vol. iv, p. 342. In the vignette to chapter 93, to illustrate the protection afforded, a buckle with human hands seizes the arm of the deceased and prevents him from going toward the East, the inauspicious direction for departed souls, pl. xxv (Papyrus, Louvre iii, 93).

[2] Budge, "The Mummy," Cambridge, 1894, p. 259.

eternal youth it was hoped the deceased would enjoy in the realm of the dead. In the 159th chapter of the Book of the Dead, we read of an *uat* of matrix-emerald; it was believed to be the gift of Thoth, serving to protect the limbs of the deceased.[3]

The amulet representing the pillow, *urs*, was generally made of hematite. The 166th chapter of the Book of the Dead is sometimes engraved thereon. Dr. Budge renders this as follows:

> Rise up from non-existence, O prostrate one! They watch over thy head at the exalted horizon. Thou overthrowest thine enemies; thou triumphest over what they do against thee, as Horus, the avenger of his father, this Osiris[4] has commanded to be done for thee. Thou cuttest off the heads of thine enemies; never shall they carry off from thee thy head (?). Verily Osiris maketh slaughter at the coming forth of the heads of his enemies; may they never remove his head from him.

Of all these amulets, the type most frequently encountered has the shape of a heart, *ab*. These are found of carnelian, green jasper, basalt, lapis-lazuli, and other hard materials. The heart, regarded in ancient Egypt as the seat of life, was the object of especial care after death. Enclosed in a special receptacle it was buried with the mummy, and the belief was that only after it had been weighed in the balance of the underworld, against the symbol of law, could it regain its place in the body of the deceased. The heart was symbolically represented by the scarab.[5]

A fine example of a heart amulet shows on one side the figure of the goddess Neith with the pennu bird or

---

[3] Budge, "The Mummy," Cambridge, 1894, p. 261.

[4] The deceased was identified with Osiris.

[5] Budge, "The Mummy," Cambridge, 1894, p. 263.

phœnix, an emblem of the resurrection, and bears inscribed the chapter of the heart.[6]

The following extract from the Book of the Dead treats of the formula to be recited over a funeral scarab cut from a hard stone, perhaps the lapis-lazuli. Egyptian tradition assigned this chapter to the reign of Semti, the fifth king of the 1st Dynasty, about 4400 B.C.[7]

Chapter of not allowing a man's heart to oppose him in the divine regions of the nether world.

My heart which came from my mother, my heart necessary for my existence on earth, do not rise up against me, do not testify as an adversary against me among the divine chiefs in regard to what I have done before the gods; do not separate from me before the great lord of Amenti. Hail to thee, O heart of Osiris, dwelling in the West! Hail to you, gods of the braided beard, august by your sceptre! Speak well of the Osiris N; make him prosper by Nehbka. I am reunited with the earth, I am not dead in Amenti. There I am a pure spirit for eternity.

To be said over a scarabæus fashioned from a hard stone, coated with gold, and placed on the heart of the man after he has been anointed with oil. The following words should be said over him as a magic charm: "My heart which came from my mother, my heart is necessary for me in my transformations."

Take your aliments, pass around the turquoise basin, and go to him who is in his temple and from whom the gods proceed.

The most ancient inscription of this especially favorite text is on the plinth of a scarab in the British Museum bearing the cartouche of Sebak-em-saf, a king of the XIV Dynasty, 2300 B.C. It is made from an exceptionally fine piece of green jasper, the body and head of the beetle being carefully carved out of the stone, while the legs are of gold, carved in relief. The scarab is inserted into

---

[6] Birch, Catalogue of Egyptian Antiquities in Alnwick Castle, London, 1880, p. 224.

[7] Pierret, "Le livre des Morts," Paris, 1882, p. 138.

AMBER HEART-SHAPED AMULET.
Italian, seventeenth century.

AN INSCRIBED SCARAB (GREEN STONE) OF THE TYPE KNOWN AS A HEART-SCARAB. DATE ABOUT 1300 B.C.

The Scribe Pa-bak: Let him say: "O Heart that I received from my mother (to be said twice), O Heart that belongs to my spirit, rise not against me as witness, oppose me not before the judges, contradict me not in the presence of the Guardian of the Scales. Thou art the spirit that is in my body, Khnum that makest sound my limbs. When thou comest to the place of judgment whither we go, cause not my name to be rejected by the assessors, but let the pronouncement of judgment be favorable, and such as causes joy to the heart."

a gold base of tabloid form, and was found at Kurna (Thebes) by Mr. Salt. As green jasper was believed to possess altogether exceptional virtues as an amulet, this particular scarab was probably regarded as especially sacred.

It appears to have been the rule to engrave certain special chapters of the Book of the Dead, among those referring to the heart, upon particular stones. Thus, for instance, the 26th chapter was engraved on lapis-lazuli, the 27th upon feldspar, the 30th upon serpentine, and the 29th upon carnelian.[8] This may perhaps have been originally due to some association of the god principally invoked in the text with the precious substance upon which the text was engraved.

The form of an eye, fashioned out of lapis-lazuli and ornamented with gold, constituted an amulet of great power; it was inscribed with the 140th chapter of the Book of the Dead. On the last day of the month Mechir, an offering "of all things good and holy" was to be made before this symbolic eye, for on that day the supreme god Ra was believed to place such an image upon his head. Sometimes these eyes were made of jasper, and could then be laid upon any of the limbs of a mummy.[9]

Of the image of Truth, made from a lapis-lazuli and worn by the Egyptian high-priest, Ælian aptly says that he would prefer the judge should not bear Truth about with him, fashioned and expressed in an image, but rather in his very soul.[10]

---

[8] "Life Work of Sir Peter le Page Renouf," Paris, 1907, vol. iv, p. 76, note.

[9] Ibid., Paris, 1907, vol. iv, p. 295.

[10] Æliani, "Varia historia," lib. xiv, cap. xxxiv, Lug. Bat., 1731, Pars altera, p. 977.

Among the Assyrian texts giving the formulæ for incantations and various magical operations, there is one which treats of an ornament composed of seven brilliant stones, to be worn on the breast of the king as an amulet; indeed, so great was the virtue of these stones that they were supposed to constitute an ornament for the gods also. The text, as rendered by Fossey, is as follows: [11]

Incantation. The splendid stones! The splendid stones! The stones of abundance and of joy.

Made resplendent for the flesh of the gods.

The *ḥulalini* stone, the *sirgarru* stone, the *ḥulalu* stone, the *sându* stone, the *uknû* stone.

The *dushu* stone, the precious stone *elmêshu*, perfect in celestial beauty.

The stone of which the *pingu* is set in gold.

Placed upon the shining breast of the king as an ornament.

Azagsud, high-priest of Bêl, make them shine, make them sparkle!

Let the evil one keep aloof from the dwelling!

The names of two of these gems, the *hulalu* and the *ḥulalini*, suggest that they were of similar class. As the fundamental meaning of the root whence the names are formed is "to perforate," it is barely possible that we have here the long-sought Assyrian designation for the pearl, which was commonly regarded in ancient times as a stone. In Arabic the perforated pearl has a special name to distinguish it from the unperforated, or "virgin pearl." All we know of the *sându* is that it must have been a dark-colored stone. The *uknû,* however, is almost certainly the lapis-lazuli. It is often mentioned in the Tel el Amarna tablets as having been among the gifts sent by the kings of Babylonia and Assyria to the Pharaohs of Egypt, and also by the latter to friendly Asiatic

---

[11] Fossey, "La Magie Assyrienne," Paris, 1902, p. 301; see Rawlinson, "Cun. insc. of West. Asia," vol. iv, 18, No. 3.

monarchs. Of the *sirgarru* and *dushu* stones nothing is known, but the *elmêshu,* the seventh in the list, was evidently regarded as the most brilliant and splendid of all; indeed, Prof. Friedrich Delitzsch hazards the conjecture that it is the diamond. In any case this stone must have been set in rings and considered very valuable, for in an Assyrian text occurs the following passage: "Like an *elmêshu* ring may I be precious in thine eyes." [12] The fact that this stone is described as having "a celestial beauty" might incline us to believe that it was a sapphire.

The idea of this mystic ornament, composed of seven gems, probably originated in Babylonia, where the number seven was looked upon as especially sacred. As we shall see, there is some reason to attribute a Hindu origin to the nine gems, "the covering" of the King of Tyre, enumerated by Ezekiel, while the breastplate on the ephod of the Hebrew high-priest, with its twelve stones, symbolizing the twelve months of the year, appears to be of later date, and seems to belong to the time of the return from the Babylonian Captivity and the building of the second temple. Certainly, the historic and prophetic books of the Old Testament know nothing of it, although the Urim and Thummim are mentioned and the elaborate description given in Exodus is generally regarded by Biblical scholars as belonging to the so-called "Priestly Codex," the latest part of the Pentateuch, gradually evolved during the Exile and given its final form in the fifth century B.C.

In the very ancient Assyrio-Babylonian epic narrative of the descent of the goddess Ishtar to Hades, the guar-

---

[12] Delitzsch, "Assyrisches Wörterbuch," Leipzig, 1896, p. 74, s. v. *elmêshu.*

dian of the infernal regions obliges the goddess to lay aside some part of her clothing and ornaments at each of the seven gates through which she passes. At the fifth, we are told that she stripped off her girdle of *aban alâdi,* or stones which aided parturition.[13] It has been asserted, and perhaps with some reason, that of the many mineral substances supposed to possess this virtue, jade (nephrite) or jadeite was the earliest known.

The Babylonian legends also tell of trees on which grow precious stones. In the Gilgamesh epic a mystic cedar tree is described. This grew in the Elamite sanctuary of Irnina and was under the guardianship of the Elamite king Humbaba. Of this tree an inscription relates:

> It produces *samtu*-stones as fruit;
> Its boughs hang with them, glorious to behold;
> The crown of it produces lapis-lazuli;
> Its fruit is costly to gaze upon.

Another tree bearing precious stones was seen by the hero Gilgamesh, after he had passed through darkness for the space of twelve hours. This must have been a most resplendent object, to judge from the following description on a cuneiform tablet:[14]

> It bore precious stones for fruits;
> Its branches were glorious to the sight;
> The twigs were crystals;
> It bore fruit costly to the sight.

One of the rarest and most significant specimens illustrating the use of valuable stones for religious cere-

---

[13] Jansen, "Assyrisch-Babylonische Mythen und Epen," Berlin, 1900.

[14] Ward, "Seal Cylinders of Western Asia," Carnegie Institution Pub., Washington, D. C., 1910, pp. 232, 234.

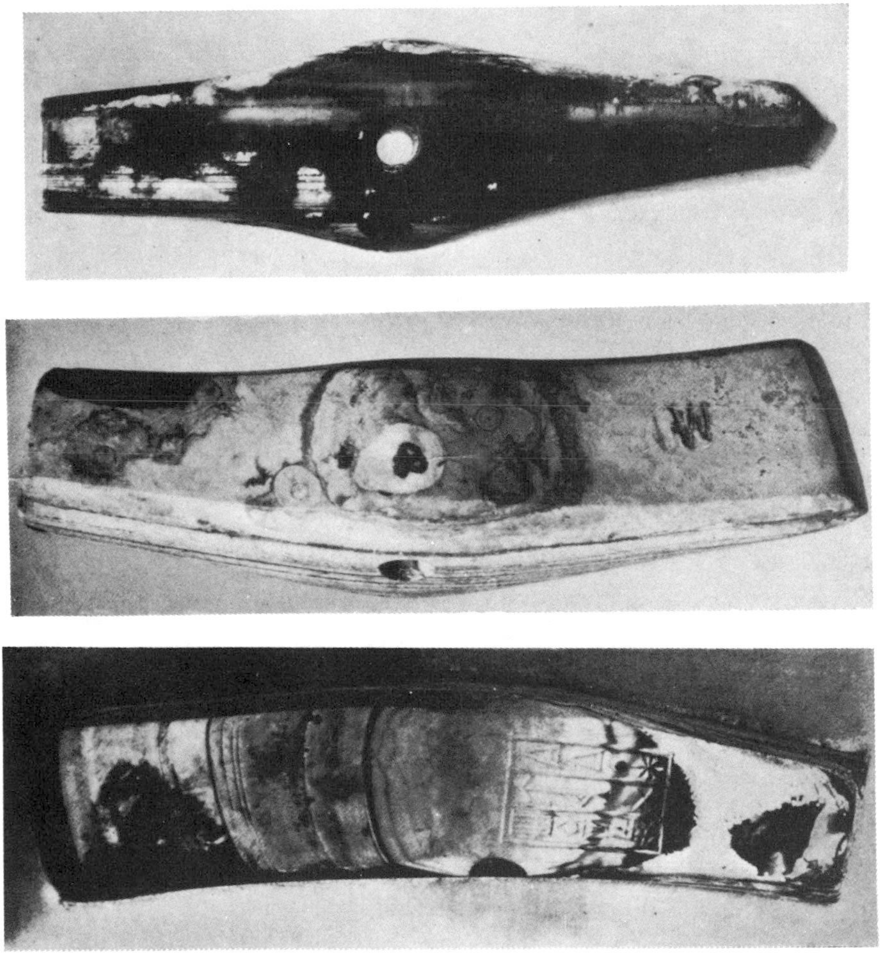

BABYLONIAN AXE HEAD

Agate, with inscription. Morgan collection, American Museum of Natural History, New York

monial purposes in the pagan world is in the Morgan-Tiffany collection. It is an ancient Babylonian axe-head made of banded agate. So regular, indeed, is the disposition of the layers in this agate that one might be justified in denominating it an onyx. Its prevailing hue is what may be called a "deer-brown"; some white splotches now apparent are evidently due to the action of fire or that of some alkali. This axe-head bears an inscription in archaic cuneiform characters, and presumably in the so-called Sumerian tongue, that believed to have been spoken by the founders of the Babylonian civilization. The form of the inscription indicates that the object dates from an earlier period than 2000 B.C.

While the characters are clearly cut and can be easily deciphered, the inscription is nevertheless exceedingly difficult to translate. It is evident that the axe-head was a votive offering to a divinity, probably on the part of a certain governor named Adduggish; but whether the divinity in question was Shamash (the sun-god), or the god Adad, or some other member of the Babylonian pantheon, cannot be determined with any finality. The French assyriologist, François Lenormant, who first described this axe-head in 1879, and Prof. Ira Maurice Price, of the Semitic Department of Chicago University, both admit that it may have been consecrated to Adad. As the weather-god, the thunderer, the axe-symbol would have been more especially appropriate to him in view of the usage, almost universal among primitive peoples, of associating stone axe-heads or axe-shaped stones with the thunderbolt, and hence with the divinity who was believed to have launched it toward the earth.

This Sumerian axe-head measures 134.5 mm. in length (5.3 inches), 35.5 mm. in width (1.4 inches), and 31 mm. in thickness (1.22 inches). It was originally secured by

Cardinal Stefano Borgia (1731–1804), for some time secretary of the College of the Propaganda in Rome, who probably acquired it from some missionary to the East. From the cardinal's family it passed for 15,000 lire ($3000) to the Tyszkiewicz Collection, and when the objects therein comprised were disposed of at public sale, the writer purchased it for the American Museum of Natural History in New York, April 16, 1902.[15]

At Alicante, in Spain, cut upon the pedestal of an ancient statue, supposed to have been that of Isis, was found an inscription giving a list of the offerings dedicated by divine command, by a certain Fabia Fabiana in honor of her granddaughter. Evidently the fond grandmother had given of her best and choicest jewels which were used to adorn the statue. They consisted of a diadem set with a "unio" (a large round pearl) and six smaller pearls, two emeralds, seven beryls, two rubies, and a hyacinth. In each ear of the statue was inserted an ear-ring bearing a pearl and an emerald; about the neck was hung a necklace consisting of four rows of emeralds and pearls, eighteen of the former and thirty-six of the latter. Two circlets bound around the ankles contained eleven beryls and two emeralds, while two bracelets were set with eight emeralds and eight pearls. The adornment was completed by four rings, two bearing emeralds, while two, placed on the little finger, were set with diamonds. On the sandals were eight beryls.[16]

---

[15] For a fuller description of this valuable relic, and a discussion of the meaning of the inscription, see "On the ancient inscribed Sumerian (Babylonian) axe-head for the Morgan Collection in the American Museum of Natural History," by George Frederick Kunz, with translation by Prof. Ira Maurice Price and discussion by Dr. William Hayes Ward. Bulletin of the Museum, vol. xxi, pp. 37–47, April 6, 1905.

[16] Montfaucon, "L'antiquité expliquée," vol. ii, Pt. II, 1719, pp. 324, 325; Plate 136.

A notable instance of an antique votive offering is the necklace of valuable precious stones dedicated to the statue of Vesta. The Byzantine historian Zosimus attributes the tragic end of Stilicho's widow, Serena, to her having despoiled the image of Vesta of this costly ornament, and finds a sort of poetic justice in the manner of her death, since she was strangled by a cord which encircled her neck.

It is not only in the works of the Fathers of the Christian Church that we find precious stones used as similes of religious virtue, in Buddhist writings also we have examples of this. In the "Questions of King Milinda," composed perhaps as early as the third century of our era, occur the following passages:[17]

Just, O King, as the diamond is pure throughout; just so, O King, should the strenuous Bhikshu, earnest in effort, be perfectly pure in his means of livelihood. This, O King, is the first quality of the diamond he ought to have.

And again, O King, as the diamond cannot be alloyed with other substance; just so, O King, should the strenuous Bhikshu, earnest in effort, never mix with wicked men as friends. This, O King, is the second quality of the diamond he ought to have.

And again, O King, just as the diamond is set together with the most costly gems; just so, O King, should the strenuous Bhikshu, earnest in effort, associate with those of the highest excellence, with men who have entered the first or second or third stage of the Noble Path, with the jewel treasures of the Arahats, of the recluses of the threefold wisdom, or of the sixfold insight. This, O King, is the third quality of the diamond he ought to have. For it was said, O King, by the Blessed one,[18] the god over all gods, in the Sutta Nipâta:

Let the pure associate with the pure,
Ever in recollection firm;
Dwelling harmoniously wise,
Thus shall ye put an end to griefs.

---

[17] "The Questions of King Milinda," tr. from the Pâli by T. W. Rhys Davids, vol. ii, Oxford, 1894, p. 128.

[18] Buddha.

The description of the New Jerusalem in the book of Revelations finds a curious parallel in the Hindu Puranas. Here we are told that the divine Krishna, the eighth incarnation of Vishnu, took up his abode in the wonderful city Devârakâ, and was visited there by the various orders of gods and geniuses.[19]

Gods, Asuras, Gandharas, Kinnaras began to pour into Dwáraká, to see Krishna and Valaráma.

Some descended from the sky, some from their cars—and alighting underneath the banyan tree, looked on Dwáraká, the matchless.

The city was square,—it measured a hundred *yojonas,* and over all, was decked in pearls, rubies, diamonds, and other gems.

The city was high,—it was ornamented with gems; and it was furnished with cupolas of rubies and diamonds,—with emerald pillars, and with court-yards of rubies. It contained endless temples. It had cross-roads decked with sapphires, and highways blazing with gems. It blazed like the meridian sun in summer.

As compared with the description in Revelations we cannot fail to note the lack of definiteness. Instead of the well-ordered scheme of color as represented by the twelve precious stones dedicated to the twelve tribes of Israel, the mystic Hindu city is simply a gorgeous mass of the most brilliant gems known in India.

The poetic description of the royal city Kusavati, given in the Maha Sudassana Suttanta, may perhaps have originated in some tradition regarding Ecbatana or Babylon. Seven ramparts surrounded Kusavati, the materials being respectively gold, silver, beryl, crystal, agate, coral and (for the last) "all kinds of gems." In these ramparts were four gates—one of gold, one of silver, one of crystal and one of jade—and at each gate seven pillars were fixed, each three or four times the

---

[19] Surindro Mohun Tagore, "Mani Málá," Pt. II, Calcutta, 1881, pp. 715, 717.

height of a man and composed of the seven precious substances that constituted the ramparts. Beyond the ramparts were seven rows of palm trees, the fourth row having trunks of silver and leaves and fruit of gold; then followed palms of beryl, with leaves and fruit of beryl; agate palms, whose fruit and leaves were of coral, and coral palms, with leaves and fruit of agate; lastly, the palms whose trunks were composed of "all kinds of gems," had leaves and fruits of the same description, "and when these rows of palm trees were shaken by the wind, arose a sound sweet and pleasant, and charming and intoxicating." [20]

In Greek literature also there is a "gem-city,"—namely, the city of the Islands of the Blessed, described by Lucian in his Vera Historia.[21] The walls of this city were of emerald, the temples of the gods were formed of beryl, and the altars therein of single amethysts of enormous size. The city itself was all of gold as a fit setting for these marvellous gems.

Hindu mythology tells of a wonderful tank formed of crystal, the work of the god Maya. Its bottom and sides were encrusted with beautiful pearls and in the centre was a raised platform blazing with the most gorgeous precious stones. Although it contained no water, the transparent crystal produced the illusion of water, and those who approached the tank were tempted to plunge into it and take a refreshing bath in what appeared to be clear, fresh water.[22]

---

[20] Bhuddist Suttas, trans. from Pali by T. W. Rhys Davids; "Sacred Books of the East," vol. xi, Oxford, 1881.

[21] Lib. ii, cap. 11. Luciani Opera, ex recog. C. Jacobitz, vol. i, Leipzig, 1884, p. 56.

[22] Surindro Mohun Tagore, "Mani Málá," Pt. II, Calcutta, 1881, p. 79.

The Kalpa Tree of Hindu religion, a symbolical offering to the gods, is described by Hindu poets as a glowing mass of precious stones. Pearls hung from its boughs and beautiful emeralds from its shoots; the tender young leaves were corals, and the ripe fruit consisted of rubies. The roots were of sapphire; the base of the trunk of diamond, the uppermost part of cat's-eye, while the section between was of topaz. The foliage (except the young leaves) was entirely formed of zircons.[23]

The Chinese Buddhist pilgrim Heuen Tsang, who visited India between 629 and 645 A.D., tells of the wonderful "Diamond Throne" which, according to the legend, had once stood near the Tree of Knowledge, beneath whose spreading branches Gautama Buddha is said to have received his supreme revelation of truth. This throne had been constructed in the age called the "Kalpa of the Sages"; its origin was contemporaneous with that of the earth, and its foundations were at the centre of all things; it measured one hundred feet in circumference, and was made of a single diamond. When the whole earth was convulsed by storm or earthquake this resplendent throne remained immovable. Upon it the thousand Buddhas of the Kalpa had reposed and had fallen into the "ecstasy of the diamond." However, since the world has passed into the present and last age, sand and earth have completely covered the "Diamond Throne," so that it can no longer be seen by human eye.[24]

In the Kalpa Sutra, written in Prakrit, one of the sacred books of the Jains, the rivals of the Buddhists, it is said that Harinegamesi, the divine commander of the

---

[23] Surindro Mohun Tagore, "Mani Málá," Pt. II, Calcutta, 1881, pp. 645, 647.

[24] Heuen Tsang, "Mémoires sur les contrées occidentales," French trans. by Stanislas Julien, Paris, 1857, vol. i, p. 461.

foot troops, seized fourteen precious stones, the chief of which was *vajra,* the diamond, and rejecting their grosser particles, retained only the finer essence to aid him in his transformations. In the same sutra the following glowing description is given of the adornment of the surpassingly beautiful goddess Sri:[25]

> On all parts of her body shone ornaments and trinkets, composed of many jewels and precious stones, yellow and red gold. The pure cup-like pair of her breasts sparkled, encircled by a garland of Kunda flowers in which glittered a string of pearls. She wore strings of pearls made by clever and diligent artists, strung with wonderful strings, a necklace of jewels with a string of Dinaras, and a trembling pair of earrings, touching her shoulders, diffused a brilliancy; but the united beauties and charms of these ornaments were only subservient to the loveliness of her face.

As engraved decoration of a fine Chinese vase of white jade with delicate crown markings, appear eight storks, each of which bears in its beak an attribute of one of the Eight Taoist Immortals. Thus we have the double gourd as attribute of the most powerful of these demi-gods known as "Li with the Iron Crutch," whose aid is sought by magicians and astrologers; the magic sword, with which Lu T'ung-pin vanquished the spirits of evil that roamed through the Chinese Empire in the form of terrible dragons; the basket of flowers, attribute of Lan Ts'ai-ho, the patron of gardeners and florists; the royal fan used by Han Chung-li, of the Chow Dynasty (1122-220 B.C.), to call again to life the spirits of the departed; the lotus flower, emblematic of the virgin Ho Hsien-Ku, venerated somewhat as a patron saint by Chinese housewives, and who acquired the gift of immortal life by the help of a powder of pulverized jade and mother-of-pearl;

[25] Gaina Sutras, trans. from Prakrit by Hermann Jacobi; "Sacred Books of the East," vol.xxii, Oxford, 1884, pp. 227, 233.

the bamboo tubes and rods with which the mighty necromancer Chang Kuo, patron of artists, evoked the souls of the dead; the flute of the musicians' patron, Han Hsiang-tzu, who owed his immortality to his craft in stealthily entering the Taoist paradise and securing a peach from the sacred tree of life; and, lastly, the castanets of Tsao Kuo-chin, especially revered by Chinese actors.

The prevailing belief in India, that treasures offered to the images or shrines of the gods will bring good fortune to the generous donor, finds expression in many ancient and modern Hindu writings. In the Rig Veda it is said that "by giving gold the giver receives a life of light and glory." In the Samaveda Upanishad we read: "Givers are high in Heaven. Those who give horses live conjointly with the sun; givers of gold enjoy eternal life; givers of clothes live in the moon." Another text (Hâiti Smriti) reads:[26]

> Coral in worship will subdue all the three worlds. He who worships Krishna with rubies will be reborn as a powerful emperor; if with a small ruby, he will be born a king. Offering emeralds will produce Gyana or Knowledge of the Soul and of the Eternal. If he worships with a diamond, even the impossible, or Nirvâna, that is Eternal Life in the highest Heaven, will be secured. If with a flower of gold a man worships for a month, he will get as much wealth as Kuvera, the Lord of Rubies, and will hereafter attain to Nirvâna and to Muskwa, or Salvation.

At Multan, one of the most ancient cities of India, situated in the Punjab, 164 miles southwest of Lahore, there was in the Hindu temple an idol having for eyes two great pearls. The eyes of the rude image of Jagganath at Puri, in Bengal (Orissa), are said to have

[26] Hendley, "Indian Jewellery," London, 1909, p. 33.

at one time been formed of precious stones, as were also those of the idols of Vishnu at Chandernagore and in the great seven-walled temple at Srirangam, whence appears to have come the Orloff diamond.

In ceremonial worship the Hindus recognize sixteen offerings, the ninth consisting of gems and jewelry, and a divine assurance of adequate return to the giver appears in the Bhagavat Purana, where Krishna says, "Whatever is best and most valued in this world and that which is most dear to you should be offered to me, and it will be received back in immense and endless quantity." On certain appointed days the holy images are decorated with the choicest garments and the richest jewelry in the temple treasury; this is especially the case on the day celebrated as the birthday of the respective divinity. However, the gifts are believed to retain their sacred character as dedicated objects only for a comparatively brief period, varying from a month or more for garments and vestments, to ten or twelve years for jewels, such as the naoratna or the panchratna, the prized and revered jewels, composed respectively of nine and five gems. The panchratna usually consists of gold, diamond, sapphire, ruby, and pearl. After the gifts have ceased to be worthy of use in the temples, they may be disposed of to defray the expenses of the foundation, including the cost of supporting the numerous priests and attendants. As the objects still retain their sacred associations, they are eagerly bought by pious Hindus, who undoubtedly regard them as valuable talismans. Thus they not only serve to bring blessings upon the donors, but also constitute one of the chief sources of income for the temples.[27]

[27] Hendley, "Indian Jewellery," London, 1909, pp. 33, 34.

One of the oldest and perhaps the most interesting talismanic jewel is that known as the naoratna or nararatna, the "nine-gem" jewel. It is mentioned in the old Hindu ratnaçastras, or treatises on gems, for example, in the Nararatnaparîkshâ, where it is described as follows:[28]

Manner of composing the setting of a ring:

In the centre	The Sun	The Ruby
To the East	Venus	The Diamond
To the Southeast	The Moon	The Pearl
To the South	Mars	The Coral
To the Southwest	Râhu	The Jacinth
To the West	Saturn	The Sapphire
To the Northeast	Jupiter	The Topaz
To the North	The descending node	The Cat's-eye
To the Northwest	Mercury	The Emerald

Such is the planetary setting.

From this description we learn that the jewel was designed to combine all the powerful astrological influences. The gems chosen to correspond with the various heavenly bodies, and with the aspects known as the ascending and descending nodes, differ in some cases from those selected in the West. For instance, the emerald is here assigned to Mercury, whereas in Western tradition this stone was usually the representative of Venus, although it is sometimes associated with Mercury also.[29] On the other hand, the diamond is dedicated to Venus, instead of to the Sun as in the Western world.

In the naoratna the five gems known to the Hindus as the mahâratnâni, or "great gems,"—the diamond, pearl,

[28] Finot, "Les lapidaries indiens," Paris, 1896, p. 175.

[29] Morales, "De las piedras preciosas," Valladolid, 1604 (fol. 16 verso).

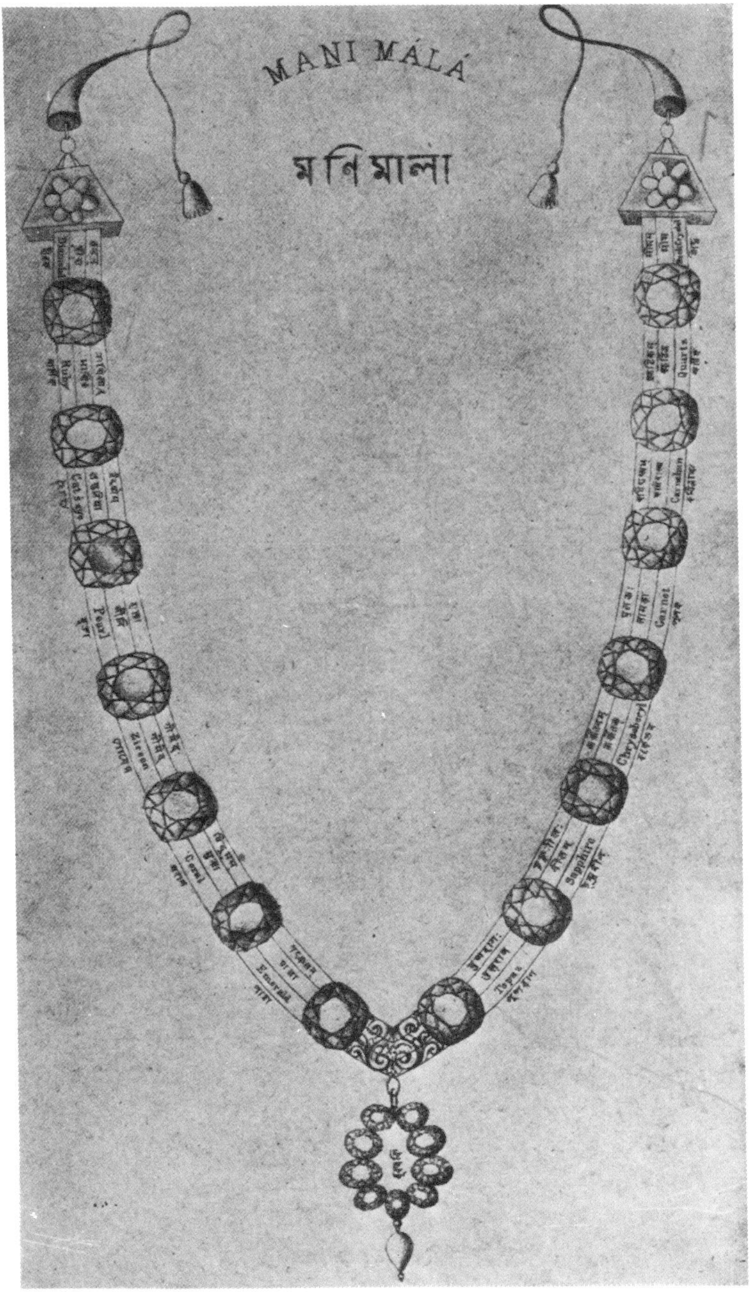

MANI MÁLÁ, OR CHAIN OF GEMS.

Comprising diamond, ruby, cat's-eye, pearl, zircon, coral, emerald, topaz, sapphire, chrysoberyl, garnet, carnelian, quartz and rock-crystal. A pendant is the nao-ratna, or "nine-gem" ornament, suspended from which is a pear-shaped pearl.

In possession of the late Rajah Sir Surindro Mohun Tagore, of Calcutta. From his "Mani Málá," Calcutta, 1879, Vol. I, iv–506 pp., 2 plates, portrait and plate; Vol. II, xiv + ii 507–1046 pp. Contains 49 figures on 10 plates.

ruby, sapphire, and emerald,—were, as we see, associated with the Sun and Moon, Venus, Mercury, and Saturn, while the four lesser gems (uparatnâni)—namely, the jacinth, topaz, cat's-eye, and coral—represent Mars, Jupiter, Râhu, and the descending node. The two last named are very important factors in astrological calculations and are often called the Dragon's Head and the Dragon's Tail. These designations signify the ascending and descending nodes, indicating the passage of the ecliptic by the Moon in her ascent above and descent below this arbitrary plane.

In three somewhat obscure passages of the Rig Veda there are references to the seven ratnas. Whether these were gems cannot be determined, since the primary meaning of the word ratna is "a precious object," not necessarily a precious stone; but it is possible that we may have here an allusion to some earlier form of talisman, in which only the Sun, Moon, and the five planets were represented.

It is easy to understand that such a talisman as the naoratna, combining the favorable influences of all the celestial bodies supposed to govern the destinies of man, must have been highly prized, and we may well assume that only the rich and powerful could own this talisman in a form ensuring its greatest efficacy. For the Hindus believed that the virtue of every gem depended upon its perfection, and they regarded a poor or defective stone as a source of unhappiness and misfortune.

In modern times this talisman is sometimes differently composed. A specimen shown in the Indian Court of the Paris Exposition of 1878 consisted of the following stones: coral, topaz, sapphire, ruby, flat diamond, cut diamond, emerald, amethyst, and carbuncle. Here the cut diamond, amethyst, and carbuncle take the place of the jacinth, pearl, and cat's-eye.

Instead of uniting the different planetary gems in a single ring, they have sometimes been set separately in a series of rings to be worn successively on the days originally named after the celestial bodies. We read in the life of Apollonius of Tyana (first century A.D.) by Philostratus: "Damis also relates that Iarchas gave to Apollonius seven rings named after the planets, and the latter wore these, one by one, in the order of the week-days."[30] Although it is not expressly stated that the appropriate stones were set in the rings, the custom of the time makes it probable that this was the case.

NINE GEMS.

English	Sanskrit	Burmese	Chinese (Canton)	Arabic
Diamond	Vajra	Chein	Chun-syak	Mâs
Ruby	Manikya	Budmiya	Se-fla-yu-syak	Yâkût bihar
Cat's-eye	Vaidûrya	Châno	Mâu-ji gan	Ain al-hirr
Zircon	Gomeda	Gomok	Pi-si	Hajar yamânî
Pearl	Muktâ	Pa-le	Chun-ti	Lûlû
Coral	Pravâla	Tadâ	Sau-ho-chi	Murjân
Emerald	Marakata	Mujâ	Luk-syak	Zumurrud
Topaz	Pushyaraga	Outfiyâ	Si-lang-syak	Yâkût al-azrak
Sapphire	Nîla	Nîlâ	Chang-syak	Yâkût al-açfar

Among the Burmese the value for occult purposes of the nine gems composing the naoratna, or nararatna, is strictly determined in the following order: first, the ruby; second, the diamond, or rock-crystal; third, the pearl; fourth, the coral; fifth, the topaz; sixth, the sapphire; seventh, the cat's-eye; eighth, the amethyst; and ninth, the emerald.[31] That the ruby, diamond and pearl should occupy places of honor is quite natural, but the rele-

[30] Philostrati, "De Vita Apollonii," lib. iii, cap. 36.

[31] Personal communication from Taw Sein Ko.

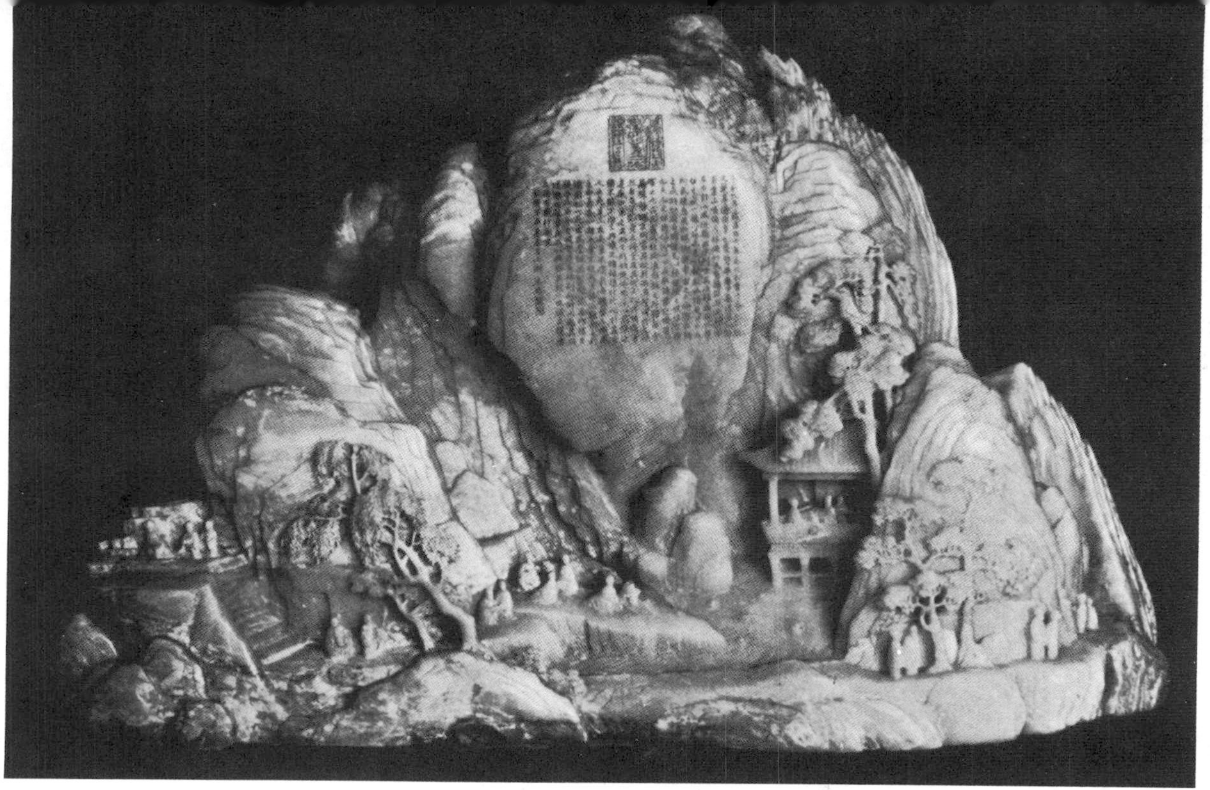

SCULPTURED JADE MOUNTAIN.

Probably the largest mass of sculptured jade in existence. The design commemorates the meetings of a literary club of the fourth century. The Chinese characters (colored red) in the side of the cliff express the famous *Lan Ting Hsu*, or "Epidendron Pavilion Essay," by Wang Hi-che (A.D. 321–379), ever since used by the Chinese as a model of elegant caligraphy, and were engraved directly from the autograph of the Emperor Ch'ien-lung, written by him in 1784. Height 23 inches, width 38½ x 18½ inches; weight 640 pounds. From the Summer Palace, west of Peking. Collection of T. D. Walker, of Minneapolis, Minn.

gation of the sapphire to sixth place, after coral and topaz, seems to be a rather unfair treatment of this beautiful stone.

The yellow girdles worn by the Chinese emperors of the Manchu dynasty were variously ornamented with precious stones according to the different ceremonial observances at which the emperor presided. For the services in the Temple of Heaven, the very appropriate choice of lapis-lazuli ornaments was made; for the Altar of Earth, yellow jade was favored; for a sacrifice on the Altar of the Sun, the gems were red corals, while white jade was selected for the ceremonies before the Altar of the Moon. Jade of different colors was used for the six precious tablets employed in the worship of heaven and earth and the four cardinal points. For the worship of Heaven there was the dark-green round tablet; for that of Earth, an octagonal tablet of yellow jade. The East was worshipped with a green pointed tablet; the West was worshipped with the white "tiger-tablet"; the North with a black, semi-circular tablet, and the South with a tablet of red jade.[32]

Of all the Chinese works on jade the most interesting and remarkable is the *Ku yü t'ou pu* or "Illustrated Description of Ancient Jade," a catalogue divided into a hundred books and embellished with upward of seven hundred figures. It was published in 1176, and lists the magnificent collection of jade objects belonging to the first emperor of the Southern Sung dynasty. One of the treasures here described was a four-sided plaque of pure white jade over two feet in height and breadth, and it was

[32] The Bishop Collection: "Investigations and Studies in Jade," New York, 1906, vol. i, p. 54, The "Yushuo" of T'ang Jing-tso, trans. by Stephen W. Bushnell.

regarded as of altogether exceptional value, for on it was a design miraculously engraven. This was a figure, seated on a mat, with a flower-vase on its left and an alms-bowl on the right, in the midst of rocks enveloped in clouds. The figure was an image of the Buddhist saint, Samantabahadra, and the plaque is said to have been washed out of a sacred cave in the year 1068, by a violent and mysterious current.[33]

Jade talismans are very popular at the present day in the Mohammedan world, and among the Turks they are so highly prized as heirlooms that it is difficult to secure any of them. There is an orthodox Mohammedan sect, whose members call themselves Pekdash, and who during their whole lifetime carry about with them a flat piece of jade as a protection against injury or annoyance of every kind.[34]

The four rain-making gods are shown wearing necklaces of coral and turquoise in the ceremonial sand-paintings of the Navajos. These four gods are respectively colored to denote the four cardinal points; black for North, blue for South, yellow for West, and white for East. The whole painting, measuring nine by thirteen feet, is guarded on three sides by magic wands; toward the East it is left unprotected, as only good spirits are believed to dwell in this direction. Each of the rain-gods carries suspended from his right wrist an elaborately decorated tobacco pouch, bearing the figure of a stone pipe. The Navajos believe that in this pouch the god places a ray of sunlight with which he lights his pipe;

[33] The Bishop Collection: "Investigations and Studies in Jade," New York, 1906, vol. i, p. 36.

[34] Kobert, "Ein Edelstein der Vorzeit," Stuttgart, 1910, p. 26.

when he smokes, clouds form in the sky and the rain descends. In the sand-picture representing the God of the Whirlwind this divinity also wears ear-pendants and a necklace of turquoise.[35]

Of the turquoise in Aztec times we have the testimony of the missionary Bernardino de Sahagun that one variety, presumably that regarded as the finest and most attractive, bore the name *teuxivitl,* which signified "turquoise of the gods." No one was allowed either to own or wear this as it was exclusively devoted to the service of the gods, whether as a temple offering, or for the decoration of the divine images. Sahagun describes this turquoise as "fine, unspotted and very clear. It was very rare and was brought to Mexico from afar. Some specimens were of rounded shape, like a hazel-nut cut in half; others were broad and flat, and some were pitted as though in a state of decomposition."[36]

The god of fire, Xiuhtecutli, or Ixçocauhqui, presided over the ceremony of piercing the ears of the young boys and girls. The image of this god was decorated with ear-rings encrusted with a mosaic of turquoise. He held in his left hand a buckler on which were five large green stones called *chalchiuitl* (jadeite), placed in the form of a cross on a plate of gold almost covering the shield.[37]

At the time of the Spanish Conquest an immense emerald, almost as large as an ostrich egg, was adored by the Peruvians in the city of Manta. This "emerald goddess" bore the name of Umiña, and, like some of

[35] Alfred Marston Tozzer, "Navajo Religious Ceremonials," Putnam Anniversary Volume, New York, 1909, pp. 323–326, 329, Plate II.

[36] Sahagun, "Historia general de las cosas de Nueva España," Mexico, 1830, vol. iii, p. 297.

[37] Sahagun, l. c., 1829, vol. i, p. 18; lib. i, cap. xiii.

the precious relics of the Christian world, was only exhibited on high feast days, when the Indians flocked to the shrine from far and near, bringing gifts to the goddess. The wily priests especially recommended the donation of emeralds, saying that these were the daughters of the goddess, who would be well pleased to see her offspring. In this way an immense store of emeralds rewarded the efforts of the priests, and on the conquest of Peru all these fine stones fell into the hands of Pedro de Alvarado,[38] Garcilasso de la Vega, and their companions. The mother emerald, however, had been so cleverly concealed by the priests of the shrine that the Spaniards never succeeded in gaining possession of it. Many of the other emeralds were destroyed because of the ignorance and stupidity of some of their new owners, who, supposing that the test of a true emerald was its ability to withstand hard blows, laid the stones on an anvil and hammered them to pieces. The old and entirely false notion that the genuine diamond could endure this treatment may have suggested the unfortunate test.

Garcilasso likens the growth of the emerald in its mine to that of a fruit on a tree, and he believed that it gradually acquired its beautiful green hue, that part of the crystal nearest the sun being the first to acquire color. He notes an interesting specimen found in Peru, half of which was colorless like glass, while the other half was a brilliant green; this he compares with a half-ripened fruit.[39]

The remarkable jade adze, generally known as the

---

[38] Garcilasso de la Vega, "Histoire des Incas." Fr. trans. by Jean Baudoin, Amsterdam, 1715, vol. ii, pp. 255–257.

[39] Ibid., p. 347.

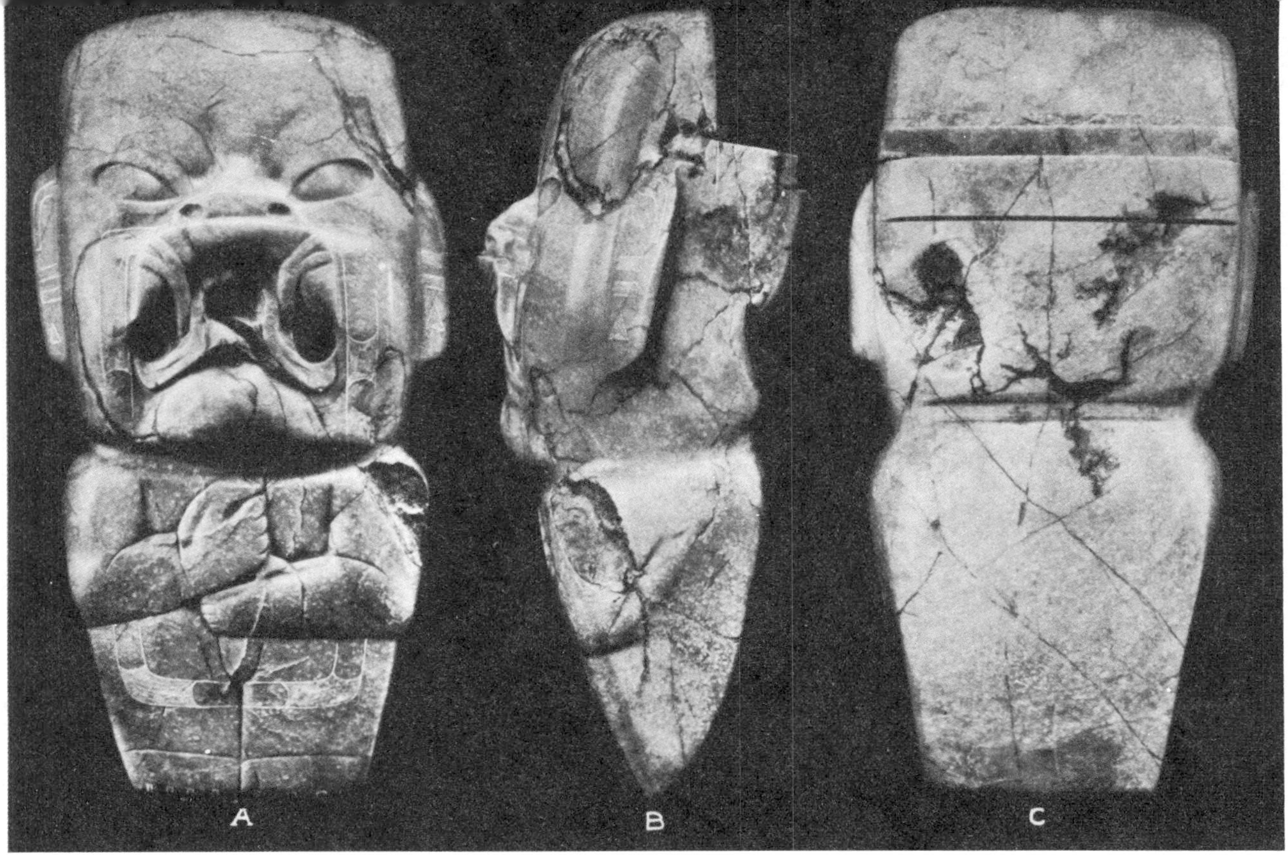

VOTIVE ADZE OF JADEITE FROM MEXICO.

a, Front view. b, Side view. c, Rear view. Kunz Collection, American Museum of Natural History, New York. $10\frac{13}{16} \times 6 \times 4\frac{5}{8}$ inches. Weight, 19 pounds Troy.

"Kunz adze," was found in Oaxaca, Mexico, brought to the United States about 1890, and is now in the American Museum of Natural History, New York. Of a light greenish-gray hue, with a slight tinge of blue, this jade artefact is 272 mm. long ($10^{13}/_{16}$ inches), 153 mm. wide (6 inches) and 118 mm. thick ($4\frac{5}{8}$ inches); its weight is 229.3 Troy ounces, nearly sixteen pounds avoirdupois. Rudely, but not unskilfully, carved upon its face is a grotesque human figure. Four small, shallow depressions, one under each eye and one near each hand, may have served to hold in place small gold films, but no trace of gold decoration is now extant. In its mechanical execution this adze offers evidence of considerable skill on the part of the Aztec lapidary, the polish equalling that of modern workers. In the fact that a large piece, which must apparently have weighed at least two pounds, has evidently been cut out of this implement by some one of its Indian owners, we can see a proof of the talismanic power ascribed to jadeite in Aztec times, for there can be little doubt that nothing less than a belief in the great virtue of jadeite coupled with the rarity of the material could have induced the mutilation of what must have been regarded in its time as a remarkable work of art.[40]

The source of the prehistoric jade (nephrite and jadeite) found in Europe, and also of that worked into ornaments by the Indians before the Spanish Conquest of America, was long the subject of contention among mineralogists and archæologists. In Germany this question was denominated the *Nephritfrage,* and the most notable contribution to the discussion was the great scien-

[40] "A Remarkable Jadeite Adze," American Association for the Advancement of Science. Kunz, "Gems and Precious Stones of North America," New York, 1890, pp. 278–280.

tific and scholarly work issued by Heinrich Fischer.[41] His conclusion was that as there was no evidence of the existence of these minerals outside of a few localities in Asia, the European and American supply must have been brought to these parts of the world from Asia, and that hence the presence of these jade artefacts in America clearly pointed to commercial intercourse at an early period between the American continent and Asia, and might be regarded as offering a strong argument in favor of an Asiatic origin for an American civilization. According to this theory the prehistoric jade objects found in Europe must have had a similar source, and would constitute a proof of the existence of traffic with remote points in Asia at a date long previous to that commonly accepted.

This view was strongly opposed by Prof. A. B. Meyer, of Dresden, and recent discoveries have effectively disproved the theory in the case of Europe at least, for nephrite has been found there *in situ* in several places. The largest mass of this material that has been taken from a European deposit is that found by the writer at Jordansmühl in Silesia, in April, 1899, and which weighed 4704 pounds.[42] The origin of American jade in the forms of nephrite and jadeite has not yet been definitely determined, but we have every reason to suppose that deposits of these minerals will eventually be discovered in various parts of the American continent, as they have already

---

[41] "Nephrit und Jadite," Stuttgart, 1880.

[42] The Bishop Collection, "Investigations and Studies in Jade," New York, 1906, vol. i, pt. iii, "Jade as a Mineral," by George Frederick Kunz, p. 177. This immense mass of nephrite which forms part of the Heber Bishop Collection loan of jade is now in the American Museum of Natural History, New York.

been in Europe. Indeed, the existence of nephrite in Alaska is already well attested.

The peculiar and characteristic qualities of these substances have made them favorite materials for ornamental objects from the earliest ages down to our own day, and in almost all parts of the world. A most important element contributing to the popularity of jade has been its supposed possession of wonderful talismanic and therapeutic virtues, and while the Western world has not the same belief in these matters as the Eastern world, a more or less definite appreciation of what jade still signifies for many in the Orient, continues to exercise an influence over both Americans and Europeans, making objects of nephrite or jadeite highly prized everywhere at the present time.

The term *chalchihuitl* was indifferently applied by the ancient Mexicans to a number of green or greenish-white stones; *quetzal chalchihuitl,* which was regarded as the most precious variety, may perhaps have more exclusively denoted jadeite. This is somewhat indefinitely described by Sahagun as being "white, with much transparency, and with a slight greenish tinge, something like jasper." Of eight ornamental objects of green stone examined some years ago by the writer, four were of jadeite, one of serpentine, another of green quartz, and the remaining two of a mixture of white feldspar and green hornblende. An inferior kind of *chalchihuitl,* said by Sahagun to have come from quarries in the vicinity of Tecalco, appears to have been identical with the so-called "Mexican onyx" which is found in veins in that place and is an aragonite stalagmite. This material, from which figures, ornaments and beads were made by the ancient Mexicans, is to-day greatly valued as an ornamental stone.

The greater number of ancient Mexican jadeite beads appear to have been rounded pebbles of this material, assorted as to size and drilled for use in making necklaces. Other green stones used at this time in Mexico were green jasper, green plasma, serpentine and also the "Tecalco onyx" or "marble" above mentioned. In many cases these substances are of such rich green that they might easily be mistaken for jadeite by those who lacked the tests or the experience at the command of modern mineralogists. Should jadeite ever be found *in situ* in Mexico, it seems probable that the discovery will be made in the State of Oaxaca, whence came the finest ancient specimens, including the splendid votive adze. Moreover, one of the few materials by which jadeite can be worked is furnished by the streams of this region, whence have been taken several rolled pebbles which the writer has identified as yellow and blue corundum, the quality being equal to that of specimens from Ceylon.[43]

Gesner describes one of the lip ornaments worn by the aborigines of South America in the following words: [44]

> A green stone or gem which the inhabitants of the West Indies use. They pierce their lips and insert this stone so that the thicker part adheres to the hole and the rest protrudes. We might call these ornaments *oripenduli* [mouth-pendants]. This stone was given me by a learned Piedmontese, Johannes Ferrerius, and he wrote of it as follows: "I send a cylindrical green stone, as long as a man's middle finger, and having at one extremity two ridges. It is stated that the Brazilians of high rank wore these, from their youth, in their pierced lips; one or more being worn according to the dignity of the wearer. While eating, or whenever they so wish for any other reason, these ornaments are removed from the lips."

---

[43] Kunz, "Chalchiuitl: a note on the jadeite discussion," Science, vol. xii, No. 298.

[44] Gesneri, "De figuris lapidum," Tiguri, 1565, fol. 107 verso, 108 recto.

Similar ornaments, made of a green quartz and of beryl, are in the Kunz collection in the Field Museum of Chicago.

The reason for these strange mutilations, which often cause serious discomfort to those who practice them, is not at all easy to determine. Some have conjectured that by the insertion of bright, colored objects in the ears, nose and lips, members of the same tribe were enabled to recognize each other at a distance; each tribe having selected a particular color. However, although certain local preferences are shown in the matter of color or material, there is no hard and fast rule in this matter, and frequently neighboring tribes will employ stones or shells of the same or similar hue and appearance. Others find in this custom a religious significance and suppose that the mutilation represents a form of sacrifice to the spirits, good or bad, who must be rendered favorable to man by some act on his part showing his unconditional submission to them. Originating in this way the idea of adornment was a secondary impulse. It is a fact that ancient peoples regarded the wearing of ear-rings as a badge of slavery, and, according to a Rabbinical legend, Eve's ears were pierced as a punishment for her disobedience, when she was driven from the Garden of Eden.

A curious theory was advanced by Knopf.[45] He calls attention to the habit children have of thrusting small bright objects into their noses and ears, and suggests that this indicates a natural propensity which, coupled with the early-developed love of adornment, induced primitive man to affix ornamental objects on or in the nose, ear, or mouth. There may be more in this than we are willing to admit, but on the whole it seems

---

[45] "De ornatu oris, nasi et aurium," Gottingæ, 1832, p. 43.

most probable that ceremonial and religious considerations gave rise to the custom.

One of the largest masses of sculptured Chinese jade is in the collection of T. B. Walker, Esq., of Minneapolis. This shows a jade mountain, with groups of figures artistically placed at its base, and winding pathways up to its summit. On the face of the rock is inscribed in beautiful Chinese characters the Epidendron Pavilion Essay of Wang Hi-che, a masterpiece of Chinese calligraphy.

An enormous mass of New Zealand jade (punamu, "green stone") weighing 7000 pounds, found in South Island in 1902, is to be seen in the Museum of Natural History, New York; it was secured by the writer and was donated to the Museum by the late J. Pierpont Morgan. This is the largest mass of jade known, or of which we have any record. On it is placed a remarkable and, in its own peculiar way, an artistic decoration, serving as a type of old Maori life, and at the same time designating the geographic source of the jade in a striking and unmistakable manner calculated to appeal to the least intelligent visitor. This is a statue of a Maori warrior of the old days, executing a war dance, characteristics of which were a distortion of the features and a thrusting out of the tongue intended to express defiance and contempt of the enemy; the time or cadence of the dance was marked by slapping the thigh with the flat of the left hand. This figure was executed from life by Sigurd Neandross; indeed it was actually cast from the model, so that there can be no doubt as to its fidelity.

Rock-crystal is included among the various objects used as fetiches by the Cherokee Indians. This stone is believed to have great power to give aid in hunting and also in divining. One owner of such a crystal kept his magic stone wrapped up in buckskin and hid it in a sacred cave; at stated intervals he would take it out of its re-

STATUE OF A MAORI WARRIOR, BY SIGURD NEANDROSS.

The base is a block of New Zealand jade from South Island, weighing three tons. It was donated by Mr. J. Pierpont Morgan to the American Museum of Natural History.

pository and "feed" it by rubbing over it the blood of a deer. This goes to prove that the stone, as a fetich, was considered to be a living entity and as such to require nourishment.[46]

Precious stones have been everywhere regarded as especially appropriate offerings at the shrine of a divinity, for the worshipper naturally thought that what was most valuable and beautiful in his eyes must also be most pleasing to the divinity he worshipped. However, we rarely find the usage which was remarked by Francisco Lopez de Gomara among the Indians of New Granada about the time of the Spanish Conquest.[47] These natives "burned gold and emeralds" before the images of the sun and moon, which were regarded as the highest divinities. Certainly to use precious stones for a "burnt offering" was an original and curious idea, although we have abundant proof that pearls were offered in this way by the mound-builders of the Mississippi Valley. In this case great quantities of pearls were burned at the obsequies of the chiefs of the tribes, or at those of any one belonging to the family of a chief.

In ancient Mexico the lapidaries adored the four following divinities as their tutelary gods: Chiconaui Itzcuintli ("nine dogs"), Naualpilli ("noble necromancer"), Macuilcalli ("five horses"), and Cintectl ("the god of harvest"). A festival was celebrated in honor of the three last-named divinities when the zodiacal sign called *chiconaui itzcuintli* was in the ascendant. A feminine divinity represented this sign and to her was attributed the invention of the garments and the orna-

[46] "Handbook of American Indians North of Mexico," ed. by Frederick Webb Hodge; Smithsonian Inst., Bur. of Am. Ethn. Bull. 30. Pt. I, p. 458; Washington, 1910.

[47] "Historia de las Indias," in "Bib. de autores españoles," vol. xxii, Madrid, 1852, p. 202.

ments worn by women. The four gods of the lapidaries were looked upon as the discoverers and teachers of the art of cutting precious stones and of piercing and polishing them, as well as of the making of labrets and ear-flaps of obsidian, rock-crystal, or amber. They also were the inventors of necklaces and bracelets.[48]

The stones worn by Chinese mandarins as a designation of their rank were undoubtedly determined originally by religious or ceremonial considerations. They are as follows; it will be noticed that red stones are given the preference:

Red or pink tourmaline, ruby (and rubellite)	1st rank.
Coral or an inferior red stone (garnet)	2d rank.
Blue stone (beryl or lapis-lazuli)	3d rank.
Rock-crystal	4th rank.
Other white stones	5th rank.

The knowledge of classical mythology was so slight among the ecclesiastics of the Middle Ages that some very queer attributions of the subjects engraved on Greek and Roman gems were made during this period. A reliquary containing a tooth of the Apostle Peter, preserved in the Cathedral of Troyes, was set with antique gems which had been plundered by French and Venetian crusaders from the treasure-house of the Greek Emperor in Constantinople, when that city was sacked in 1204 during the Fourth Crusade. Among these gems was one representing Leda and the Swan—certainly a curious subject for the adornment of a Christian reliquary. Another Greek or Roman gem, long preserved in a church, was furnished by its Christian owners with an inscription

---

[48] Sahagun, "Historia general de las cosas de Nueva España," Mexico, 1829, vol. ii, pp. 389–391, lib. ix, cap. xvii.

indicating that the figure engraved upon it was that of St. Michael, while in reality it was a representation of the god Mercury. Still another gem was provided with an inscription signifying that the subject was the temptation of Mother Eve in the Garden of Eden, but the Greek gem engraver's intent had been to carve the figures of Zeus and Athena, standing before an olive tree, a design which appears on some Athenian coins; at the feet of the divinities appears a serpent. In a similar way the grain-measure crowning the head of Jupiter-Serapis led to the attribution of a gem so engraved to the patriarch Joseph.[49]

An engraved amethyst bearing the figure of a little Cupid is said to have been worn in a ring by St. Valentine. While this may be somewhat doubtful, it is by no means impossible, for many pagan gems were worn by pious Christians, who reconciled their consciences to the use of these beautiful but scarcely religious ornaments by giving to the pagan symbols a Christian meaning. Certainly, in view of the time-honored customs connected with St. Valentine's Day, there seems something peculiarly appropriate in the design of the ring supposed to have been worn by St. Valentine.

That precious stones had sense and feeling was quite generally believed in medieval times, and a legend told of St. Martial illustrates this idea. The gloves worn by this saint were studded with precious stones, and when on a certain occasion a sacrilegious act was committed in his presence, the gems, horrified at the sight, sprang out of their settings and fell to the ground before the eyes of the onlookers.

---

[49] Klot, "Ueber den Nützen und Gebrauch der alten geschnittenen Steine," Altenburg, 1768, p. 57.

The St. Sylvester or St. James stone is a banded agate in two colors, the one dark and the other light, with a cat's-eye effect so that both colors are equally visible. The light side represents the old year, with its known occurrences, and the opaque side represents the new year, which is dark like futurity. This is a typical stone for a New Year's present or for one born on St. Sylvester's Day, the last day of the year. The popular tradition is that the member of a family or a household who is last to arise on that day will be the last to arise all the year around.

TITLE PAGE OF A GROUP OF TREATISES BY VARIOUS AUTHORS, COLLECTED AND EDITED BY CONRAD GESNER AT ZÜRICH IN 1565.

The upper one of the two rings figured is set with a natural pointed diamond, the lower one with a piece of amber enclosing an insect; grouped around are twelve stones representing those of the Breastplate.

The famous "Sacro Catino" preserved in Genoa was long believed to be made of a single immense emerald, but careful investigation proved that it was of no more valuable material than green glass. A legend still current in the early part of the sixteenth century represented this cup, or dish, as having been used by Christ at the Last Supper, and stated that it was one of the utensils which King Herod ordered to be brought from Galilee to Jerusalem for the celebration of the paschal feast; but his purpose having been changed by Divine Providence, he made other use of it.[50]

A queer story has been told regarding the Genoese emerald. At one time when the government was hard pressed for money, the Sacro Catino was offered to a rich Jew of Metz as pledge for a loan of 100,000 crowns. He was loath to take it, as he probably recognized its spurious character, and when his Christian clients forced him to accept it under threats of dire vengeance in case of refusal, he protested that they were taking a base advantage of the unpopularity of his faith, since they could not find a Christian who would make the loan. However, when some years later the Genoese were ready to redeem this precious relic, they were much puzzled to learn that a half-dozen different persons claimed to have it in their possession, the fact being that the Jew had fabricated a number of copies which he had succeeded in pawning for large sums, assuring the lender in each case that the redemption of the pledge was certain.

Among the celebrated emeralds noted by George Agricola[51] (1490–1555) was a large one preserved in a monastery near Lyons, France. This is also mentioned by Gesner, who states that it was shaped as a dish, or shallow cup, and was held to be the Holy Grail, like its rival at Genoa.[52] Another of Agricola's emeralds was somewhat smaller, but nevertheless measured nine inches in diameter and was in the chapel of St. Wenceslaus, at Prague; this may have been a chrysoprase, as at the present day many fine specimens of this stone can be seen in St. Wenceslaus, where the walls are inlaid with the

[50] Erasmi Stellæ, "Interpretamentum Gemmarum," 3d ed., Erfurti et Lipsiæ, 1736, p. 27.

[51] Agricolæ, "De natura fossilium," lib. vi, Basileæ, 1546, p. 289.

[52] Gesner, "De figuris lapidum," Tiguri, 1565, ff. 112v, 113r.

golden green gem-stone. Still another, larger than the last named, was set in the gold monstrance in Magdeburg, and was believed to have been the handle of Emperor Otho I's knife, since it was perforated. Possibly, however, the emerald, if genuine, was an Oriental stone, for it was customary to pierce rubies, sapphires, emeralds, etc., in the East so as to string them for necklaces or attach them as pendants to a jewel.

In the convent-church of St. Stephan, in Persian Armenia, erected about the middle of the seventeenth century, it is related by the French traveller Tavernier that there was preserved a cross said to be made out of the basin in which Christ washed the feet of the Apostles. Set in this cross was a white stone, and the priests asserted that when the cross was laid upon the body of one seriously ill, this stone would turn black if he were about to die, but would regain its white hue after his death.[53]

No jewelled sacred image has been the object of greater reverence than has been accorded to the rude little wooden carving popularly known as the "Sacro Bambino" or "Sacred Baby," in the old church of Ara Coeli in Rome. This figure was carved, in 1847, by a monk, out of a piece of olive-wood from one of the ancient trees growing on the Mount of Olives near Jerusalem. The carving was executed in the Holy Land and was sent thence to Italy, and although the ship bearing it was shipwrecked, this precious freight was miraculously preserved and is supposed to have been conveyed to its destination in some mysterious way. The reverence of the thousands of pilgrims who in the course of time have

---

[53] "Les six voyages de Jean Baptiste Tavernier," La Haye, 1718, vol. i, p. 48; Voyages en Perse, liv. i, chap. iv.

gazed with veneration upon this quaint and curious work of art, has found expression in the bestowal of a wealth of gems and jewels, including necklaces, brooches, rings, etc., with which the silken dress of the image is studded. A crown of gold adorned with precious stones rests upon the head of the olive-wood figure, which is jealously guarded by the priests and only shown to the faithful as a particular favor, except on the occasion of certain religious festivals.

One of the most renowned emeralds in the world surmounted the elaborately jewelled imperial crown that was placed upon the head of the venerated image of the Virgen del Sagrario in the Cathedral of Toledo. This emerald, of a rich green color, was cut as a perfect sphere and measured about 40 millimetres, or 1½ inches, in diameter. The crown itself was the work of the Toledan goldsmith, Don Diego Alejo de Montoya, who began his task in 1574 and devoted twelve years to its completion. It is described as being of almost pure gold and executed in the Renaissance style. Curiously chased in arabesque designs and enamelled in various colors, the framework of the crown served as a magnificent background for the gems constituting its adornment, which comprised rubies, emeralds, and Oriental pearls; a row of angels and cherubs sustained the arches which bore at their summit the allegorical figures of Faith, Hope, and Charity; upon that representing Faith rested the splendid emerald. This precious ornament was still preserved in the Cathedral in 1865, but was so carelessly guarded that it was stolen in 1869.[54]

If we are to believe the following anecdote, the em-

[54] José Ignacio Miró, "Estudio de las piedras preciosas," Madrid, 1870, pp. 135, 136.

erald disappeared at an earlier date: It is said that in 1809, during the French occupation of Spain, Marshal Junot visited this cathedral, and the emerald was pointed out to him as one of the chief glories of the shrine. As soon as the marshal's covetous glance rested upon the gem, he plucked it from its setting, remarking, coolly, to the astonished and horrified bystanders, "This belongs to me." Then, smiling and bowing, he left the cathedral with the emerald safely ensconced in his waistcoat pocket. Later, it was replaced by an imitation in glass.

The famous collection of jewels gathered together in the treasury of the Santa Casa, at Loreto, Italy, was plundered during the French occupation in 1797, and all trace of most of the magnificent ornaments has been lost. These represented the gifts of many crowned heads and titled personages; among the former was the unfortunate Henrietta Maria, wife of Charles I, who donated a golden heart-shaped jewel with the words "Jesus Maria" incrusted in diamonds. This jewel is described as being "as big as both a man's hands, opened onto two leaves, on one of which was the figure of the Blessed Virgin and on the other a portrait of the queen herself.[55] Of the many rich vestments for decorating the statue of the Virgin in the sanctuary, the most splendid was the gift of the Infanta Isabel of Flanders, and was valued at 40,000 crowns. In a seventeenth-century account by an English traveller it is thus described:[56]

> Its set thick with six rows of diamonds downe before, to the number of three thousand, and its all wrought over with a kinde of embroidery of little pearle set thick everywhere within the flowers with great round pearle, to the number twenty thousand pearles in all.

---

[55] Lassels, "The Voyage of Italy," Paris, 1670, Pt. II, p. 344.

[56] Lassels, l. c., p. 339.

The same writer tells us the niche in which the statue was placed was bordered with a row of precious stones of great number, size, and value, the colors being so varied that this bordering formed "a rich Iris of several colors." There is also said to have been a great pearl, set in gold, and engraved with the image of the Virgin and Child.[57] It seems probable that this was a jewel made of a baroque pearl, or pearls, completed by enamel-work so as to represent the sacred figures.

The pectoral cross worn in solemn processions by the prior of the monastery of San Lorenzo del Escorial was adorned with eight perfect emeralds, five diamonds, and five pearls. From it hung a splendid pear-shaped pearl, the gift of Philip II in 1595, and one of the finest of those acquired by this monarch. In 1740 the cross was valued at 50,000 crowns, Philip's great pearl not being included in this valuation.[58]

The monastery of Streoneshalh, later Whitby Abbey, was founded about 656 A.D. by Oswy, King of Northumbria, in fulfilment of a vow made before his victory over the pagan king Penda, at the battle of Winwidfield, fought in November, 654. St. Hilda was made abbess of this monastery, and Oswy's daughter Aelfleda took the veil and eventually, in 680, succeeded Hilda as abbess; she died in 713.[59] Tradition relates that at this early date crosses and rosaries were made for the inmates of the monastery from the jet found in the neighborhood. The "Whitby jet," so popular and fashionable in the eighteenth century, was largely derived from the same

[57] Scotto, "Itinerario d'Italia," Roma, 1747, p. 314.

[58] José Ignacio Miró, "Estudio de las piedras preciosas," Madrid, 1870, pp. 136, 137, 229.

[59] Cartularium abbathiæ de Whiteby, Surtees Soc. Pub., vol. lxix, pp. xvi–xx.

source, and since then has had several revivals, until replaced by black-stained chalcedony, the so-called onyx, and, later still, by steel carved with glass and glass itself.

In the sixteenth century jet was popularly called "black amber," and Cardano states that in his time beads of this material were made up into rosaries. He also says that curious figures made of jet were brought from Spain to Italy.[60]

Many are unaware of the fact that a number of ornamental objects made of nephrite and jadeite—unquestionably of European origin—are to be seen in the quiet little town of Perugia. These objects, collected principally in central and southern Italy, constitute the Belucci Collection, in that city. This collection also contains other specimens of worked jadeite, which must have been brought to Europe at the time of the Spanish conquest of Mexico and Peru. A very interesting example shows us the utilization of a pagan celt to form a Christian emblem. By the removal of a rectangular piece from each of the four corners of the jadeite celt, a perfect cross has been made, the back and front of which still offer the original polish given to the material centuries ago by the native American worker. The superstitious belief propagated in Europe by the returning Spanish sailors, very probably an invention of their own to enhance the value of their jade and jadeite, that these minerals were worn by the natives as a cure for diseases of the kidneys, whence the name *lapis nephriticus,* rendered the material exceptionally precious in the eyes of many, and quite possibly it may have been thought that, by transforming this object into the sacred form of the cross, a talisman would be produced that would not only effect the

---

[60] Cardani, "De subtilitate," lib. v, Basileæ, 1560, p. 370.

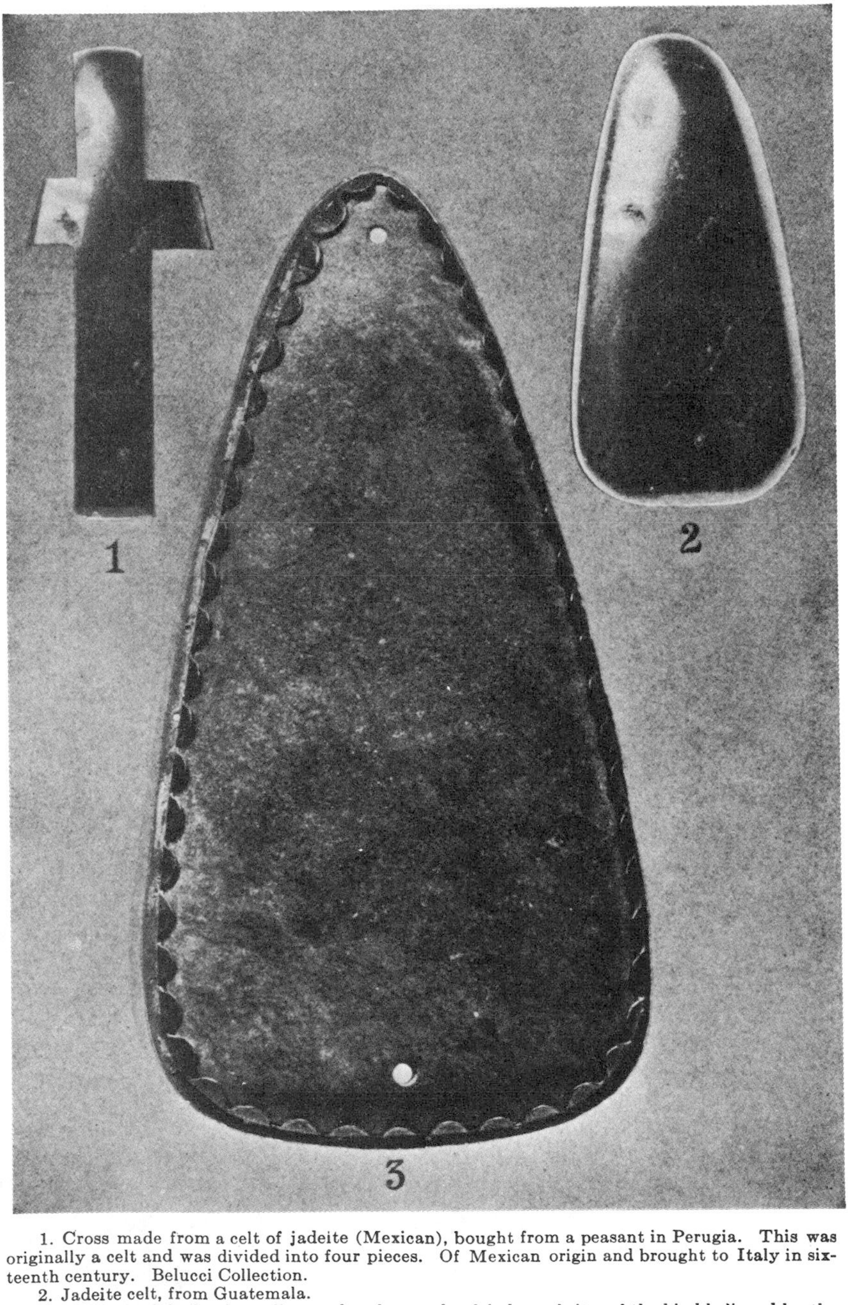

1. Cross made from a celt of jadeite (Mexican), bought from a peasant in Perugia. This was originally a celt and was divided into four pieces. Of Mexican origin and brought to Italy in sixteenth century. Belucci Collection.
2. Jadeite celt, from Guatemala.
3. Celt, Aboriginal. A small stone hatchet made of jade nephrite, of the kind believed by the peasants to be thunderbolts. Mounted in silver to be worn as a charm. This specimen, tied over the loins, is said to have been worn as recently as fifty years ago by a Scottish gentleman as a cure for kidney disease. British Museum.

cure of a special disease, but would also by its superior virtue guard the wearer from harm and danger of all kinds. Here may also be seen some celts of European jade sewed up in little bags to be worn on the loins.

Certain curious amulets called *magatama* (crooked jewels) have been found in Japanese graves of the iron age;[61] they are formed of various materials, among others of steatite, jasper, carnelian, agate, rock crystal, chrysoprase and nephrite (jade). In the shell heaps of a period preceding the iron age, the *magatama* are frequently made of horn, or of boar's or wolf's teeth, and their peculiar form, which is variously explained as a symbol, may have been conditioned by the shape of the materials originally used. The magatama were evidently regarded as amulets. "They are generally perforated at the thick end, and were worn on a string, together with beads and bugles of the same material." These peculiar ornaments were used to adorn the statues of the gods and were also employed as imperial insignia and distinctive marks of high rank. At the present day they are numbered among the three emblems of sovereignty in Japan.

A green and a red magatama are combined in the national emblem of Korea and a similar figure is used in China to symbolize the union of the masculine and feminine principles (Yang and Yin) in nature. Dr. Baelz believes that the swastika emblem, encountered in so many different parts of the world, belongs to the same order of ideas.

The Bghai tribes of Burma have many superstitions in regard to stones, such as garnets, rock-crystal, chalcedony, carnelian, agate, onyx and others of less value,

[61] Dr. Baelz, of the Imperial University of Tokyo, in Report of the Smithsonian Institution for 1904, pp. 523–547.

their repute not depending entirely or principally upon their quality as gem-stones. In almost every household is installed a stone fetish, and blood offerings are on occasion made to this. A question as to the reason for this offering elicited the following reply: "If we do not give it blood to eat it will eat us." A common belief was that spirits good or bad dwelt in the stones, and in case a great misfortune befell a family, this was sometimes laid to the charge of such a spirit. The father of a family having died, his widow commanded her son to throw away their magic stone. This he did, but the spirit was not to be denied, for shortly afterward this very stone was found to have returned to its accustomed place, and had even brought two companion stones with it![62]

Ruy Gonzalez de Clavijo, who travelled in the East during the years 1403–1406, gives a description of a slab of stone bearing the outlines of a "natural picture," and placed in the church of St. Sophia, in Constantinople:[63]

> In the wall, on the left-hand side, there is a very large white slab, on which, among many other figures, was drawn, very naturally, without any human artifice of sculpture or painting, the most sacred and blessed Virgin Mary, with our Lord Jesus Christ in her most holy arms, with his most glorious forerunner, St. John the Baptist, on one side. These images, as I said before, are not drawn or painted with any color, or inlaid, but the stone itself gave birth to this picture, with its veins, which may be clearly seen; and they say that when this stone was cut, to be placed in this most holy place, the workman saw these most wonderful and fortunate images on it, and, as this church was the most important one in the city, that stone was deposited in it. The said images appear as if they were in the clouds of heaven, and as if there was a thin veil before them.

[62] Mason, "Burmah, its People and Natural Productions," Rangoon, 1860, pp. 109, 110.

[63] Narrative of the Embassy of Ruy Gonzalez de Clavijo to the Court of Timour, trans. by Clement R. Markham, London, 1859, p. 38, Hakluyt Soc. Pub.

Many other examples of these "natural gems" are noted by early writers. Among them was an agate gem in the treasury of the Basilica of St. Mark, in Venice. Upon this gem appeared the head of a king, adorned with a diadem, the whole design being figured naturally by the veining of the agate, and not owing anything to artifice. In the same city, upon a column in the church of San Georgio Maggiore, could be seen the likeness of Our Lord, hanging from the Cross.[64]

Such stones, with peculiar markings indicating the form of human heads and figures, were regarded as the work of higher powers.

Another remarkable example is described by Kircher as follows: [65]

> In Rome, in the Chapel of the Sacred Virgin, near the organ to the right hand of those who enter the Church of St. Peter, an image may be seen in which the Blessed Virgin of Loreto is so artistically depicted by Nature that it appears to be the work of an artist's hand. She is attired in a triple garment, divided by a zone, and holds in her arms the child, who is distinguished by a crown, as is the mother. Around may be seen the figures of angels.

The red spots upon the bloodstone were said in Christian legend to represent the blood of Christ. This idea has been beautifully utilized in some gems cut from this stone, whereon the thorn-crowned head of Christ is so placed that the red spots of the bloodstone figure the drops of blood trickling down the hair and face of the Saviour. Such a gem might well be looked upon as a Christian amulet and one that could be reverently worn by any believer.

---

[64] Chiocci, "Museum Calceolarium," Veronæ, 1622, p. 251.

[65] Kircher, "Mundus Subterraneus," Amstelodami, 1665, p. 36; Tabula IV, Fig. 6.

The ignorance in the Middle Ages of the art of gem-engraving often induced the belief that engraved stones were the work of nature. A striking instance of this was the celebrated stone over the figure of the Mother of Jesus, on the tomb of St. Elizabeth of Marburg. On this gem appeared two heads touching each other, and it was, according to tradition, not a work of art, but a freak of the sculptress Nature. An oft-repeated legend tells us that a former Elector of Mainz offered the whole district of Amöneberg for this costly stone, which robber hands removed at Cassel. It is in reality a fine onyx engraved with the heads of Castor and Pollux.[66]

We might be disposed to regard rather sceptically the tales regarding wonderful stones bearing the image of Christ, or that of the Virgin Mary, and we may be inclined to believe that the old accounts are exaggerated or distorted by the pious imaginations of the writers. Nevertheless, in our own time we have a well-attested case of the discovery of such a stone.

In 1880, while visiting the village of Oberammergau, Bavaria, to witness the Passion Play, Mrs. Eugenia Jones-Bacon, of Atlanta, Georgia, found on Mount Kopfel, which overlooks the village, a small stone composed of chert and limestone, and having on its surface excrescences so disposed that, when the stone was held at a certain angle, the shadows cast by them formed a striking likeness of the head of Christ as depicted in Christian art. This peculiar freak specimen has been carefully examined by experts and has been pronounced to be entirely a work of nature. The mineralogist is not disposed to see here anything more than coincidence, and

---

[66] Creuzer, "Antik geschnittene Steine vom Grabmahl der heiligen Elizabeth," Leipsic and Darmstadt, 1834, p. 25.

yet the most sceptical cannot fail to be impressed by the fact that such a stone was found at the time and place of the Passion Play. As Max Müller said, in commenting on this strange discovery: "The chapter of accidents is much larger than we imagine," and the present writer feels disposed to add that it is remarkable how often we find what we are looking for, especially if we are only looking or thinking of one object or subject.

The religious symbolism of the diamond was a favorite theme with the thirteenth century "lapidaria," or rhymed treatises on precious stones. Just as it could only be discovered by night—an old fancy—so was the Incarnation a hidden mystery; it gave forth a great light, just as Jesus illumined the depths of Hades when he descended thither; it was unconquerably hard, and who can resist the might of God? [67]

The mediæval Italians who were fond of seeking some hidden and significant meaning in the names of precious stones, in the case of the diamond (diamante), read the phrase *amante di Dio,* or "lover of God." [68] This was a reason for regarding the brilliant gem as a sacred stone and one especially suitable for religious use.

The Rosicrucians, who sought to combine pagan with Christian types and figures, saw in the amethyst and the amethystine color a symbol of the divine male sacrifice, since the stone and the color were typical of love, truth, passion, suffering, and hope. The love of Christ led him to make the supreme sacrifice and suffer the agony of the Cross, and the Crucifixion was followed by

---

[67] Barbier de Montault, "Le Trésor de l'Abbaye de Sainte-Croix de Poitier"; in Mém. de la Soc. d'Antiq. de l'Ouest, Sec. Ser., vol. lv, 1881, pp. 105, 106; Poitiers, 1882.

[68] Italian MS. of the fourteenth century in the author's library; fol. 41 b.

the Resurrection, whence came the hope of mankind to enjoy eternal happiness in heaven.

The chiastolite, or macle, shows the representation of a cross on its surface, this effect being produced by the regular arrangement of carbonaceous impurities along the axes of the crystal. The name signifies a marking resembling the Greek letter Χ (chi). This marking is often very striking in appearance, and the crystal was naturally regarded as having a mystical and religious significance. It was said to stanch the flow of blood from any part of the body if worn so as to touch the skin, and it was also believed to increase the secretion of milk. All kinds of fevers were cured by this mineral if it were worn suspended from the neck, and the divine symbol it bore served to drive away evil spirits from the neighborhood of the wearer.

This very interesting mineral occurs very frequently in mica schists. When found, it appears about the thickness of a small finger, tapering slightly at each edge. If broken near one end, it often shows a white cross with a veined outline of black, making a distinct cross with black markings. The crystals frequently measure from two to four inches in length, and are found in Massachusetts, California, and other places. If small segments are broken off, it will be found that the black outline will become stronger, and the white less marked, until finally a black cross will appear, with white markings. The white material is the result of two white wedges pushed point onward until the ends meet, the narrow end of one wedge being crossed by the broad end of the second wedge, and the black filling in the balance of the square. No two of these square crosses can thus ever be exactly alike, and, when polished, the crystals naturally form an

STAUROLITE CRYSTALS (FAIRY STONES).
Patrick County, Virginia.

interesting stone that was known as *lapis crucifer,* or cross-stone by the ancients.

The peculiar form of the mineral known as staurolite (from the Greek σταυρός cross) is due to the twinning of two crystals at right angles. In Cronstedt's treatise on mineralogy, published in Stockholm in 1758, we are told that the staurolite was sometimes called Baseler Taufstein (baptismal-stone) or *lapis crucifer,* the former name being used in Basel, where the stone was employed as an amulet at baptisms. However, the *lapis crucifer* of De Boot appears from his description to have been the chiastolite. In Brittany these twin crystals were worn as charms, and local legends state that they had dropped from the heavens.

Fine crystals of staurolite have been found in Patrick County, Virginia, and there is said to be a beautiful local legend in regard to their origin. Near where they are found there wells up a spring of limpid water, and the story goes that one day, long, long ago, when the fairies were dancing and playing around this spring, an elfin messenger winged his way through the air and alighted among them. He bore to them the sad tidings of the crucifixion of Christ in a far-off city. So mournful was his recital of the sufferings of the Saviour that the fairies burst into tears, and these fairy tear-drops, as they fell to earth, crystallized into the form of the cross. These natural crosses are in great demand as charms, and ex-President Roosevelt is said to wear one of them mounted as a watch-charm.

There has been found in the southern part of New Mexico, and in northern Mexico, a blue variety of calamine, a hydrous silicate of zinc, colored blue by an admixture of copper. This stone has been cut into gem form and has been sold to a certain extent as a cheap gem. It is translucent and is sometimes veined with white wavy lines. The Mexican Indians employed in the mines often set up a cross and a candle near where they are working, so that they may pay their devotions at this improvised shrine. In Sonora and Western Chihuahua the Indians frequently place a piece of the stone to which we have alluded alongside the cross. They may be attracted by its beautiful blue color, or they may believe that it is a turquoise, although it does not resemble this latter stone, which is more opaque, of a different shade of blue and of a different composition.

In some epitaphs the hope of the resurrection finds expression in likening the body enclosed in its narrow coffin to a precious jewel in its casket. The following lines from a tombstone erected in 1655 to the memory of Mary Courtney, at Fowell, Cornwall, England, give a good example of this class of inscription:[69]

Near this a rare jewell's set,
Clos'd up in a cabinet.
Let no sacrilegious hand
Breake through—'tis y[e] strickte comaund
Of the jeweller: who hath sayd
(And 'tis fit he be obey'd)
I'll require it safe and sound
Both above and under ground.

In a churchyard at Prittlewell, Essex, England, a rather whimsical treatment of the same idea is offered by some verses engraved on the stone marking the graves of two wives of a certain Freeborne, the first of whom died in 1641 and the second in 1658. The bereaved hus-

---

[69] Ravenshaw, "Antiente Epitaphs," London, 1878, p. 110.

band seems to have been perfectly willing to await the Day of Judgment for the return of his lost spouses:[70]

Under this stone two precious gems do ly
Equall in weight, worth, lustre, sanctity:
Yet perhaps one of them do excell;
Which was't who knows? ask him y^t^ knew y^em^ well
By long enjoyment. If he thus be prest,
He'el pause, then answere: truly both were best:
Were't in my choice that either of ye twain
Might be returned to me to enjoy agayne,
Which should I chuse? Well, since I know not whether;
Ile mourne for the losse of both, but wish for neither,
Yet here's my comfort, herein lyes my hope,
The tyme a comeinge cabinets shall ope
Which are lockt fast: then shall I see
My Jewells to my joy, my Jewells mee.

The Christian symbolism of colors has in many cases determined the use of certain colored gems for religious ornaments, and therefore the following summary of their principal significance is of interest here:[71]

WHITE is regarded as the first of the canonical colors, and as emblematic of purity, innocence, virginity, faith, life, and light. For this reason it is used in the ceremonies of Easter and Christmas, as in those of the Circumcision and Epiphany of Our Lord. As the color of virginity it is especially appropriate for the festival of the Virgin Mary, and as that of faith not sealed with blood, for the festivals of the saints who were not martyred. The heavenly host of angels and saints wear white robes, and in pictures of the Assumption of the Virgin she is frequently clad in white.

RED is used at the feasts of the Exaltation and Invention of the Cross, at Pentecost, and at the Feast of Martyrs. It suggests and symbolizes suffering and martyrdom for the faith, and the supreme

---

70 Ravenshaw, "Antiente Epitaphs," London, 1878, p. 113.

71 See Audsley, "Handbook of Christian Symbolism," London, 1865, pp. 135–137.

sacrifice of Christ upon the Cross. Divine love and majesty are also typified by this color.

Blue is an emblem of the celestial regions and of the celestial virtues. Nevertheless, as this is not one of the five canonical colors, it is not employed for the decoration of churches or for ecclesiastical vestments. In Christian art, however, the Virgin and the saints and angels are often robed in blue.

Yellow of a golden hue is emblematic of God's goodness and of faith and good works, but it is not a canonical color. A dull yellow, however, has the opposite signification, and is a type of treachery and envy. Hence Judas is garbed in yellow of a dull hue, and heretics wore garments of this shade when they were condemned to the stake.

Green is the canonical color for use on Sundays, week-days, and ordinary festivals. Hope and joy and the bright promises of youth are signified by green.

Violet, another canonical color, is appropriate for use on Septuagesima and Quinquagesima Sundays, during Lent, and on Advent Sunday. The chastening and purifying effects of suffering find expression in this color.

Black, also a canonical color, is a symbol of death and of the mourning and sorrow inspired by death. Therefore it is only used in the Church on Good Friday, to symbolize the sorrow and despair of the Christian community at the death of Christ, a sorrow soon to be turned to joy by His glorious resurrection.

FRONTISPIECE OF THE "VESTITUS SACERDOTUM HEBRÆORUM," OF JOHANN BRAUN, AMSTERDAM, 1680.

The vignettes at the top illustrate the source of the materials of the vestments, etc.; as the nopal, source of the cochineal insect; gold-thread; linen; a sheep for wool; Tyre and the purple murex. The other vignettes show separate parts of the high-priest's attire, and in the centre appear two figures of the high-priest, each garbed in different sets of vestments.

VIII

On the High-Priest's Breastplate

VERY early, and very naturally, the religious nature of man led to the use of precious stones in connection with worship—the most valuable and elegant objects being chosen for sacred purposes. Of this mode of thought, we have a striking instance in the accounts given, in the book of Exodus, of the breastplate of the High-priest, and the gems contributed for the tabernacle by the Israelites in the wilderness. Another religious association of such objects is their use to symbolize ideas of the Divine glory, as illustrated in the visions of the prophet Ezekiel and in the description of the New Jerusalem in the book of Revelation. Apart from such legitimate uses, however, gems have become associated with all manner of religious fancies and superstitions, traces of which appear in the Talmud, the Koran, and similar writings; they have also been dedicated to various heathen deities. Even in modern times, some trace of the same ideas remains in the ecclesiastical jewelry and its supposed symbolism.

In the vision of Ezekiel i, 26, and in a brief allusion to the similar appearance of the God of Israel in Exodus xxiv, the throne of Jehovah, or the pavement beneath his feet, is compared to a sapphire, and the Apostle John, in the Apocalypse, describes the Great White Throne as surrounded by a rainbow like an emerald.

The Rabbinical writings, instead of the simple grandeur of these biblical comparisons, give us many fanciful ideas. The stones of the breastplate are here represented as sacred to twelve mighty angels who guard the gates of

Paradise, and wondrous tales are told of the luminous gems in the tent of Abraham and the ark of Noah. Mohammedan legend represents the different heavens as composed of different precious stones, and in the Middle Ages these religious ideas became interwoven with a host of astrological, alchemistic, and medical superstitions.

The following is the description of the breastplate given in Exodus (xxviii, 15–30):

And thou shalt make the breastplate of judgment with cunning work; after the work of the ephod thou shalt make it; of gold, of blue, and of purple, and of scarlet, and of fine twined linen shalt thou make it.

Foursquare it shall be being doubled; a span shall be the length thereof, and a span shall be the breadth thereof.

And thou shalt set in it settings of stones, even four rows of stones: the first row shall be a sardius, a topaz, and a carbuncle: this shall be the first row.

And the second row shall be an emerald, a sapphire, and a diamond.

And the third row a ligure, an agate, and an amethyst.

And the fourth row a beryl, and an onyx, and a jasper; they shall be set in gold in their enclosings.

And the stones shall be with the names of the children of Israel, twelve, according to their names, like the engravings of a signet; every one with his name shall they be according to the twelve tribes.

And thou shalt make upon the breastplate chains at the ends of wreathen work of pure gold.

And thou shalt make upon the breastplate two rings of gold, and shalt put the two rings on the two ends of the breastplate.

And thou shalt put the two wreathen chains of gold in the two rings which are on the ends of the breastplate.

And the other two ends of the two wreathen chains thou shalt fasten in the two ouches, and put them on the shoulder-pieces of the ephod before it.

And thou shalt make two rings of gold, and thou shalt put them upon the two ends of the breastplate in the border thereof, which is in the side of the ephod inward.

And two other rings of gold thou shalt make, and shalt put them on the two sides of the ephod underneath, toward the forepart thereof, over against the other coupling thereof, above the curious girdle of the ephod.

And they shall bind the breastplate by the rings thereof unto the rings of the ephod with a lace of blue, that it may be above the curious girdle of the ephod, and that the breastplate be not loosed from the ephod.

And Aaron shall bear the names of the children of Israel in the breastplate of judgment upon his heart, when he goeth in unto the holy place, for a memorial before the Lord continually.

And thou shalt put in the breastplate of Judgment the Urim and the Thummim; and they shall be upon Aaron's heart, when he goeth in before the Lord: and Aaron shall bear the judgment of the children of Israel upon his heart before the Lord continually.

Of the miraculous quality of the stones worn by the high-priest, the Jewish historian Josephus (37–95 A.D.) says:[1]

From the stones which the high-priest wore (these were sardonyxes, and I hold it superfluous to describe their nature, since it is known to all), there emanated a light, as often as God was present at the sacrifices; that which was worn on the right shoulder instead of a clasp emitting a radiance sufficient to give light even to those far away, although the stone previously lacked this splendor. And certainly this in itself merits the wonder of all those who do not, out of contempt for religion, allow themselves to be led away by a pretence of wisdom. However, I am about to relate something still more wonderful, namely, that God announced victory in battle by means of the twelve stones worn by the high-priest on his breast, set in the pectoral. For such a splendor shone from them when the army was not yet in motion, that all the people knew that God himself was present to aid them. For this reason the Greeks who reverence our solemnities, since they could not deny this, called the pectoral *λόγιον* or oracle. However, the pectoral and the onyxes ceased to emit this radiance two hundred years before the time when I write this, because God was displeased at the transgressions of the Law.

[1] Flavii Josephi, "De Antiq. Jud.," lib. iii, cap. viii, 9; Opera, ed. Dindorf, Parisiis, 1845, vol. i, pp. 100, 101.

This writer, who must have seen the high-priest wearing his elaborate vestments, says that the breastplate was adorned "with twelve stones of exceptional size and beauty, a decoration not easily to be acquired, on account of its enormous value."[2] However these gems were not merely rare and costly; they also possessed wonderful and miraculous powers. Writing about 400 A.D., St. Epiphanius, Bishop of Constantia, tells of a marvellous *adamas* which was worn on the breast of the high-priest, who showed himself to the people, arrayed in all his gorgeous vestments, at the feasts of Pascha, Pentecost, and Tabernacles. This adamas was termed the δήλωσις or "Declaration," because, by its appearance, it announced to the people the fate that God had in store for them. If the people were sinful and disobedient, the stone assumed a dusky hue, which portended death by disease, or else it became the color of blood, signifying that the people would be slain by the sword. If, however, the stone shone like the driven snow, then the people recognized that they had not sinned, and hastened to celebrate the festival.[3]

There seems to be little doubt that this account is nothing more than an elaboration and modification of the passage in Josephus. Evidently the λόγιον (oracle) of Josephus has become the δήλωσις (declaration).

When Moses wished to engrave on the stones of the breastplate the names of the twelve tribes of Israel, he is said to have had recourse to the miraculous *shamir*. The names were first traced in ink on the stones, and the *shamir* was then passed over them, the result being that

[2] "Ant. Jud.," lib. iii, cap. vii, 5, Flavii Josephi Opera, Basileæ, 1544, p. 75.

[3] Sancti Patri Epiphanii, "De XII Gemmis," Tiguri, 1566, ff. 12–14. Edited by Conrad Gesner from a unique MS. in his possession.

THE HEBREW HIGH-PRIEST ATTIRED WITH HIS VESTMENTS.
(From Johann Braun's "Vestitus Sacerdotum Hebraeorum," Amsterdam, 1680, opp. p. 822.)

the traced inscriptions became graven on the stones. In proof of the magical character of this operation, no particles of the gems were removed in the process.[4] The name really designates "emery."

An argument against the use of especially rare and costly stones in the decoration of the breastplate has been found in its probable size.[5] We are told that when folded it measured a span in each direction, and this would indicate that its length and breadth were each from eight to nine inches. In this case the stones themselves might have measured two by two and a half inches, and, in view of the number of characters required to express some of the tribal names, these dimensions do not seem excessive. It is highly improbable that in the time of Moses precious stones like the ruby, the emerald, or the sapphire would have been available in these dimensions. The difficulty of engraving very hard stones with the appliances at the command of the Hebrews of this period must also be taken into consideration. As we shall see, however, there is good reason to believe that after the Babylonian Captivity a new breastplate was made, and at that time it may have been easier to secure and work precious stones of great value and a high degree of hardness. We must also bear in mind that in those periods perfection was not so great a requisite as rich color.

In his commentary on Exodus xxviii, Cornelius à Lapide (Cornelius Van den Steen) discusses the question of the diamond in the high-priest's breastplate. In the first place, he notes that the diamond was very costly, and

---

[4] Ginsburg, "Legends of the Jews," Eng. trans., Phila., 1909, vol. i, p. 34.

[5] See J. L. Myers in the "Encyclopædia Biblica," vol. iv, pp. 4799–4812.

that a large stone would have been needed to bear the name of Judah or that of any other tribe. He considers that a stone of the requisite size would have cost a hundred thousand gold crowns, and he asks, "Whence could

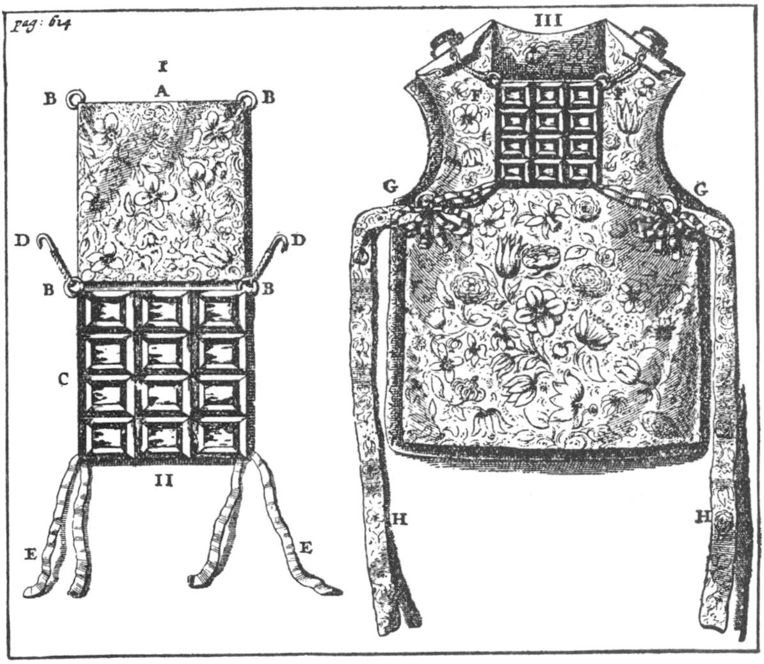

I, II, THE BREASTPLATE UNFOLDED.

A, lower fold; *B, B, B, B*, rings for attachment to Ephod; *C*, the twelve gems in their settings; *D, D*, hooks for attachment to shoulder; *E, E*, bands to pass through rings in Ephod.

III. EPHOD WITH BREASTPLATE FOLDED AND ATTACHED.

G, G, rings through which pass bands of Breastplate; *H, H*, bands of Ephod. From Johann Braun's "Vestitus Sacerdotum Hebraeorum," Amsterdam, 1680.

the poor Hebrews have obtained such a sum of money, and where could they have found such a diamond?" He proceeds to give still another reason for doubting that the diamond was in the breastplate,—namely, that it would have marked too great a distinction between the tribes, the result being that the tribe to which the diamond was assigned would have been puffed up with pride, while the others would have been filled with hatred and envy, "for the diamond is the Queen Gem of all the gems."[6]

The use of the breastplate to reveal the guilt of an offender is testified to in a Samaritan version of the book of Joshua, which has been discovered by Dr. Moses Gaster, chief rabbi of the Spanish and Portuguese Jews in England. According to this version, Achan steals a golden image from a heathen temple in Jericho. The high-priest's breastplate reveals his guilt, for the stones lose their light and grow dim when his name is pronounced.

Many conjectures have been made as to the origin of the breastplate with the mystic Urim and Thummim enclosed within it. That an Egyptian origin should be sought seems most probable. A breast-ornament worn by the high-priest of Memphis, as figured in an Egyptian relief, consists of twelve small balls, or crosses, intended to represent Egyptian hieroglyphics. As it cannot be determined that these figures were cut from precious stones, the only definite connection with the Hebrew ornament is the number of the figures; this suggests, but fails to prove, a common origin. The monuments show that the high-priest of Memphis wore this ornament as early as the fourth Dynasty, or, approximately, 4000 B.C.[7]

Of the Urim and Thummim, the mysterious oracle of the ancient Hebrews, St. Augustine (354–450 A.D.), after acknowledging the great difficulty of interpreting the

---

[6] See Gimma, "Della storia naturale delle gemme," Napoli, 1730, vol. i, pp. 208, 209.

[7] Hommel, "Altisraelitische Ueberlieferung," pp. 281, sqq.; Erman, "Aegypten," Tübingen, 1885, p. 402.

meaning of the words and the character of the oracle, adds that some believed the words to signify a single stone which changed color according as the answer was favorable or unfavorable, while the priest was entering the sanctuary; still he thought it possible that merely the letters of the words Urim and Thummin were inscribed upon the breastplate.[8]

After the capture of Jerusalem by Titus in 70 A.D., the treasures of the temple were carried off to Rome, and we learn from Josephus that the breastplate was deposited in the Temple of Concord, which had been erected by Vespasian. Here it is believed to have been at the time of the sacking of Rome by the Vandals under Genseric, in 455, although Rev. C. W. King thinks it is not improbable that Alaric, king of the Visigoths, when he sacked Rome in 410 A.D., might have secured this treasure.[9] However, the express statement of Procopius that "the vessels of the Jews" were carried through the streets of Constantinople, on the occasion of the Vandalic triumph of Belisarius, in 534, may be taken as a confirmation of the conjecture that the Vandals had secured possession of the breastplate and its jewels.[10]

It must, however, be carefully noted that Procopius nowhere mentions the breastplate and that it need not have been included among "the vessels of the Jews." It appears that this part of the spoils of Belisarius was placed by Justinian (483–565) in the sacristy of the church of St. Sophia. Some time later, the emperor is said to have heard of the saying of a certain Jew to the

[8] Aureli Augustini, "Opera Omnia," vol. iii, Part I, col. 637; Patrologiæ Latinæ, ed. Migne, vol. xxxviii, Paris, 1864.

[9] "Natural History of Precious Stones," London, 1870, p. 333.

[10] Procopius, ed. Dindorf, Bonnae. 1833, vol. i, p. 445; "De bello Vandalico," lib. ii, cap. 9.

effect that, until the treasures of the Temple were restored to Jerusalem, they would bring misfortune upon any place where they might be kept.[11] If this story be true, Justinian may have felt that the fate of Rome was a lesson for him, and that Constantinople must be saved from a like disaster. Moved by such considerations, he is said to have sent the "sacred vessels" to Jerusalem, and they were placed in the Church of the Holy Sepulchre.

This brings us to the last two events which can be even plausibly connected with the mystic twelve gems,—namely, the capture and sack of Jerusalem by the Sassanian Persian king, Khusrau II, in 615, and the overthrow of the Sassanian Empire by the Mohammedan Arabs, and the capture and sack of Ctesiphon, in 637.[12] If we admit that Khusrau took the sacred relics of the Temple with him to Persia, we may be reasonably sure that they were included among the spoils secured by the Arab conquerors, although King, who has ingeniously endeavored to trace out the history of the breastplate jewels after the fall of Jerusalem in 70 A.D., believes that they may be still "buried in some unknown treasure-chamber of one of the old Persian capitals."

A fact which has generally been overlooked by those who have embarked on the sea of conjecture relative to the fate of the breastplate stones is that a large Jewish contingent, numbering some twenty-six thousand men, formed part of the force with which the Sassanian Persians captured Jerusalem, and they might well lay claim

---

[11] Procopius, ed. Dindorf, Bonnae, 1833, vol. i, p. 445; "De bello Vandalico," lib. ii, cap. 9.

[12] For an account of the immense booty taken by the Arabs, under Sa'ad, on this occasion, see Rawlinson, "Seventh Great Oriental Monarchy," London, 1876, pp. 564–566. The total value has been placed as high as $125,000,000.

to any Jewish vessels or jewels that may have been secured by the conquerors. In this case, however, it is still probable that these precious objects fell into the hands of the Mohammedans who captured Jerusalem in the same year in which they took Ctesiphon.

One circumstance which may have contributed to the preservation of these gems in their original form after they fell into the hands of the Romans is the fact that each one was engraved with the name of one of the Jewish tribes, the inscription being probably in the older form of Hebrew writing, which was used in the coinage even as late as the last revolt in 137 A.D. Hence, recutting would have been necessary to fit them for use as ornaments, a process not easily accomplished, and involving a great loss of size. We must also bear in mind that the intrinsic value of the gems may not have been so great as many suppose, since all of them were probably of the less perfect forms of the precious and semi-precious varieties. It is very likely that the enthusiastic statements of Josephus in this connection were dictated by national pride, or arose from the tendency to exaggeration so common among the Oriental writers. Certainly, if the breastplate known to Josephus was made not long after the return of the Jews from the Babylonian Captivity, their financial resources at the time of its fabrication were quite restricted.

Admitting as a possibility that the Arabs may have secured possession of the breastplate, how would they have regarded it? The heroes of the Old Testament, and especially Moses, were such sacred personalities in the eyes of Mohammedans that this relic would have been as precious for them as for us. However, the victorious Arabs who overran the Sassanian Empire, although filled with religious zeal, were no students of archæology, and would have been quite unable to decipher the strange characters engraved on the stones. They would most probably have supposed them to be Persian characters, and would, therefore, have valued these stones no higher than others in the Persian treasure. This can serve as an explanation of the fact that no allusion to the breastplate with its adornment can be found in the works of those Mohammedan writers, such as Tabari, who treat of the overthrow of the Sassanian Empire. We may be sure that the Persians themselves would have accorded no special honor to objects connected with the Hebrew religion, since their own Zoroastrian faith had no connection with it.

In 628, not long before the date of the Arab invasion, the most precious relic of Christendom, the cross discovered by Helena, mother of Constantine the Great, and believed to be the very cross on which Christ died, was surrendered to the Greek Emperor Heraclius by Kobad II, son of Khusrau II, on the conclusion of a treaty of peace between the Eastern and Sassanian Empires. This cross was one of the sacred objects borne away to Persia from Jerusalem by Khusrau in 615 A.D. It is said to have been guarded carefully through the influence of Sira, Khusrau's Christian wife. There is a bare possibility that other objects of religious veneration, taken from Jerusalem, may have been given up by the Persians at the same time, and that the unique character of the most important relic so overshadowed all others that historians have failed to note the fact. The cross was restored to Jerusalem by Heraclius in 629, only to fall into the hands of the Mohammedans when that city was taken by the Arabs under Omar, in 637. Hence, if the jewelled breastplate had also been surrendered by Kobad, it would probably have shared the same fate.

SILVER CROSS WITH QUARTZ CAT'S EYE.
Russian, sixteenth century. Collection of Mrs. Henry Draper.

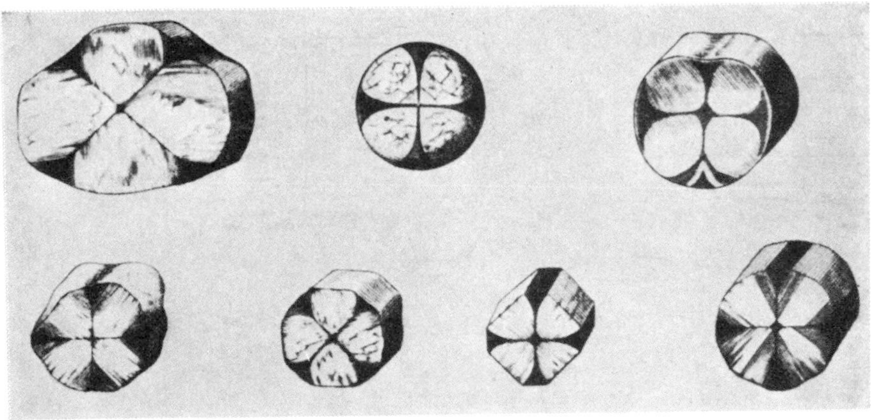

SPECIMENS OF CHIASTOLITE (LAPIS CRUCIFER). (See page 271.)
From the "Metallotheca Vaticana" of Mercatus, Rome, 1719, p. 238. In the author's library.

We have here a wide field for conjecture,—but, unfortunately, nothing more. Still, in the absence of any definite and trustworthy information, there is a kind of romantic interest in viewing the various possible relations of the mystery surrounding the fate of the most precious gems, historically at least, that have ever existed. More especially is this interest justified in the case of all who are disposed to prize gems and jewels for their symbolic significance, for, as we have shown, this significance, as far as concerns natal stones and the spiritual interpretation of the qualities of the heart and soul symbolized by the color and character of the principal precious and semi-precious stones, has its root in the veneration felt by early Christian writers, beginning with the author of the Apocalypse, for the unforgotten and unforgettable gems that were worn by the Hebrew high-priest.

A rather ingenious utilization of the reputed powers of Aaron's breastplate comes to us in a book printed in Portland, Maine.[13] The writer assumes that the Urim and Thummim enclosed in the folds of the breastplate consisted of twelve stones, duplicates of those engraved with the names of the tribes, and so disposed that, when they were shaken to and fro and then allowed to come to rest, three of them would become visible through an aperture in the ephod just beneath the rows of set stones. The signification of the oracle is given by the various combinations of color offered by the three stones that reveal themselves; to each combination a prearranged meaning is given. That anything of the kind could have been true of the original Urim and Thummim is scarcely worthy the trouble of refutation, but the

[13] C. H. Emerson, "Psychocraft" [Portland, Me., 1911].

practical result of this modern experiment is a clever oracle which will probably enjoy a certain vogue.

For those who, with the late lamented Lieutenant Totten, see in the tribes of Manasseh and Ephraim the Anglo-Saxons of England and the United States, and who look upon George V as the king who sits upon the throne of David, these symbolical stones of the breastplate acquire an added significance. While not pretending to be able to follow all the intricate and certainly most ingenious and interesting speculations of this school of Biblical exegesis, we cannot help expressing some astonishment that Ephraim should be thought to prefigure England and Manasseh the United States, instead of *vice versa.* In Gen. xlviii, 17–20, the text more especially referred to in these speculations, Jacob's blessing is bestowed upon Ephraim, in spite of Joseph's protest that it should go to the *eldest* son, Manasseh. To this protest Jacob answers: "I know it, my son, I know it: he also [Manasseh] shall become a people, and he also shall be great: but truly his younger brother shall be greater than he, and his seed shall become a multitude of nations." Certainly the very composite population of the United States perfectly merits this description. As a general rule, the Hebrews, when using the names Ephraim and Manasseh as tribal designations, maintained the twelve-fold division of the people, by substituting these tribes for Joseph and by dropping the name of Levi from the list, the tribe of Levi being assigned as priests to the care of the sanctuary, and not participating in the division of the Land of Promise.

In the Midrash Bemidbar, the Rabbinical commentary on Numbers, the tribes are given in their order, with the stone appropriate to each and the color of the tribal standard pitched in the desert camp, this color corresponding in each case with that of the tribal stone. This list represents a tradition dating back to at least the twelfth century and possibly much earlier than that; hence its value should not be underestimated, although we may not accept it without some reserves.[14]

Odem	Reuben	Red
Pitdah	Simeon	Green
Bareketh	Levi	White, black and red
Jophek	Judah	Sky-blue
Sappir	Issachar	Black (like stibium)
Yahalom	Zebulun	White
Leshem	Dan	Sapphire-color
Shebo	Gad	Gray
Ahlamah	Naphtali	Wine-color
Tarshish	Assher	Pearl-color (?)
Shoham	Joseph	Very black
Yashpheh	Benjamin	Colors of all the stones

In the attempt to determine the identity of the stones enumerated in Exodus xxviii and xxxix, as adorning the breastplate of the high-priest, we must bear in mind that this "breastplate of Aaron" and the one described by Josephus, and brought by Titus to Rome after the capture of Jerusalem in 70 A.D., are in all probability entirely distinct objects. The former, if it ever existed, except in the ideal world of the authors of the Priestly Codex, must have been composed of the stones known to and used by the Egyptians of the thirteenth or fourteenth century, B.C., some of them being, perhaps, set in the "jewels of gold and jewels of silver" borrowed by the Israelites from the Egyptians just before the Exodus; on the other hand, the most trustworthy indications re-

[14] "Der Midrasch Bemidbar Rabba," German transl. by Dr. Aug. Wünsche, Leipzig, 1885, pp. 15, 16. Parasha II. Of the *tarshish* it is said the color resembled that of "the costly stone with which women adorn themselves," possibly the pearl is signified. Hebrew text in "Sepher Midrash Rabba," Vilna, 1845, pt. iii, "Sepher Bemidbar," p. 23.

garding the stones of the breastplate of the Second Temple, made perhaps in the fifth century B.C., should be sought in the early Greek and Latin versions of the Old Testament, and in the treatise on precious stones by Theophrastus, who wrote about 300 B.C. The Natural History of Pliny, that great storehouse of ancient knowledge, and other early writers, may also be used with profit.

TITLE PAGE OF THE EDITION OF MARBODUS ON PRECIOUS STONES, PUBLISHED IN COLOGNE, 1539.

Shows the figure of the High-priest and the names and tribal attributions of twelve stones of the Breastplate.

I. Odem. [אֹדֶם.] The etymology of this word clearly indicates that we have to do with a red stone, most probably the carnelian. We know that in ancient Egypt hieroglyphic texts from the Book of the Dead were engraved upon amulets made from this stone, and it was also used for early Babylonian cylinders. Fine specimens of carnelian were obtained from Arabia. The Greek Septuagint and the Latin Vulgate, as well as Josephus, in the "Wars of the Jews" (V, 5, 7), and Epiphanius, all translate *sardius,* the ancient designation of carnelian; in his "Antiquities," however, Josephus renders *odem* by "sardonyx." The Egyptian word *chenem* was used to designate red stones, and seems to have been applied indifferently to red jasper and red feldspar as well as to carnelian; indeed, the first-named material was more freely used in early Egyptian work than the carnelian. It is, therefore, probable that in Mosaic times *odem* signified red jasper, while for the fifth century B.C. "carnelian" would be the better rendering. This modern name of the *sardius,* signifying the "flesh-colored" stone, first appears in the Latin translation of a treatise by Luca ben Costa, who wrote in the tenth century A.D. The name of Reuben is said to have been engraved on the *odem* stone, which occupied the first place on the breastplate.

II. Piṭdah. [פִּטְדָה.] There seems to be little doubt that this is the *topazius* of ancient writers, which usually signified our chrysolite, or peridot, not our topaz; for Pliny and his successors describe the topazius as a stone of a greenish hue. A legend related by Pliny gives as the place of origin an island in the Red Sea, called Topazos, from *topazein,* "to conjecture," because it was difficult to find. However, the Hebrew *piṭdah* appears to have been derived from the Sanskrit *pîta,* "yellow," and should, therefore, have originally signified a yellow stone, perhaps our topaz. W. M. Flinders Petrie, probably influenced by this Sanskrit etymology, sees in it the yellow serpentine used in ancient Egypt. If, nevertheless, we admit that a light green stone occupied the second place on the Mosaic breastplate, it was perhaps the light green serpentine. This was called *meh* in Egyptian, and was often used for amulets. In the case of the later breastplate we may substitute the peridot. On this second stone was engraved the name Simeon.

III. Bareketh. [בָּרֶקֶת.] Here the Septuagint, Josephus, and the Vulgate agree in translating *smaragdus,* and as we know that emerald mines were worked at

Mount Zabarah, in Nubia, before the beginning of our era, and that the emerald was known and used in Egypt, there does not seem to be any reason for rejecting the usual translation "emerald." Still it must be admitted that *smaragdus* often designates other green stones than the emerald. The suggestion has been made (by Myers and Petrie) that the passage in Revelation iv, 3, where the rainbow is likened to the *smaragdus,* indicates that the writer used this name for rock-crystal; but this conjecture is scarcely satisfactory, since it confuses the prismatic effects of light which has traversed the crystal with the crystal itself. There can be little doubt that a stone of brilliant coloration, like the emerald, not a colorless one, like rock-crystal, would be used as a simile of the rainbow. Whether the Mosaic breastplate already contained the emerald is another question, and it seems rather more likely that green feldspar, freely used in ancient Egypt for amulets, and known as *uat,* was the third stone of the proto-breastplate. The Authorized Version makes "the carbuncle" the third instead of the fourth stone. Upon the *bereketh* was engraved the name Levi.

IV. Nophek. [נֹפֶךְ.] This name is rendered ἄνθραξ by the Septuagint and Josephus, and "carbunculus" by the Vulgate. This designation, signifying literally "a glowing coal," was used for certain stones distinguished by their peculiarly brilliant red color, such as the ruby and certain fine garnets. While it is quite possible that the Oriental ruby may have been in the breastplate seen by Josephus, it is almost certain that it could not have been in the original breastplate of Mosaic times, since there is absolutely no proof that this stone was known in ancient Egypt. Hence we are inclined to believe that in the thirteenth century B.C. the name *nophek* designated the almandine garnet, or some similar variety of that stone. The Authorized Version has "emerald" here instead of in the third place. On this fourth stone of the breastplate was engraved the tribal name of Judah.

V. Sappir. [ספיר.] This is rendered *sapphirus* in all the old versions.[15] The stone cannot have been our sapphire, for both Theophrastus and Pliny describe the *sapphirus* as a stone with golden spots, thus showing that they meant the lapis-lazuli, which is often spotted with particles of pyrites having a golden sheen. This stone was named *chesbet* by the Egyptians, and was highly prized by them, a quantity of lapis-lazuli often appearing as an important item in the lists of tribute paid to Egypt and among the gifts sent by Babylonia to the Egyptian monarchs, and obtained from the oldest mines in the world. These were worked at a period 4000 B.C. and still are worked to this day. From this material amulets and figures were made, many of which have been preserved for us, and the Egyptian high-priest is said to have worn, suspended from his neck, an image of Mat, the Goddess of Truth, made of lapis-lazuli. The name is composed of the Latin *lapis,* "a stone," and *lajuward,* the name of the stone in Persian. From this latter word is also derived our "azure." In ancient times the lapis-lazuli was the blue stone *par excellence,* because of its beautiful color and the valuable ultramarine dye derived from it. Although Pliny writes (xxxvii, 39) that this stone was too soft for engraving, this fact need not have prevented its use in the breastplate, since the stones set therein were not intended for use as seals and hence were not subjected to any wear. In this connec-

[15] There are two evident transpositions in the text of Josephus between the fifth and sixth and the eighth and ninth stones respectively.

tion, however, it is somewhat strange that the Hebrew word *sappir* appears to indicate a stone especially adapted to receive inscriptions. The fact that the lapis-lazuli was greatly esteemed in ancient Egypt, and was still much used as an ornamental stone in Greek and Roman times, renders it probable that it was set not only in the original breastplate, but also in that of a later age. Upon this fifth stone the name Issachar was inscribed.

VI. Yahalom. [יַהֲלֹם] The sixth stone of the Septuagint version and of Josephus is the *ἴασπις*, probably green jasper, or jade, and this has been assumed to show that in the original Hebrew text *yashpheh* was the sixth stone, in place of *yahalom*. The twelfth stone of the Greek version is the *ὀνύχιον* or "onyx," and this seems to be the most probable equivalent of the Hebrew *yahalom*. Some Hebrew sources, however, render it "diamond," and Luther in his German version of the Bible, as well as our own Authorized Version, translates it thus. This rendering is based upon the derivation of the word *yahalom* from a verb meaning "to smite," thus making the name of the stone signify "the smiter," a designation not inappropriate for the diamond, which, because of its extreme hardness, has the power to cut, or "smite," all other stones. However, for this purpose the emery corundum, or smiris-point *shamir*, mentioned in Zechariah, was most likely used. The diamond was certainly not used in this way in very early times, although it is possible that the stone was employed in engraving in the fifth century B.C. These considerations induce us to prefer the traditional interpretation of *yahalom*, and translate it "onyx." In this case "the smiter" could be explained as denoting the use of the engraved onyx for sealing, as the engraved figure or letters were struck upon some soft material to make an impression. Zebulun was the tribal name inscribed on the *yahalom.*

VII. Leshem. [לֶשֶׁם.] No stone in the breastplate is more difficult to determine than this one. The Septuagint, Josephus, and the Vulgate all translate *ligurius,* an appellation sometimes applied to amber, a substance quite unfitted for use in the breastplate among the other engraved stones. Probably the original significance of *ligurius* was amber, this name being used because Liguria, in northern Italy, was the chief source of supply for Greece and the Orient; amber which had been gathered on the shores of the Baltic being brought by traders to Liguria and forwarded thence to other lands. As, however, the Greeks had another name for amber, *electron,* the name *ligurion* appears to have been applied later to a variety of the jacinth somewhat resembling amber in color, and then to other varieties of the same stone. The original form of the name was evidently *ligurion,* which was later changed to *lyncurion,* and was then explained as meaning the urine of the lynx (from λύγξ, and οὖρον, urine). This fanciful etymology gave rise to the story that the *ligurios,* or rather *lyncurius,* was the solidified urine of the lynx. The term *lyncurion,* as used by Theophrastus, may possibly have included the sapphire as well as the jacinth, since he lays especial stress upon the coldness of this substance, a quality characteristic of the sapphire, and also of the still denser jacinth. Hence, it appears that we have, even in the name *ligurius,* some justification for accepting the rendering *hyacinthus,* suggested by the list of foundation stones in Revelation xxi, 20, and already proposed by Epiphanius, Bishop of Constantia, about 400 A.D. Whether *hyacinthus* should be rendered "sapphire" or "jacinth" is not easy to determine, as this name seems to have been used indifferently for both stones; with the Arabs, under the form *yakut,* it became a generic term for all the varieties of the corundum gems. The sapphire was engraved in Greek and Roman times and is, perhaps, the *leshem* stone of the Second Temple. For the Mosaic breastplate we are forced to seek for some stone known in ancient Egypt, where the sapphire does not seem to have been introduced at an early date. If we could accept the suggestion of Brugsch that the Egyptian *neshem* stone, reputed to have wonderful magic virtues, was the same as the Hebrew *leshem,* a brown agate would have been the seventh stone in the original breastplate, as Wendel gives very strong reasons for rendering *neshem* in this way. The color designations were very freely used in Egyptian, and therefore a reddish or a yellowish brown agate may have been used. The *leshem* bore the tribal name Joseph.

VIII. Shebo. [שְׁבוֹ.] This is uniformly rendered in the ancient versions and in Josephus by "agate," a composite stone highly esteemed in very ancient times, and hence worthy of a place among the stones of the breastplate; at a later period, as Pliny notes (xxxvii, 54), it became so common that it was but little regarded. Nevertheless the fact that the various kinds of agates were believed to have many talismanic and therapeutic virtues, the great variety of coloration observable in these stones, and the curious figures and markings displayed by many of them, served to make them favorite objects. The etymology of the word *shebo* suggests that it designated more especially a banded agate, and that set in the proto-breastplate was most probably one with gray and white bands, as this variety often appears in Egyptian work. There would have been no lack of contrast between this stone and the reddish or yellowish-brown agate, of uniform color, which may have occupied the seventh place. For the later breastplate we may choose any one of the many kinds of banded agate. This stone had engraved upon it the name Benjamin.

IX. Aḥlamah. [אַחְלָמָה.] As to this stone also, all the authorities are in agreement, and render *aḥlamah* by "amethyst." This was not, however, the Oriental amethyst, a variety of corundum, but a dark blue or purple variety of quartz. Both Arabia and Syria furnished a supply of amethysts. The Hebrew name shows that this stone was believed to possess the virtue of inducing dreams and visions (cf. *halom*—"dream"), while, as is well known, the Greek name characterizes it as an enemy or preventive of inebriety. The amethyst was known in ancient Egypt and probably was named *hemag*. In the Book of the Dead a heart made of *hemag* is mentioned, and two such heart-shaped amulets of amethyst are preserved in the Boulaq Museum. As the amethyst retained its repute as a stone of beauty and power through the Greek and Roman periods, we may safely assert that it was set in both the first and second breastplates. Upon the *aḥlamah* was engraved the name Dan.

X. Tarshish. [תַּרְשִׁישׁ] The Septuagint renders this word "chrysolite," where it is used in the description of the breastplate, as does Josephus also. In the Authorized Version, "beryl" is the rendering. We have already stated that the topaz of the ancients was usually our chrysolite, or peridot, and the name "chrysolite" appears to have been used to designate our topaz. This is indeed indicated by the literal meaning of the word, "golden-stone." The *tarshish* received its name from Tartessus, in Spain, an important commercial station of the Phœnicians. The stone derived from this source was not, of course, our Oriental topaz, a variety of corundum, nor was it the true topaz; neither is it at all likely that the name *tarshish* signified, at least originally, the genuine topaz; most probably it denoted a variety of quartz which occurs in Spain. This is originally black, but is decolorized by heating to a deep brown, and if the heating be prolonged the stone becomes paler and eventually entirely transparent. The ancients were familiar with this property. In ancient Egyptian records a stone called *thehen* is frequently mentioned as a material from which amulets were made. This Egyptian name signified primarily a "yellow stone," and might designate either the topaz or the yellow jasper, known and used in Egypt at a very early date; the topaz was probably not known there earlier than 500 or 600 B.C. Hence, in spite of the unquestionable difficulty offered by the geographical name *tarshish,* which might seem to confine us to a Spanish origin for the stone, the probabilities favor the selection of the yellow jasper as the tenth gem in Aaron's breastplate. For that made with pious zeal by those who labored to renew the glories of the Old Jerusalem, we choose the topaz,—possibly, indeed, a fine specimen of the genuine topaz,—for whatever the quality of the yellow stone originally brought from Tartessus, the name may well have been applied to the genuine topaz when that stone became known to the Jews, either in Babylonia, or after their return to Palestine. The *tarshish* was engraved with the name Naphtali.

XI. Shoham. [שֹׁהַם] The Septuagint translates "beryl," but in our Authorized Version and in that used by Roman Catholics, the so-called Douai Version, the word is invariably rendered "onyx." Diodorus Siculus and Dionysius Periegetes, writing in the first century B.C., are the first classical authors who use the name beryl.

While this name does not appear in the treatise of Theophrastus, he evidently includes the beryl among his *smaragdi;* indeed, the true emerald is simply a variety of the beryl, and owes its beautiful coloration to a slight admixture of chromium. The finest beryls were brought from India. Besides the specimen set in the breastplate, the high-priest wore on his shoulders two *shoham* stones, each engraved with the names of six of the tribes. After carefully weighing the evidence, we believe that the stones worn by the high-priest of the Second Temple were aquamarines (beryls). In our endeavor to determine the *shoham* stones used in Mosaic times, we have no very definite information to guide us; on the whole, the conjecture of J. L. Myers, that they were malachites, seems to have much in its favor, for this material was known to the ancient Egyptians and appears to have been often used for amulets. The Egyptian name for malachite, as well as for other green stones, was *mafek,* and a ring of *mafek* is mentioned in an Egyptian text; undoubtedly, at a later period in Egyptian history, *mafek* may also have denoted the beryl. In view of the fact that the turquoise was unquestionably known to the Egyptians at a very early date, the supply being derived from mines in the Sinai Peninsula, which were rediscovered by Macdonald, we might be tempted to suggest that the *shoham* stones were turquoises. The light blue or blue-green of the specimens of this stone found on Mt. Sinai would make an even better contrast with the neighboring jade than would the bright green malachite. On the *shoham* of the breastplate the name Gad was engraved.

XII. Yashpheh. [יָשְׁפֵה.] If, as appears almost certain, this name originally occupied the sixth place in the original Hebrew text, all the ancient versions agree in translating it "jasper." An Assyrian form of the name was *yashpu,* as is shown by the Tell el Amarna letters in the cuneiform writing dating from not long before the Exodus. Of all the so-called jaspers none were so highly valued as those of a green color. The talismanic and therapeutic qualities of the "green jaspers" are often noted by ancient writers, and, according to Galen, these stones were recommended for remedial use by Egyptian writers on medicine. Abel Remusat, the great French Orientalist, writing in 1820, was one of the first to see in the *yashpheh* of the Hebrews and in the green jasper of the Greeks and Romans, the material jade (nephrite or jadeite), the Chinese *yu*-stone. These minerals were used both in the Old and the New World, and were everywhere believed to possess wonderful virtues. Very likely the powers supposed to characterize jade were later attributed to green jasper, but there is every reason to suppose that the true jade was always more highly prized than its jasper substitute, for it was much rarer, and was easily distinguishable, by its translucency, from jasper of a similar color. Until quite recently only Turkestan, Burma and New Zealand have supplied jade and most of that used in other lands came from prehistoric relics or from sources unknown to us. It seems highly probable that the *yashpheh* which adorned the breastplate made for Aaron was a piece of nephrite or jadeite; possibly in the later breastplate green jasper may have been employed. This stone was inscribed with the tribal name Assher.

In the following lists of the precious and semi-precious stones contained in the earlier and later breastplates, the writer does not claim to have finally solved the problem presented by the Hebrew accounts of the high-priest's adornment, but he hopes that the distinction established here between the Mosaic breastplate and that of the Second Temple, separated from each other by an interval of eight centuries, may serve to clear up some of the difficulties encountered in the treatment of this subject.

	The Breastplate of Aaron.	The Breastplate of the Second Temple.
I	Red jasper	Carnelian
II	Light-green serpentine	Peridot
III	Green feldspar	Emerald
IV	Almandine garnet	Ruby
V	Lapis-lazuli	Lapis-lazuli
VI	Onyx	Onyx
VII	Brown agate	Sapphire or jacinth
VIII	Banded agate	Banded agate
IX	Amethyst	Amethyst
X	Yellow jasper	Topaz
XI	Malachite	Beryl
XII	Green jasper, or jade	Green jasper, or jade

The following lists show the variations of the different ancient authorities in regard to the names of the gems in the breastplate:

	Hebrew.	Septuagint. (Greek) About 250 B.C.	Josephus (Greek) About 90 A.D.	Vulgate (Latin) About 400 A.D.	Authorized Version 1611 A.D.	Revised Version 1884 A.D.
1	Odem	Sardion	Sardonyx	Sardius	Sardius	Sardius (or Ruby)
2	Pitdah	Topazion	Topazos	Topazius	Topaz	Topaz
3	Bareketh	Smaragdos	Smaragdos	Smaragdus	Carbuncle	Carbuncle (or Emerald)
4	Nophak	Anthrax	Anthrax	Carbunculus	Emerald	Emerald (or Carbuncle)
5	Sappir	Sappheiros	Iaspis	Sapphirius	Sapphire	Sapphire
6	Yahalom	Iaspis	Sappheiros	Jaspis	Diamond	Diamond (or Sardonyx)
7	Leshem	Ligurion	Liguros	Ligurius	Ligure	Jacinth (or Amber)
8	Shebo	Achatês	Amethystos	Achatês	Agate	Agate
9	Ahlamah	Amethystos	Achatês	Amethystus	Amethyst	Amethyst
10	Tarshish	Chrysolithos	Chrysolithos	Chrysolithus	Beryl	Beryl (orChalcedony)
11	Shoham	Bêryllion	Onyx	Onychinus	Onyx	Onyx (or Beryl)
12	Yashpheh	Onychion	Bêryllos	Beryllus	Jasper	Jasper

The high-priest's breastplate, as described in Hebrew tradition, was regarded by the Jews with peculiar reverence, and the stones set in it were believed to be emblematic of many things. It is, therefore, quite natural that these stones are described in the book of Revelation as the foundation stones of the New Jerusalem. The names are in some cases not identical with those given in Exodus, but this may arise from various renderings of the Hebrew names in the Targums or in the Greek versions.

The text in Revelation (xxi, 9–21) is as follows:

And there came unto me one of the seven angels which had the seven vials full of the seven last plagues, and talked with me, saying, Come hither, I will show thee the bride, the Lamb's wife:

And he carried me away in the spirit to a great and high mountain, and showed me that great city, the holy Jerusalem, descending out of heaven from God.

Having the glory of God: and her light was like unto a stone most precious, even like a jasper-stone, clear as crystal;

And had a wall great and high, and had twelve gates, and at the gates twelve angels, and names written thereon, which are the names of the twelve tribes of the children of Israel:

On the east, three gates; on the north, three gates; on the south, three gates; and on the west, three gates.

And the wall of the city had twelve foundations, and in them the names of the twelve apostles of the Lamb.

And he that talked with me had a golden reed to measure the city, and the gates thereof, and the wall thereof.

And the city lieth foursquare, and the length is as large as the breadth: and he measured the city with the reed, twelve thousand furlongs. The length and the breadth and the height of it are equal.

And he measured the wall thereof, an hundred and forty and four cubits, according to the measure of a man, that is, of the angel.

And the building of the wall of it was of jasper: and the city was pure gold, like unto clear glass.

And the foundations of the wall of the city were garnished with all manner of precious stones. The first foundation was jasper; the second, sapphire; the third, a chalcedony; the fourth, an emerald;

The fifth, sardonyx; the sixth, sardius; the seventh, chrysolite; the eighth, beryl; the ninth, a topaz; the tenth, a chrysoprasus; the eleventh, a jacinth; the twelfth, an amethyst.

And the twelve gates were twelve pearls; every several gate was of one pearl: and the street of the city was pure gold, as it were transparent glass.

It is easy to trace in this description the substitution of the twelve apostles for the twelve tribes in connection with the precious stones enumerated, and, besides this, we also have the twelve angels, associated at a later date with the months and the signs of the zodiac.

Of the twelve foundation stones the Revelation of St. John expressly states that they had "in them the names of the twelve apostles of the Lamb." The assignment of each stone to the respective apostle was made in later times according to the order given in the lists of the apostles contained in the so-called Synoptic Gospels, Matthew, Mark, and Luke. These lists are not quite identical—Andrew, for instance, being placed second in Matthew and Luke, but fourth in Mark—and the same stone was not always assigned to a given apostle. Frequently the list was modified by the addition of the apostle Paul, really the thirteenth apostle. In this case he was usually given the second place immediately after St. Peter, and to the brothers James and John, the "Sons of Thunder," was assigned a single stone; in some later arrangements St. Paul occupies the last place, after St. Matthias, who was chosen to take the place of Judas Iscariot, and whose name as an apostle first appears in the Acts.

LISTS OF THE APOSTLES.

Gospel of St. Matthew x, 2–4.	Gospel of St. Mark iii, 16–19.	Gospel of St. Luke vi, 14–16.
Peter	Peter	Peter
Andrew	James	Andrew
James	John	James
John	Andrew	John
Philip	Philip	Philip
Bartholomew	Bartholomew	Bartholomew
Thomas	Matthew	Matthew
Matthew	Thomas	Thomas
James the Less	James the Less	James the Less
Thaddeus	Thaddeus	Simon Zelotes
Simon Zelotes	Simon Zelotes	Judas
Judas Iscariot	Judas Iscariot	Judas Iscariot

The passage in Revelation xxi, 19, 20, is not the only one in that book treating of precious stones, for we read in chapter iv, 2, 3:

And immediately I was in the Spirit: and, behold, a throne was set in heaven, and *one* sat on the throne.

And he that sat was to look upon like a jasper and a sardine stone: and *there was* a rainbow round about the throne, in sight like unto an emerald.

The commentators, both ancient and modern, have given many different explanations of the symbolic meaning of the similes employed here. Some have seen in the two stones a type of the two judgments of the world, by fire and by water; others find that they signify the holiness of God and his justice. Of the rainbow "like unto an emerald," Alford says we should not think it strange that the bow is green, instead of prismatic: "the *form* is that of the covenant bow, the color even more refreshing and more directly symbolizing grace and mercy."[16]

[16] Alford, "The Greek Testament," vol. iv, Pt. 2, p. 594.

The significance of the twelve Apocalyptic gems is given by Rabanus Maurus, Archbishop of Mainz (786–856), in the following words:[17]

In the jasper is figured the truth of faith; in the sapphire, the height of celestial hope; in the chalcedony, the flame of inner charity. In the emerald is expressed the strength of faith in adversity; in the sardonyx, the humility of the saints in spite of their virtues; in the sard, the venerable blood of the martyrs. In the chrysolite, indeed, is shown true spiritual preaching accompanied by miracles; in the beryl, the perfect operation of prophecy; in the topaz, the ardent contemplation of the prophecies. Lastly, in the chrysoprase is demonstrated the work of the blessed martyrs and their reward; in the hyacinth, the celestial rapture of the learned in their high thoughts and their humble descent to human things out of regard for the weak; in the amethyst, the constant thought of the heavenly kingdom in humble souls.

The origin of the foundation stones named in Revelation xxi, 19, 20, may be found in the text, Isaiah liv, 11, 12, where we read:

O thou afflicted, tossed with tempest, *and* not comforted, behold, I will lay thy stones with fair colours, and lay thy foundations with sapphires.

And I will make thy windows of agates, and thy gates of carbuncles, and all thy borders of pleasant stones.

As we see, only three stones are mentioned by name: the sapphire, the carbuncle, and "agates." This last rendering is quite doubtful, as the Hebrew word (*kodkodim*) signifies shining or gleaming stones, and their use for windows indicates that they must have been transparent. It is easy to understand that in later times the twelve stones of the breastplate, dedicated to the twelve tribes of Israel, were used to fill out and com-

[17] Rabani Mauri, "Opera Omnia," vol. v, col. 470. Patrologiæ Lat., vol. cxi, Parisiis, 1864.

plete the picture, following the indication given by the general terms "stones with fair colours" and "pleasant stones."

In commenting on this text Rabbi Johanan is quoted in the Babylonian Talmud as saying that God would bring jewels and pearls thirty ells square (twenty ells in height and ten in width) and would place them on the gates of Jerusalem. There may be in this some reminiscence of the Apocalyptic foundation stones. A sceptical disciple said to the Rabbi, "We do not ever find a jewel as large as the egg of a dove." But not long afterward, when this same disciple was sailing in a boat on the sea, he saw angels sawing stones as immense as those described by Rabbi Johanan, and when he asked for what they were designed, the reply was, "The Holy One, blessed be He, will place them on the gates of Jerusalem." [18]

---

[18] "New Edition of the Babylonian Talmud," ed. and trans. by Michael L. Rodkinson, vol. v (xiii), New York, 1902, p. 210. Baba Batra.

IX

Birth-Stones

THE origin of the belief that to each month of the year a special stone was dedicated, and that the stone of the month was endowed with a peculiar virtue for those born in that month and was their natal stone, may be traced back to the writings of Josephus, in the first century of our era, and to those of St. Jerome, in the early part of the fifth century. Both these authors distinctly proclaim the connection between the twelve stones of the high-priest's breastplate and the twelve months of the year, as well as the twelve zodiacal signs. Strange to say, however, in spite of this early testimony, we have no instance of the usage of wearing such stones as natal stones until a comparatively late date; indeed, it appears that this custom originated in Poland some time during the eighteenth century. The reason for this seems to have been that the virtues attributed to each particular stone, more especially the therapeutic virtues, rendered it necessary to recommend the wearing of one or the other, according to the disease from which the person was suffering, for his natal stone might not have the power to cure his particular ailment, or might not bring about the fulfilment of his dearest wish. In other words, the belief in the special virtues of the stone was paramount, and it was long before the mystic bond between the stone of the month and the person born in that month was fully realized.

The order in which the foundation stones of the New Jerusalem are given in the book of Revelation determined the succssion of natal stones for the months. The first stone was assigned to St. Peter and to the month of March, to the leader of the apostles and to the month of the spring equinox; the second to the month of April; the third to May, etc. When, however, many centuries later,—probably in Poland, as we have stated,—with the aid of the rabbis or the Hebrew gem traders, the wearing of natal stones became usual, certain changes had been made in this order and some stones not mentioned among those of the breastplate, or of the New Jerusalem, were substituted for certain of these,—notably the turquoise for the month of December, the ruby for July, and the diamond for April. In modern times the turquoise has become the stone for July while the ruby has been assigned to December.

There is some evidence in favor of the theory that at the outset all twelve stones were acquired by the same person and worn in turn, each one during the respective month to which it was assigned, or during the ascendancy of its zodiacal sign. The stone of the month was believed to exercise its therapeutic or talismanic virtue to the fullest extent at that period. Perhaps the fact that this entailed a monthly change of ornaments may rather have been a recommendation of the usage than the reverse.[1]

It seems highly probable that the development of the belief in natal stones that took place in Poland was due to the influence of the Jews who settled in that country shortly before we have historic notice of the use of the twelve stones for those born in the respective months. The lively interest always felt by the Jews regarding the gems of the breastplate, the many and various commen-

[1] Brückmann, "Abhandlung von Edelsteinen," 2d ed., Braunschweig, 1773, p. 358.

taries their learned men have written upon this subject, and the fact that the well-to-do among the chosen people have always carried with them in their wanderings many precious stones, all this seems to make it likely that to the Jews should be attributed the fashion of wearing natal stones.

However, whether this conjecture be correct or erroneous, the fashion once started became soon quite general and has as many votaries to-day as ever before. There can be no doubt that the owner of a ring or ornament set with a natal stone is impressed with the idea of possessing something more intimately associated with his or her personality than any other stone, however beautiful or costly it may be. If it be objected that this is nothing but imagination due to sentiment, we must bear in mind that imagination is one of the most potent factors in our life; indeed, the great Napoleon is quoted as saying that it ruled the world.

Probably the very earliest text we have in which the stones of the breastplate are positively associated with the months of the year is to be found in the "Antiquities of the Jews," by Flavius Josephus.[2] This runs as follows:

> Moreover, the vestments of the high-priest being made of linen signifies the earth, the blue denotes the sky, being like lightning in its pomegranates, and resembling thunder in the noise of the bells. And as for the ephod, it showed that God had made the universe of four elements, and as for the gold interwoven in it, I suppose it related to the splendor by which all things are to be enlightened. He also appointed the breastplate to be placed in the middle of the ephod to resemble the earth, for that occupies the middle place in the world; and the girdle,

---

[2] Flavii Josephi, ed. Dindorf, Parisii, 1847, vol. ii, p. 97; "Antiq. Jud," lib. iii, cap. 7, paragraph 7. In the second century, Clemens Alexandrinus (lib. v, cap. 6) repeats this idea of Josephus, adding that the four rows in which the gems were disposed signified the four seasons of the year.

which encompassed the high priest about, signifies the ocean, for that goes about everything. And the two sardonyxes that were in the clasps on the high-priest's shoulders, indicate to us the sun and the moon. And for the twelve stones, whether we understand by them the months, or the twelve signs of what the Greeks call the zodiac, we shall not be mistaken in their meaning. And for the cap, which was of a blue color, it seems to me to mean heaven, for otherwise the name of God would not have been inscribed upon it. That it was also adorned with a crown, and that of gold also, is because of the splendor with which God is pleased.

This passage was adapted by St. Jerome, three hundred years later, in his letter to Fabiola,[3] and undoubtedly laid the foundation for the later custom of wearing one of these stones as a natal or birth-stone for a person born in a given month, or for an astral or zodiacal stone for one born under a given zodiacal sign. As we see, both uses are indicated by the passage of Josephus. In the later centuries, as the book of Revelation, which was generally less favored at the outset than the other parts of the New Testament, became a subject of devout study, and a mine of mystical suggestions, the twelve foundation stones (Rev. xxi, 19) of the New Jerusalem largely took the place of the stones of the breastplate. While this list of foundation stones is unquestionably based upon the much earlier list of the stones adorning Aaron's breastplate, the ordering differs considerably and there are some changes in the material; possibly many, if not all, of these differences may be due to textual errors or to a transcription from memory.

That the foundation stones were inscribed with the names of the apostles is expressly stated (Rev. xxi, 14), but it was not until the eighth or ninth century that the commentators on Revelation busied themselves with

[3] Sancti Hieronymi, "Opera Omnia," ed. Migne, Parisiis, 1877, vol. i, col. 616; Epistola lxiv, paragraph 16.

finding analogies between these stones and the apostles. At the outset, the symbolism of the stones was looked upon from a purely religious standpoint. Few of the early fathers—we may except Epiphanius—thought or cared much for the stones themselves, or knew much of them; but, in time, their natural beauty became more and more highly developed as the lapidarian art demanded better cut and choicer material, their supposed virtues came to the fore, and the symbolism was strengthened and emphasized by a reference to their innate qualities and also to their peculiar powers. The fact that this part of the tradition was rather of pagan than of Christian origin probably contributed to render it less attractive to the early Christians, so that it was not until Christianity had become practically universal in the Greek and Roman world and the opposition to pagan traditions, as such, was weakened and, indeed, largely forgotten, that the virtues of the stones were made prominent, and certain parts of these superstitions were retained, as were some of the pagan ceremonies in the Christian religion.

One of the earliest writers to associate directly with the apostles the symbolism of the gems given as foundation stones of the New Jerusalem by St. John in Revelation xxi, 19, is Andreas, bishop of Cæsarea. This author was at one time assigned by critics to the fifth century A.D.,[4] but more recent investigation has shown that he probably belonged to the last half of the tenth century. His exposition reads as follows:[5]

The jasper, which like the emerald is of a greenish hue, probably signifies St. Peter, chief of the apostles, as one who so bore Christ's death in his inmost nature that his love for Him was always vigorous

---

[4] Lücke, "Versuch einer Einleitung in die Offenbarung Johannes," Bonn, 1852, p. 964.

[5] Patrologiæ Græcæ, ed. Migne, vol. cvi, Parisiis, 1863, cols. 433–438.

and fresh. By his fervent faith he has become our shepherd and leader.

As the sapphire is likened to the heavens (from this stone is made a color popularly called lazur), I conceive it to mean St. Paul, since he was caught up to the third heaven, where his soul was firmly fixed. Thither he seeks to draw all those who may be obedient to him.

The chalcedony was not inserted in the high-priest's breastplate, but instead the carbuncle, of which no mention is made here. It may well be, however, that the author designated the carbuncle by the name chalcedony. Andrew, then, can be likened to the carbuncle, since he was splendidly illumined by the fire of the Spirit.

The emerald, which is of a green color, is nourished with oil, that its transparency and beauty may not change; we conceive this stone to signify John the Evangelist. He, indeed, soothed the souls dejected by sin with a divine oil, and by the grace of his excellent doctrine lends constant strength to our faith.

By the sardonyx, showing with a certain transparency and purity the color of the human nail, we believe that James is denoted, seeing that he bore death for Christ before all others. This the nail by its color indicates, for it may be cut off without any sensible pain.

The sardius with its tawny and translucent coloring suggests fire, and it possesses the virtue of healing tumors and wounds inflicted by iron; hence I consider that it designates the beauty of virtue characterizing the apostle Philip, for his virtue, animated by the fire of the Holy Spirit, cured the soul of the wounds inflicted by the wiles of the devil, and revived it.

The chrysolite, gleaming with the splendor of gold, may symbolize Bartholomew, since he was illustrious for his divine preaching and his store of virtues.

The beryl, imitating the colors of the sea and of the air, and not unlike the jacinth, seems to suggest the admirable Thomas, especially as he made a long journey by sea, and even reached the Indies, sent by God to preach salvation to the peoples of that region.

The topaz, which is of a ruddy color, resembling somewhat the carbuncle, stops the discharge of the milky fluid with which those having eye-disease suffer. This seems to denote Matthew, for he was animated by a divine zeal, and, his blood being fired because of Christ, he was found worthy to enlighten by his Gospel those whose heart was blinded, that they might like new-born children drink of the milk of the faith.

The chrysoprase, more brightly tinged with a golden hue than gold itself, symbolizes St. Thaddæus; the gold (*chrysos*) symbolizing the kingdom of Christ, and the *prassius,* Christ's death, both of which he preached to Abgar, King of Edessa.

The jacinth, which is of a celestial hue, signifies Simon Zelotes, zealous for the gifts and grace of Christ and endowed with a celestial prudence.

By the amethyst, which shows to the onlooker a fiery aspect, is signified Matthias, who in the gift of tongues was so filled with celestial fire and with fervent zeal to serve and please God, who had chosen him, that he was found worthy to take the place of the apostate Judas.

Some theologians were opposed to the assignment of the foundation stones to the apostles, for they held that only Christ himself could be regarded as the foundation of his Church. Hence the symbolism of these stones was made to apply to Christ alone, the color of the stone often guiding the commentator in his choice of ideas denoted by the different gems. Thus, one writer, applying all the meanings to Christ, finds that the greenish Jasper denotes satisfaction; the sky-blue Sapphire, the soul; the bright-red Chalcedony, zeal for truth; the transparent green Emerald, kindness and goodness; the nail-colored Sardonyx, the strength of spiritual life; the red Sardius, readiness to shed His blood for the Church; the yellow Chrysolite, the excellence of His divine nature; the sea-green Beryl, moderation and the control of the passions; the glass-green Topaz (chrysolite?), uprightness; the harsh-colored Chrysoprase, sternness towards sinners; the violet or purple Jacinth, royal dignity, and, lastly, the purple Amethyst, with a touch of red, perfection.[6]

Andreas of Cæsarea freely recognizes his indebtedness to the much more ancient source, St. Epiphanius, bishop of Constantia in Cyprus, who died in 402 A.D., and who wrote a short but very valuable treatise on the

---

[6] Georgius Vitringa, "Nauwkeurige onderzoek van de goddelyke Openbaring der H. Apostels Johannes," Dutch trans. of Latin by M. Gargon, Amsterdam, 1728, vol. ii, p. 681.

stones of the breastplate, noting in several cases the therapeutic and talismanic virtues of these stones and giving his opinion as to the order in which the names of the tribes were inscribed upon them.[7] As the foundation stones of Revelation are rightly called "apostolic stones," so those of the breastplate merit the designation of "tribal stones," as well as that of astral stones; indeed, the Jews of medieval times definitely associated the tribes with the zodiacal signs in the following order:

Judah	Aries
Issachar	Taurus
Zebulun	Gemini
Reuben	Cancer
Simeon	Leo
Gad	Virgo
Ephraim	Libra
Manasseh	Scorpio
Benjamin	Sagittarius
Dan	Capricorn
Naphtali	Aquarius
Asher	Pisces

For Rabanus Maurus the nine gems of the king of Tyre named in Ezekiel xxxviii, 13, are types of the nine orders of angels, just as the twelve foundation stones of Revelation signify the twelve apostles.[8]

It is evident, from early and later usage, that, at the place and time where and when these stones were first utilized for birth-stones, the year must have begun with the month of March. This will be apparent when we compare the following eight lists, carefully gathered from various sources:

[7] Sancti Patris Epiphanii episcopi Cypri ad Diodorum Tyri episcopum, "De XII. Gemmis, quæ erant in veste Aaronis," ed. Gesner, Tiguri, 1565.

[8] Rabani Mauri, "Opera Omnia," vol. v, col. 465; in Patrologiæ Latinæ, ed. Migne, vol. xvi, Paris, 1864.

Month	Jews	Romans	Isidore Bishop of Seville	Arabians	Poles	Russians	Italians	15th to 20th Century
January	Garnet	Garnet	Hyacinth	Garnet	Garnet	Garnet Hyacinth	Jacinth Garnet	Garnet
February	Amethyst	Amethyst	Amethyst	Amethyst	Amethyst	Amethyst	Amethyst	Amethyst Hyacinth Pearl
March	Jasper	Bloodstone	Jasper	Bloodstone	Bloodstone	Jasper	Jasper	Jasper Bloodstone
April	Sapphire	Sapphire	Sapphire	Sapphire	Diamond	Sapphire	Sapphire	Diamond Sapphire
May	Chalcedony Carnelian Agate	Agate	Agate	Emerald	Emerald	Emerald	Agate	Emerald Agate
June	Emerald	Emerald	Emerald	Agate Chalcedony Pearl	Agate Chalcedony	Agate Chalcedony	Emerald	Cat's-eye Turquoise Agate
July	Onyx	Onyx	Onyx	Carnelian	Ruby	Ruby Sardonyx	Onyx	Turquoise Onyx
August	Carnelian	Carnelian	Carnelian	Sardonyx	Sardonyx	Alexandrite	Carnelian	Sardonyx Carnelian Moonstone Topas
September	Chrysolite	Sardonyx	Chrysolite	Chrysolite	Sardonyx	Chrysolite	Chrysolite	Chrysolite
October	Aquamarine	Aquamarine	Aquamarine	Aquamarine	Aquamarine	Beryl	Beryl	Beryl Opal
November	Topas	Topas	Topas	Topas	Topas	Topas	Topas	Topas Pearl
December	Ruby	Ruby	Ruby	Ruby	Turquoise	Turquoise Chrysoprase	Ruby	Ruby Bloodstone

It may be interesting to show in these eight lists the stones which are most favored in each month in the following way, the numerals indicating the number of lists in which the stones appear (including the alternate stones):

January..... Garnet 7, hyacinth 2.
February.... Amethyst 8, hyacinth 1, pearl 1.
March....... Jasper 5, bloodstone 4.
April........ Sapphire 7, diamond 2.
May.......... Agate 5, emerald 4, chalcedony 1, carnelian 1.
June .... Emerald 4, agate 4, chalcedony 3, turquoise 1, pearl 1, cat's-eye 1.
July......... Onyx 5, sardonyx 1, carnelian 1, ruby 1, turquoise 1.
August ...... Carnelian 5, sardonyx 3, moonstone 1, topaz 1, alexandrite 1.
September ... Chrysolite 6, sardonyx 2.
October...... Beryl, 8, aquamarine 5, opal 1.
November ... Topaz 8, pearl 1.
December.... Ruby 6, turquoise 2, chrysoprase 1, bloodstone 1.

With the exception of January, where we have the garnet instead of the jacinth, and of December, which gives us the ruby instead of the chrysoprase, the first choices are practically identical with the foundation stones, bearing in mind that the eleventh stone is that for January, the twelfth that for February, the first that for March and so on.

Of the assignment of the natal stones to the different months of the year or to the zodiacal signs, Poujet fils, writing in 1762, states that in his opinion this fashion started in Germany—others say in Poland—some two centuries before his time, and he adds that, though this arrangement was purely imaginary, and unknown to ancient writers, it soon became popular, and many, more especially of the fair sex, seeing in it an element of mystery, wished to wear rings set with the stone appropriate

FACSIMILE OF THE BETROTHAL RING OF THE VIRGIN IN THE CATHEDRAL OF PERUGIA.

The original ring, which is of chalcedony, is shown on St. Agatha's Day, July 29, to cure ailments of mothers. This cord and facsimile of ring acquired by the author at Perugia, May 6, 1902.

to the month of their birth, the stone being engraved with the appropriate zodiacal sign.[9] However correct Poujet may be regarding the period at which the fashion of wearing natal rings was introduced, he is, as we have already shown, quite wrong in believing that the serial arrangement of the stones and their assignment to months or signs was purely imaginary, for it is unquestionably based on the list in Revelation, which in its turn goes back to the twelve stones of the high-priest's breastplate.

The fashion of wearing a series of twelve stones denoting (or bearing) the zodiacal signs seems to have existed in the sixteenth century, for Catherine de' Medici is said to have worn a girdle set with twelve stones, among which were certain onyxes as large as crown-pieces, upon which talismanic designs had been engraved. Two hundred years later this girdle is stated to have been in the possession of a M. d'Ennery, whose collection of antique medals was regarded as the finest in Paris at the time.[10] It is not, however, certain that the twelve stones of Catherine's girdle were those attributed to the zodiacal signs both at an earlier and later period.

Though the substitution of a new schedule for the time-honored list of birth-stones has received the approval of the National Association of Jewellers at the meeting in Kansas City August, 1912, it can scarcely be said to offer a satisfactory solution of the question, which has its importance not only from a commercial point of view, but also because the idea that birth-stones possess a certain indefinable, but none the less real significance, has long been present and still exercises a spell over the

---

[9] Poujet fils, "Traité des pierres précieuses," Paris, 1762, p. 4.
[10] Poujet fils, l.c.

minds of all who are gifted with a touch of imagination, or romance, if you will. The longing for something that appeals to this sense is much more general than is commonly supposed, and is a not unnatural reaction against the progress of materialism, against the assertion that there is nothing in heaven or earth but what we can definitely apprehend through our senses.

It is this persuasion that should be chiefly considered in any attempt to tamper with the traditional attribution of the stones to particular months or to the zodiacal signs. Once we allow the spirit of commercialism pure and simple to dictate the choice of such stones, according to the momentary interest of dealers, there is grave danger that the only true incentive to acquire birth-stones will be weakened and people will lose interest in them. Sentiment, true sentiment, is one of the best things in human nature. While if darkened by fear it may lead to pessimism, with all the evils which such a state of mind implies, if illumined by hope it gives to humanity a brighter forecast of the future, an optimism that helps people over difficult passages in their lives. Thus, sentiment must not be neglected, and nothing is more likely to destroy it than the conviction that it is being constantly exploited for purposes of commercialism. For this reason, the interest as well as the inclination of all who are concerned in this question of birth-stones should induce a very careful handling of the subject.

Quite true it is that there are now, and have been in the past, several lists of these stones, differing slightly from one another, but all are based essentially either upon the list of foundation stones given in Revelation (xxi, 19) or upon that of the gems adorning the breastplate of Aaron and enumerated in Exodus (xxxix, 10–13). For convenient reference, we give the latter according to the Authorized Version of the Scriptures, and also as corrected by later research, and the former according to the Authorized Version.

	Breastplate. Authorized Version.	Breastplate. Later Correction.	Foundation Stones. Authorized Version.
I	Sardius	Carnelian	Jasper
II	Topaz	Chrysolite (peridot)	Sapphire
III	Carbuncle	Emerald	Chalcedony
IV	Emerald	Ruby	Emerald
V	Sapphire	Lapis-lazuli	Sardonyx
VI	Diamond	Onyx	Sardius
VII	Ligure	Sapphire	Chrysolite
VIII	Agate	Agate	Beryl
IX	Amethyst	Amethyst	Topaz
X	Beryl	Topaz	Chrysoprasus
XI	Onyx	Beryl	Jacinth
XII	Jasper	Jasper	Amethyst

While the arrangement differs in Revelation, the stones are nearly identical. For chalcedonius, we should probably read *carchedonius,* a name of the ruby; sardonyx is the onyx of Exodus; the jacinth (sapphire) is probably the "ligure"; the sapphire was the lapis-lazuli, and sardius is equivalent to carnelian. There thus remains only the chrysoprase, which for some reason has substituted the agate. In the eventual association of the foundation stones with the months, the first, the jasper, was assigned to March, with which month the year was reckoned to begin.

The list suggested and adopted in Kansas City reads as follows:

Month.	Birth-stone.	Alternate Stone.
January	Garnet	
February	Amethyst	
March	Bloodstone	Aquamarine
April	Diamond	
May	Emerald	
June	Pearl	Moonstone
July	Ruby	

Month.	Birth-stone.	Alternate-stone.
August	Sardonyx	Peridot
September	Sapphire	
October	Opal	Tourmaline
November	Topaz	
December	Turquoise	Lapis-lazuli

Among the many changes in this list from that habitually followed, it will be noted that the ruby is transferred from December to July, changing places with the turquoise, which became the gem of December. This has been favored on the ground that the warmer-colored gem was best adapted for a July birth-stone, while the paler turquoise was best suited to a winter month, when the sun's rays are feeble. The contrary, however, is true; for it is in winter that we seek for warmth, while in the heat of summer nothing is more grateful than coolness. This transposition is, in effect, simply a return to the ordering of these stones in the Polish list, which may perhaps have become popular in Europe in the eighteenth century through Marie Leczinska, the queen of Louis XV. Another undesirable change takes the chrysolite (peridot) from the place it has always occupied as the gem of September, and makes of it an alternate for August, with the sardonyx, while the sapphire, properly the gem for April, is made the birth-stone for September. For October neither the tourmaline nor the opal is as appropriate as the beryl, while for June we should prefer the asteria to the moonstone as a substitute for the pearl.

This suggested radical change or violation cannot be permitted. The time-honored ordering is familiar now to all who are interested in the matter, and any change, even if one apparently for the better, is liable to disturb the popular confidence in those who are supposed to be familiar with the subject. Above all, there should be no duplication or triplication of birth-stones for any given month, the choice between a birth-stone or an astral or zodiacal stone or the combination of these affording all the variety that is necessary or should be desired.

As the diamond does not appear to have been known to the ancients and is not given in any of the lists of birth-stones before the last century, and as diamonds, like gold and platinum, may easily be used as accessories to other stones, would it not perhaps be better to omit the diamond from the list of the stones of the months, and rather use these gems as a bordering or other ornate addition to the stone of the month? The pearl, which is not a stone in any sense of the word, should not appear in the list at all; but it can be worn in some device suggesting a sentiment, as, for instance, an emblem of purity, etc.

The tourmaline, as a gem only known in modern times or since the eighteenth century, seems out of place in the list of birth-stones, which ought only to comprise precious or semi-precious stones which have been known and worn from ancient times.

"Astral stones" or "zodiacal stones" are terms used to designate those gems which were believed to be peculiarly and mystically related to the zodiacal signs. While these signs constitute a twelve-fold division of the year just as do the months, they do not exactly coincide with the latter as now reckoned, but overlap them, so that the sign Aquarius, for instance, covers the period from January 21 to February 20, that of Pisces from February 21 to March 20, that of Aries, the spring sign, from March 21 to April 20, and so on down to Capricornus, which begins at the winter solstice. Thus, every necessary opportunity is afforded for enlarging the selection of natal stones while preserving the traditional order of those appropriate to the months, an order which in its origin dates back to the early Christian centuries and which, from the close relation with the sacred gems of the Scriptures, it seems almost sacrilegious to violate by arbitrary changes.

CARNELIAN, ENGRAVED WITH THE ZODIACAL SIGNS, TAURUS, LEO AND CAPRICORN; IN THE CENTRE A SIX-RAYED STAR, THE FORM OF ONE OF THESE RAYS DENOTING A COMET. (See p. 341.)

Referred to the nativity of Augustus and to a comet which appeared shortly after the assassination of Julius Cæsar. From De Mairan's "Lettres au R. P. Parrenin," Paris, 1770, opp. p. 274.

Then, in addition, we have the "talismanic gems," or the stones of the twelve guardian angels, one set over all those born in each month. Here we have another time-honored list, differing from either of those mentioned above, so that, in almost if not quite every case, each person has the choice between three different stones as "birth-stones," or can have them combined in an artistic jewel so as to profit by all the favorable influences promised by the old authorities. Thus, there is absolutely no excuse for playing fast and loose with an ancient, popular, and quasi-religious belief in the special virtue of *one particular stone* for each month, and that one the gem long prescribed by usage.

As it might seem appropriate that one born in the United States should wear a gem from among those which our country furnishes, the following list was some time since prepared by the writer, not in any sense as a substitute for the real birth-stones, but as possible accessory gems (when they were not identical), gems which might be worn from a spirit of patriotism. Of course where the stone in question is really that traditionally recommended, the fact that it is at the same time an American gem-stone is an added argument in its favor.

Month.	Stones.	Where found.
January...	Garnet, rhodolite	Montana, New Mexico, Arizona, North Carolina
February..	Amethyst	North Carolina, Georgia, Virginia
March.....	Californite	California
April......	Sapphire	Montana, Idaho
May.......	Green tourmaline	Lake Superior
June......	Moss-agate	California, Montana, Wyoming, Arizona
July.......	Turquoise	New Mexico, California, Arizona
August....	Golden beryl	California, Connecticut, North Carolina
September.	Kunzite	California
October....	Aquamarine	North Carolina, Maine, California
November..	Topaz	Utah, California, Maine
December..	Rubellite	Montana

The year is divided into four seasons or cycles,—spring, summer, fall, and winter,—and each season has its particular gem. The emerald is the gem of the spring, the ruby the gem of summer, the sapphire the gem of autumn, and the diamond the gem of winter.

For spring, no precious stone is more appropriate than the emerald. Its beautiful color is that of Nature, for Nature clothes herself with green when she awakens from her long rest of winter. Having decked herself with green of the various tints and colors, she has selected a background by which a contrast is made for the flowers that come in the spring and summer and ripen into fruit and seeds of autumn. To be a seasonable gem it must be rare, and the emerald is rare. Whether found in the mines of Bogotá, whether mined in ancient times at Zabarah in Egypt, or in the past century in the Ural Mountains, it has never been found in abundance. It is softer in color than the ruby and less hard in structure.

The ruby, although as a natal stone it belongs to December, is the gem of summer. It is born in the hot climates,—the pigeon's-blood ruby in Burma, the pomegranate-red in Ceylon, and the more garnet-hued type in Siam,—these three equatorial countries produce the ruby. Those of large size are always rare, and this is the gem which Job valued more highly than any other, although "garnet" may perhaps be a better rendering. It is on an equal plane in hardness, in composition, in crystalline structure, and in every way, with the sapphire. These are sister gems, structurally alike, yet varying in complexion, due to a slight difference which some scientists think is not even dependent upon the coloring matter.

The sapphire—the gem of autumn, the blue of the autumn sky—is a symbol of truth, sincerity, and constancy. Less vivid than its sister gem, the ruby, it typifies calm and tried affection, not ardent passion; it is therefore appropriate to the autumn season, when the declining sun no longer sends forth the fiery rays of summer but shines with a tempered brilliancy.

The diamond, the gem of winter, typifying the sun, is the gem of light. Its color is that of ice, and as the dew-drop or the drop of water from a mountain stream sparkles in the light of the sun, as the icicle sparkles in winter, and the stars on a cold winter night, so the diamond sparkles, and it combines and contrasts with all known gems. Like light, it illumines them just as the sun does the plants of the earth. The diamond, the gem of light, like light itself when broken into a spectrum, gives us all known colors, and by combining all these colors it gives us white. Like gold, the diamond was made rare, so that it must be searched for, and the mines and deposits contain less of these two substances in a given area than of any other known materials. It is thirty to a hundred times more rare than gold, for if gold occurs one part in 250,000, it can scarcely be worked with profit, while the diamond can be worked to advantage when found only one part in 10,000,000,—yes, even one part in 25,000,000—and, like gold, it sometimes spurs the searcher on to wealth or to ruin. As great nuggets of gold have occasionally been found, so has a diamond been discovered large enough to make the greatest ruler pause to pay its price, and one which it took an entire country to give to that ruler who sways his sceptre over countries in which the world's greatest diamonds have been found.

When the God of the Mines called his courtiers to bring him all known gems, he found them to be of all colors and tints, and of varying hardnesses, such as the ruby, emerald, sapphire, etc., etc. He took one of each; he crushed them; he compounded them, and said: "Let this be something that will combine the beauty of all; yet it must be pure, and it must be invincible." He spoke: and lo! the diamond was born, pure as the dew-drop and invincible in hardness; but when its ray is resolved in the spectrum, it displays all the colors of the gems from which it was made. "Mine," said the god, "must be the gem of the universe; for my queen I will create one that shall be the greatest gem of the sea," and for her he created the pearl.

Gems of Spring	Gems of Summer
Amethyst	Zircon
Green diamond	Garnet (demantoid and ouvarite)
Chrysoberyl	Chrysoberyl (alexandrite)
Spinel (rubicelle)	Spinel
Pink topaz	Pink topaz
Olivine (peridot)	Ruby
Emerald	Fire opal

Gems of Autumn	Gems of Winter
Hyacinth	Diamond
Topaz	Rock-crystal
Sapphire	White sapphire
Jacinth	Turquoise
Cairngorm	Quartz
Adamantine spar	Moonstone
Tourmaline	Pearl
Oriental chrysolite	Labradorite

SENTIMENTS OF THE MONTHS

JANUARY

Natal stone .... Garnet.
Guardian angel .... Gabriel.
His talismanic gem .... Onyx
Special apostle .... Simon Peter.
His gem .... Jasper.
Zodiacal sign .... Aquarius.
Flower .... Snowdrop.

No gems save garnets should be worn
By her who in this month is born;
They will insure her constancy,
True friendship and fidelity.

The gleaming garnet holds within its sway
Faith, constancy, and truth to one alway.

FEBRUARY

Natal stone ....Amethyst.
Guardian angel ....Barchiel.
His talismanic gem ....Jasper.
Special apostle ....Andrew.
His gem .... Carbuncle.
Zodiacal sign ....Pisces.
Flower ....Primrose.

The February-born may find
Sincerity and peace of mind,
Freedom from passion and from care,
If she an amethyst will wear.

Let her an amethyst but cherish well,
And strife and care can never with her dwell.

MARCH

Natal stone ....Jasper, bloodstone.
Guardian Angel ....Malchediel.
His talismanic gem ....Ruby.
Special apostles .... James and John.
Their gem ....Emerald.
Zodiacal sign ....Aries.
Flower ....Ipomœa, violet.

Who on this world of ours her eyes
In March first opens may be wise,
In days of peril firm and brave,
Wears she a bloodstone to her grave.

Who wears a jasper, be life short or long,
Will meet all dangers brave and wise and strong.

APRIL

Natal stone ....Diamond, sapphire.
Guardian angel ....Ashmodei.
His talismanic gem ....Topaz.
Special apostle ....Philip.
His gem ....Carnelian.
Zodiacal sign ....Taurus.
Flower ....Daisy.

She who from April dates her years
Diamonds should wear, lest bitter tears
For vain repentance flow. This stone
Emblem of innocence is known.

Innocence, repentance—sun and shower—
The diamond or the sapphire is her dower.

MAY

Natal stone ....Emerald.
Guardian angel ....Amriel.
His talismanic gem ....Carbuncle.
Special apostle ....Bartholomew.
His gem ....Chrysolite.
Zodiacal sign ....Gemini.
Flower ....Hawthorn.

Who first beholds the light of day
In spring's sweet flow'ry month of May,
And wears an emerald all her life,
Shall be a loved and happy wife.

No happier wife and mother in the land
Than she with emerald shining on her hand.

JUNE

Natal stone ....Agate.
Guardian angel ....Muriel.
His talismanic gem ....Emerald.
Special apostle ....Thomas.
His gem ....Beryl.
Zodiacal sign ....Cancer.
Flower ....Honeysuckle.

Who comes with summer to this earth,
And owes to June her hour of birth,
With ring of agate on her hand
Can health, long life, and wealth command.

Thro' the moss-agate's charm, the happy years
Ne'er see June's golden sunshine turn to tears.

JULY

Natal stone .... Turquoise.
Guardian angel .... Verchiel.
His talismanic gem .... Sapphire.
Special apostle .... Matthew.
His gem .... Topaz.
Zodiacal sign .... Leo.
Flower .... Water-lily.

The heav'n-blue turquoise should adorn
All those who in July are born;
For those they'll be exempt and free
From love's doubts and anxiety.

No other gem than turquoise on her breast
Can to the loving, doubting heart bring rest.

AUGUST

Natal stone .... Carnelian.
Guardian angel .... Hamatiel.
His talismanic gem .... Diamond.
Special apostle .... James, the son of Alpheus.
His gem .... Sardonyx.
Zodiacal sign .... Virgo.
Flower .... Poppy.

Wear a carnelian or for thee
No conjugal felicity;
The August-born without this stone,
'Tis said, must live unloved, alone.

She, loving once and always, wears, if wise,
Carnelian—and her home is paradise.

SEPTEMBER

Natal stone .... Chrysolite.
Guardian angel .... Tsuriel.
His talismanic gem .... Jacinth.
Special apostle .... Lebbeus Thaddeus.
His gem .... Chrysoprase.
Zodiacal sign .... Libra.
Flower .... Morning-glory.

A maid born when September leaves
Are rustling in the autumn breeze,
A chrysolite on brow should bind—
'Twill cure diseases of the mind.

If chrysolite upon her brow is laid,
Follies and dark delusions flee afraid.

OCTOBER

Natal stone .... Beryl.
Guardian angel .... Bariel.
His talismanic gem .... Agate.
Special apostle .... Simon. (Zelotes.)
His gem .... Jacinth.
Zodiacal sign .... Scorpio.
Flower .... Hops.

October's child is born for woe,
And life's vicissitudes must know;
But lay a beryl on her breast,
And Hope will lull those woes to rest.

When fair October to her brings the beryl,
No longer need she fear misfortune's peril.

NOVEMBER

Natal stone .... Topaz.
Guardian angel .... Adnachiel.
His talismanic gem .... Amethyst.
Special apostle .... Matthias.
His gem .... Amethyst.
Zodiacal sign .... Sagittarius.
Flower .... Chrysanthemum.

1. Moss agate mocha stone. Hindoostan.
2. Moss agate, Brazil, S. A.

Who first comes to this world below
With drear November's fog and snow
Should prize the topaz's amber hue—
Emblem of friends and lovers true.

Firm friendship is November's, and she bears
True love beneath the topaz that she wears.

DECEMBER

Natal stone	Ruby.
Guardian angel	Humiel.
His talismanic gem	Beryl.
Special apostle	Paul.
His gem	Sapphire.
Zodiacal sign	Capricornus.
Flower	Holly.

If cold December give you birth—
The month of snow and ice and mirth—
Place on your hand a ruby true;
Success will bless whate'er you do.

December gives her fortune, love and fame
If amulet of rubies bear her name.

A HINDU LIST OF GEMS OF THE MONTHS [11]

April	Diamond
May	Emerald
June	Pearl
July	Sapphire
August	Ruby
September	Zircon
October	Coral
November	Cat's-eye
December	Topaz
January	Serpent-stone
February	Chandrakanta
March	The gold Siva-linga

[11] Surindro Mohun Tagore, "Mani Málá," Pt. II, Calcutta, 1881, pp. 619, 621.

When the zodiacal signs were engraved on gems to give them special virtues and render them of greater efficacy for those born under a given sign, the Hebrew characters designating the sign (or at least the initial character) were often cut upon the gem. As the letters in which the earliest of our sacred writings were written, a peculiar sanctity was often ascribed to these Hebrew characters, which were perhaps the more highly valued that they were unknown to the owners of the gems, and hence possessed a certain air of mystery for them. The subjoined list of the signs with the Hebrew equivalents may be of interest on this account.

HEBREW NAMES OF THE SIGNS OF THE ZODIAC

Libra........................	מאזנים	Moznayim
Scorpio........................	עקרב	Åkrab
Sagittarius........................	קשת	Keshet
Capricornus........................	גדי	Ġedi
Aquarius........................	דלי	Deli
Pisces........................	דגים	Dagim
Aries........................	טלה	Taleh
Taurus........................	שור	Ṡhor
Gemini........................	תאומים	Te'omim
Cancer........................	סרטן	Sartan
Leo........................	אריה	Aryėh
Virgo........................	בתולה	Betulah

GEMS OF WEEK DAYS

Sunday: Topaz—diamond.

The bairn that is born
On Sonnan's sweet day
Is blithe and is bonnie,
Is happy and gay.

Sunday's talismanic gem: the pearl.

Monday: Pearl—crystal.

The bairn that is born
Of Monan's sweet race
Is lovely in feature
And fair in the face.

Monday's talismanic gem: the emerald.

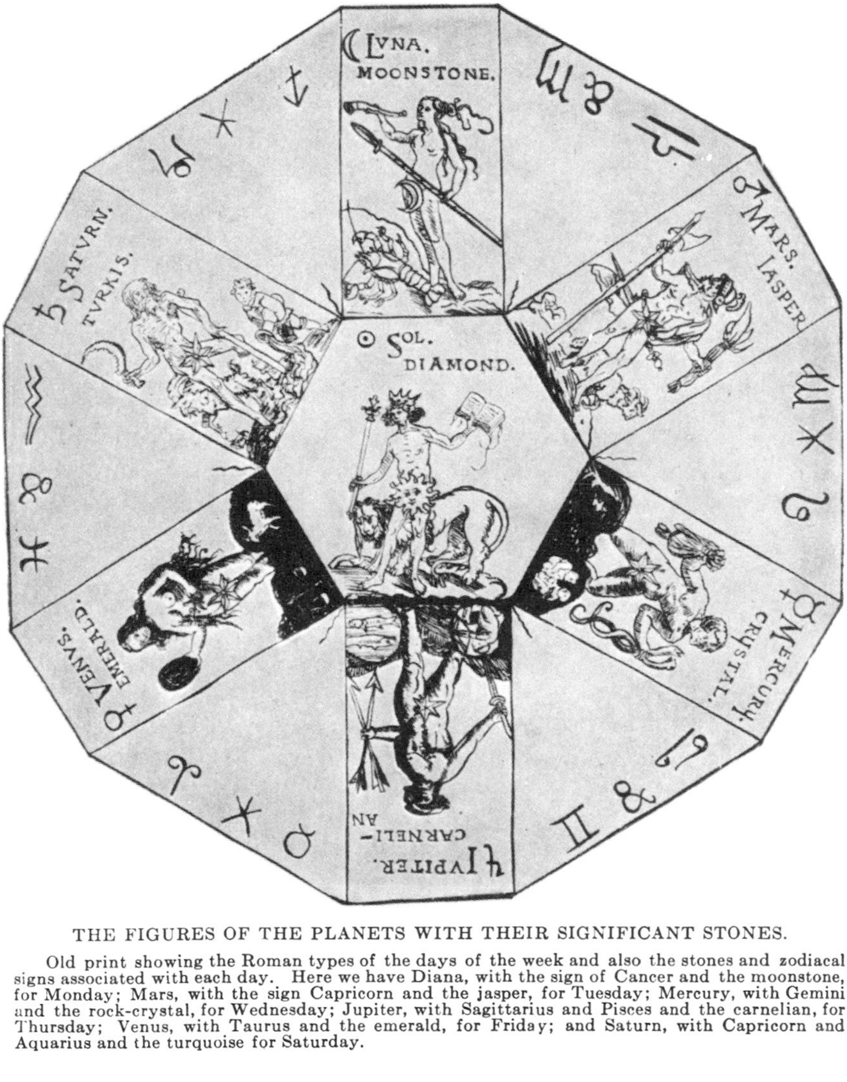

THE FIGURES OF THE PLANETS WITH THEIR SIGNIFICANT STONES.

Old print showing the Roman types of the days of the week and also the stones and zodiacal signs associated with each day. Here we have Diana, with the sign of Cancer and the moonstone, for Monday; Mars, with the sign Capricorn and the jasper, for Tuesday; Mercury, with Gemini and the rock-crystal, for Wednesday; Jupiter, with Sagittarius and Pisces and the carnelian, for Thursday; Venus, with Taurus and the emerald, for Friday; and Saturn, with Capricorn and Aquarius and the turquoise for Saturday.

Tuesday: Ruby—emerald.

If Tuisco assists
And at birth keeps apace,
The bairn will be born
With a soul full of grace.

Tuesday's talismanic gem: the topaz.

Wednesday: Amethyst—loadstone.

But if Woden be there,
Many tears will he sow,
And the bairn will be born
But for sadness and woe.

Wednesday's talismanic gem: the turquoise.

Thursday: Sapphire—carnelian.

Jove's presence at birth
Means a long swath to mow,
For if born on Thor's day
Thou hast far, far to go.

Thursday's talismanic gem: the sapphire.

Friday: Emerald—cat's-eye.

If Venus shall bless thee,
Thou shalt bless many living;
For Friga's bairn truly
Is loving and giving.

Friday's talismanic gem: the ruby.

Saturday: Turquoise—diamond.

Seater-daeg's bairn
In sweat shall be striving,
For Saturn has doomed it
To work for a living.

Saturday's talismanic gem: the amethyst.

No gems have afforded more interest to the Oriental peoples than those that are known as phenomenal gems; that is, such as exhibit a phenomenal quality, either as a moving line as in the chrysoberyl cat's-eye, or the quartz cat's-eye, or as a star, a class represented by the star-sapphire and the star-ruby, all these being considered to bring good fortune to the wearer. A splendid star-sapphire is in the hilt of the sword presented as an Easter gift to King Constantine of Greece, then Prince Constantine, by the Greeks of America, on Easter Day 1913.[12] This ornate and beautiful sword was made by Tiffany & Co. Then there is the alexandrite cat's-eye which, in addition to its chatoyant effect, changes from green to red, showing its natural color by day and glowing with a ruddy hue by artificial light. The cat's-eye effect here is caused by a twinning of the crystal; that is, when the gem is cut, with a dome, across the twinning line, this shows itself as a smooth band of white light, with a translucent or transparent space at one side, the line varying in sharpness and in breadth as the illumination becomes more intense. If the light is very bright, the line is no wider than the thinnest possible silver or platinum wire.

The quartz cat's-eye, less distinct than the chrysoberyl cat's-eye, is also found in the East, and possesses the property that when cut straight across, an apparent striation in the stone produces the cat's-eye effect, but the material is not so rich or brilliant nor is the gem as beautiful as is the true cat's-eye. The alexandrite variety of chrysoberyl is colored by chromium and is dichroitic, appearing green when viewed in one direction and red in another; in artificial light, however, the green color is lost and the red alone becomes apparent.

The moonstone, with its moonlike, silvery-white light, changes on the surface as the light varies. This is due to a chatoyancy produced by a reflection caused by certain cleavage planes present in feldspar of the variety to which the moonstone belongs.

---

[12] The star-sapphire has already been described on pp. 106, 107.

PHENOMENAL GEMS FOR THE DAYS OF THE WEEK

Sunday	Sunstone
Monday	Moonstone
Tuesday	Star sapphire
Wednesday	Star ruby
Thursday	Cat's-eye
Friday	Alexandrite
Saturday	Labradorite

Fashion in some parts of the Orient dictates the use of special colors for raiment and jewels to be worn on the different days of the week. In Siam deep red silks and rubies are appropriate for Sunday wear; white fabrics and moonstones are prescribed for Monday; light red garments and coral ornaments are favored for Tuesday; striped stuffs and jewels set with the cat's-eye are considered the proper wear for Wednesday; green materials and emeralds are decreed for Thursday; silver-blue robes and ornaments set with diamonds are chosen for Friday, and on Saturday those who obey the dictates of fashion are clad in dark blue garments and wear sapphires of a similar hue.

Our age is not satisfied with the marvellous progress of science, which has rendered possible the realization of many of the old magicians' dreams. In spite of this there seems to be a growing tendency to revive many of the old beliefs which appeared to have been definitely discarded; therefore we need not be surprised that the nineteenth century offers us a work on the magic art, written precisely in the spirit that animated an Agrippa or a Porta in the fifteenth and sixteenth centuries.[13] This work gives elaborate directions as to the manner in which the "Magus" should proceed to perform his magic rites.

---

[13] Eliphas Lévi, "Rituel de la haute magie," Paris, 1861.

Each day has its special and peculiar ritual. Sunday is the day for the "Works of Light," and on this day a purple robe should be worn and a tiara and bracelets of gold; the ring placed on the finger of the operator should be of gold and set with a chrysolite or a ruby. A white robe with silver stripes is to be worn on Monday, the day of the "Works of Divination and Mystery," and the high-priest of the mysteries wears over his robe a triple necklace of pearls, "crystals," and selenites; the tiara should be covered with yellow silk, and bear in silver characters the Hebrew monogram of Gabriel, as given by Cornelius Agrippa in his "Occult Philosophy." Tuesday is assigned to the "Works of Wrath," and on this day the robe must be red, the color of fire and blood, with a belt and bracelets of steel; the tiara should have a circlet of iron, and a sword or a stylus is to be used in place of a wand; the ring is set with an amethyst. The day for the "Works of Science" is Wednesday, when a green robe is worn and a necklace of hollow glass beads, filled with quicksilver; the ring is adorned with an agate. On Thursday, appointed for the "Works of Religion or Politics," a scarlet robe is worn; upon the forehead of the operator is bound a plate of tin, engraved with the symbol of the planet Jupiter and various mystic characters; the ring bears either an emerald or a sapphire. Friday, the day of Venus, is naturally dedicated to the "Works of Love," and the celebrant wears a sky-blue robe; his ring shows a turquoise, and his tiara is set with lapis-lazuli and beryl. The "Works of Mourning" belong to Saturday, when a black or a brown robe is worn, embroidered in orange-colored silk with mystic characters; from the neck of the operator hangs a leaden medal, bearing the symbol of the planet Saturn, and on his finger is a ring set with an onyx, upon which a double-faced Janus has been engraved while Saturn was in the ascendant.

GEMS OF THE HOURS

Hours of the Day

7	Chrysolite	1	Jacinth
8	Amethyst	2	Emerald
9	Kunzite	3	Beryl
10	Sapphire	4	Topaz
11	Garnet	5	Ruby
12	Diamond	6	Opal

Hours of the Night

7	Sardonyx	1	Morion
8	Chalcedony	2	Hematite
9	Jade	3	Malachite
10	Jasper	4	Lapis-lazuli
11	Loadstone	5	Turquoise
12	Onyx	6	Tourmaline

WEDDING ANNIVERSARIES

1	Paper	19	Hyacinth
2	Calico	20	China
3	Linen	23	Sapphire
4	Silk	25	Silver
5	Wood	26	Star sapphire, blue *
6	Candy	30	Pearl
7	Floral	35	Coral
8	Leather	39	Cat's-eye *
9	Straw	40	Ruby
10	Tin	45	Alexandrite
12	Agate	50	Gold
13	Moonstone *	52	Star ruby *
14	Moss agate	55	Emerald
15	Rock-crystal, glass	60	Diamond, yellow
16	Topaz	65	Star sapphire, gray *
17	Amethyst	67	Star sapphire, purple
18	Garnet	75	Diamond

* For this number, and for the succeeding multiples of thirteen, the gem is believed to counteract the malign influence of the number.

X

Planetary and Astral Influences of Precious Stones

THE talismanic influence of the stones associated with the planets and also with the signs of the zodiac is closely connected with the early ideas regarding the formation of precious stones. In an old work on the occult properties of gems we read:

The nature of the magnet is in the iron, and the nature of the iron is in the magnet, and the nature of both polar stars is in both iron and magnet, and hence the nature of the iron and the magnet is also in both polar stars, and since they are Martian, that is to say, their region belongs to Mars, so do both iron and magnet belong to Mars.

The author then proceeds to describe an analogous relation between a man and any natural object or product to which his imagination draws him, and shows that, if this object be one that stands in a sympathetic relation with the star beneath which the man was born, the man, the star, and the object will constitute a triplicity of great utility. As an explanation of the peculiarly intimate relation between stars and precious stones we read, on page 12:

Metals and precious stones usually lie with their first seeds deep down in the earth and require continuous moisture and a mild heat. This they obtain through a reflection of the sun and the other stars in the manifold movement of the heavens. . . . Therefore, also, the metals and precious stones are nearest related to the planets and the stars, since these influence them most potently and produce their peculiar qualities, for they are enduring and unchangeable and show therein their concordance [with the stars and the planets].[1]

Hence it is that the influence over human fortunes ascribed by astrology to the heavenly bodies is conceived to be strengthened by wearing the gem appropriate to certain planets or signs, for a subtle emanation has passed into the stone and radiates from it. A combination of several different stones, each partaking of this special quality, was believed to have an influence similar to that exercised by several planets in conjunction,—that is, grouped in the same "house" or division of the heavens.

The same is true of the stones dedicated to the guardian angels; the color and appearance of the stone was not merely emblematic of the angel, but, by its sympathetic quality, it was supposed to attract his influence and to provide a medium for the transmission of his beneficent force to the wearer. The whole theory, whether consciously or unconsciously, rested on the idea of harmony, of the accord of certain ethereal vibrations, either those of the visible light of the stars and planets or the purely psychic emanations from the spiritual "powers and principalities."

The wearing of the appropriate zodiacal gem was always believed to strengthen the influence of the zodiacal sign upon those born under it, and to afford a sympathetic medium for the transmission of the stellar influences. The gem was thus something more than a mere symbol of the sign. The same was true of the stone of the saint who ruled the month and that of the holy guar-

[1] Wilhelmus Eo, "Coronæ Gemma Nobilissima," Newheusern, 1621, pp. 38–9.

dian angel set over those born in the month. In each and every case the material form and color of the stone was believed to attract the favor and grace of the saint or angel, who would see in the selection of the appropriate gem an act of respect and veneration on the part of the wearer.

The old writers are never tired of insisting upon the idea that, while the image graven upon a stone was in itself dead and inactive, the influence of the stars during whose ascendancy the work had been executed communicated to the inert material talismanic qualities and virtues which it before lacked. In these instances the images could be regarded as outward and visible signs of the planetary or zodiacal influence. Even in the case of the bezoar stone, a generally recognized antidote for all sorts of poisons, it was held that the scorpion's bite could be most effectually healed by a bezoar upon which this creature's figure had been cut during the time when the constellation Scorpio was in the ascendancy.[2]

In the production of engraved stones to serve as amulets, the influence of the respective planet was made to enter the stone by casting upon the latter, during the process of engraving, reflections from a mirror which had been exposed to the planet's rays. In addition to this, the work was executed while the planet was in the ascendant, and the design was emblematic of it. With these combined influences the gem was believed to be thoroughly impregnated with the planetary virtue.[3]

An old writer finds in the hardness of precious stones a reason for their retaining longer the celestial virtues they receive. After they have been extracted these vir-

---

[2] Gaffarelli, "Curiositates inauditæ," Hamburgi, 1706, pp. 146, 147.

[3] Schindler, "Der Aberglaube des Mittelalters," Breslau, 1858, p. 131.

tues persist in them and they keep "the traces and gifts of mundane life which they possessed while clinging to the earth.[4] These "gifts of mundane life" signify the stored-up energy derived from the stars and planets, which penetrates the matter of the stone, and each stone is peculiarly sensitive to the emanations from a certain planet, star, or group of stars.

A fine carnelian gem engraved with a design consisting of a star surrounded by the images of a ram, a bull, and a lion, is described by M. Mairan.[5] He sees in the star the emblem of the splendid comet which appeared shortly after the assassination of Cæsar, and which, according to Suetonius, was believed to be the soul of Cæsar newly received into the sky; the ram, bull, and lion are the symbols of the zodiacal signs Aries, Taurus, and Leo, the first-named sign referring perhaps to the death of Cæsar on the Ides, or fifteenth of March; while the other two signs may allude to the position of the comet at different dates.

In the Cabinet du Roi, in Paris, there was an engraved carnelian, the design showing Jupiter enthroned, with thunderbolt and sceptre, and Mars and Mercury standing on either side of the central figure. Separated from the gods of the upper air by a bow, probably representing the arch of the sky, appears the bust of Neptune, emerging from the sea. The border of the design is formed by the twelve signs of the zodiac, Virgo being of an unusual type,—the virgin and a unicorn,—said to have been used only during the reign of Domitian (81–96 A.D.).[6]

---

[4] Reichelti, "De amuletis," Argentorati, 1676, p. 45; citing Ficini, "De vita coelit.," cap. 13.

[5] Mairan, "Lettres au R. P. Parrenin," Paris, 1770, pp. 275 sqq.

[6] Mairan, l.c., pp. 199, 211.

Some choice examples of astrological gems may be seen in the Metropolitan Museum of Art, New York; among these is a green jasper bearing symbols of Luna, Capricorn, and Taurus. This gem is from the collection of the late Rev. C. W. King, which has been acquired for the Museum, and is described as figuring the horoscope of the owner. In the same collection is a banded agate engraved with Sagittarius as a centaur, surrounded by the stars of this constellation in their proper order. King states that this was the earliest horoscopical gem known to him. Still another gem of this collection is a sard bearing the symbol of Aries carrying a long caduceus; this type appears on the coins of Antioch, because that city was founded in the month over which the sign Aries presides.[7]

The Austrian Imperial Collection in Vienna contains the celebrated Gemma Augustea, sometimes called the Apotheosis of Augustus. This commemorates the Pannonian triumph of Tiberius, 13 A.D., and above the figure of Augustus appears the sign of Capricornus, the constellation of his nativity; beneath the figure of Tiberius is engraved the sign of Scorpio, under which that emperor was born. This celebrated cameo, the work of the famous gem-engraver Dioskorides, is mentioned in an inventory of the treasury of St. Sernin, in Toulouse, dated 1246. It is said to have been offered by Francis I of France to Pope Clement VII, on the occasion of their meeting in Marseilles in 1535; however, as the gem only reached Marseilles two days after the pope's departure, Francis decided to retain possession of it. The

[7] "Collection of Engraved Gems," Metropolitan Museum of Art, Handbook No. 9, pp. 53, 54.

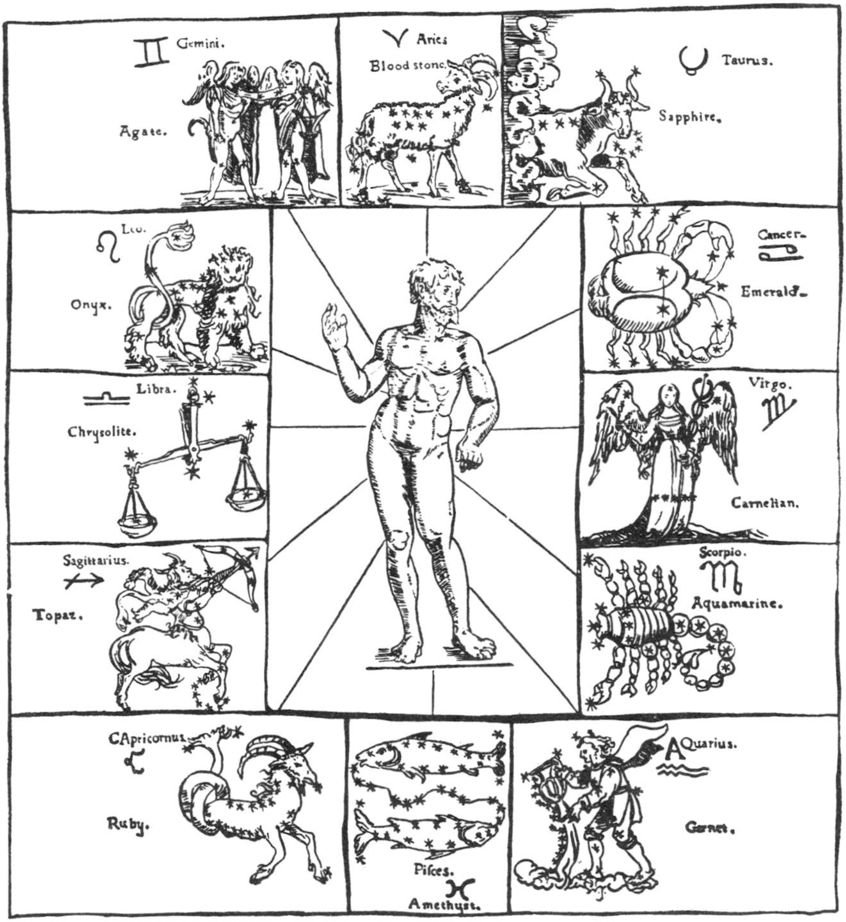

THE ZODIACAL STONES WITH THEIR SIGNS.

Old print illustrating the influence believed to be exerted on the different parts of the body by the respective zodiacal signs, and through their power by the stones associated with them. This belief often determined the administration of special precious-stone remedies by physicians of the seventeenth and earlier centuries.

royal treasure at Fontainebleau was plundered in 1590, and the stone was offered for sale, and was purchased, in 1619, by Emperor Rudolph II, for the sum of 12,000 ducats.

A ruby called sandastros is described by Pliny as containing stellated bodies which he compares to the Hyades; hence, says he, they are the objects of great devotion with the Chaldæi or Assyrian Magi. According to Morales (De las piedras preciosas), the ruby and the diamond were both under the influence of the sign of Taurus; the same writer informs us that the Hyades and the sun were supposed to have a potent effect upon the ruby or carbuncle. In ancient Babylonia the sign of Taurus was regarded as the most important, and Winckler believes that the presence in this sign of the five stars of the Hyades and the seven of the Pleiades was brought into connection with the twelve-fold division of the zodiac. The Hyades signified the five signs visible in Babylonia at the summer solstice, while the Pleiades typified the seven invisible signs. It seems probable that the Pleiades were associated with the diamond, although Morales, who was very familiar with the Moorish astrology current among the Spaniards of his time, attributed the crystal to this group. His attribution proves at least that the stone of the Pleiades was a colorless one.

In Sanskrit the diamond is called *vajra,* "thunderbolt," and also *indrâjudha,* "Indra's weapon"; another name is *açira,* "fire," or "the Sun."[8] All these designations are probably suggested by the brilliant flashes of

---

[8] Garbe, "Die indische Mineralien," Naharari's Râjanighantu, Varga XIII, Leipzig, 1882, p. 80.

light emitted by this stone. It is not easy to determine the reason that induced the Hindus to dedicate the diamond to the planet Venus rather than to the Sun or to the Moon. However, as the most brilliant of the planets, Venus was not unworthy of the honor, and if we substitute the Goddess of Love for her planet, it seems quite appropriate that she should be adorned with the most brilliant of precious stones. Certainly these sparkling gems are often enough offered at the shrine of Venus in our own day, and they often serve to win the good graces of the divinity to whom they are presented.

The Sanskrit name for the sapphire, *nîla,* signifies "blue," so that, as the topaz is the "yellow stone" *par excellence,* the sapphire is the blue stone (*nilaçman*). In both cases the name indicates a variety of corundum, distinguished merely by the coloring matter. As a talisman the Hindus believed that the sapphire rendered the planet Saturn favorable to the wearer, an important consideration from the astrological point of view, for Saturn's influence was generally supposed to be unfavorable. The Hindus distinguished four classes of sapphires, corresponding to the four castes: Brahmins, Kshatriyas, Vaisyas, and Sudras. The respective sapphires were light blue, reddish blue, yellowish blue, and dark blue. The same distinction is made in the case of the diamond, and a like rule applies to both stones, namely, that only the appropriate stone should be worn by the members of each caste, in order to profit by the virtues inherent in the sapphire or diamond.[9]

One of the Sanskrit appellations of the hyacinth (zircon) is *râhuratna,*—that is, the jewel dedicated to the mysterious "dragon," that was supposed to be the

[9] Garbe, "Die indische Mineralien," Naharari's Râjanighantu, Varga XIII, Leipzig, 1882, p. 83.

cause of the periodic eclipses of the Sun and Moon.[10] As the stone was sacred to this malevolent influence, we need not be surprised that it was believed to avert misfortune, for nothing was so effective against the lesser spirits of evil as an evil genius of great power.

According to the Hindu mystics it was very lucky to have a turquoise at hand at the time of the new moon, for whoever, after first looking at the moon on the *pratipada* (the first day after new-moon), should cast his eyes upon a turquoise, was destined to enjoy immeasurable wealth.[11]

ZODIACAL GEMS

Aquarius.
January 21 to February 21.
The Garnet.

If you would cherish friendship true,
In Aquarius well you'll do
To wear this gem of warmest hue—
The garnet.

Pisces.
February 21 to March 21.
The Amethyst.

From passion and from care kept free
Shall Pisces' children ever be
Who wear so all the world may see
The amethyst.

Aries.
March 21 to April 20.
The Bloodstone.

Who on this world of ours his eyes
In Aries opens shall be wise
If always on his hand there lies
A bloodstone.

---

[10] Garbe, "Die indische Mineralien," Naharari's Râjanighantu, Varga XIII, Leipzig, 1882, p. 84.

[11] Surindro Mohun Tagore, "Mani Málá," Pt. II, Calcutta, 1881, p. 883.

Taurus.

April 20 to May 21.

The Sapphire.

If on your hand this stone you bind,
You in Taurus born will find
'Twill cure diseases of the mind,
The sapphire.

Gemini.

May 21 to June 21.

The Agate.

Gemini's children health and wealth command,
And all the ills of age withstand,
Who wear their rings on either hand
Of agate.

Cancer.

June 21 to July 22.

The Emerald.

If born in Cancer's sign, they say,
Your life will joyful be alway,
If you take with you on your way
An emerald.

Leo.

July 22 to August 22.

The Onyx.

When youth to manhood shall have grown,
Under Leo lorn and lone
'Twill have lived but for this stone,
The onyx.

Virgo.

August 22 to September 22.

Carnelian.

Success will bless whate'er you do,
Through Virgo's sign, if only you
Place on your hand her own gem true,
Carnelian.

Libra.

September 22 to October 23.

The Chrysolite.

Through Libra's sign it is quite well
To free yourself from evil spell,
For in her gem surcease doth dwell,
The chrysolite.

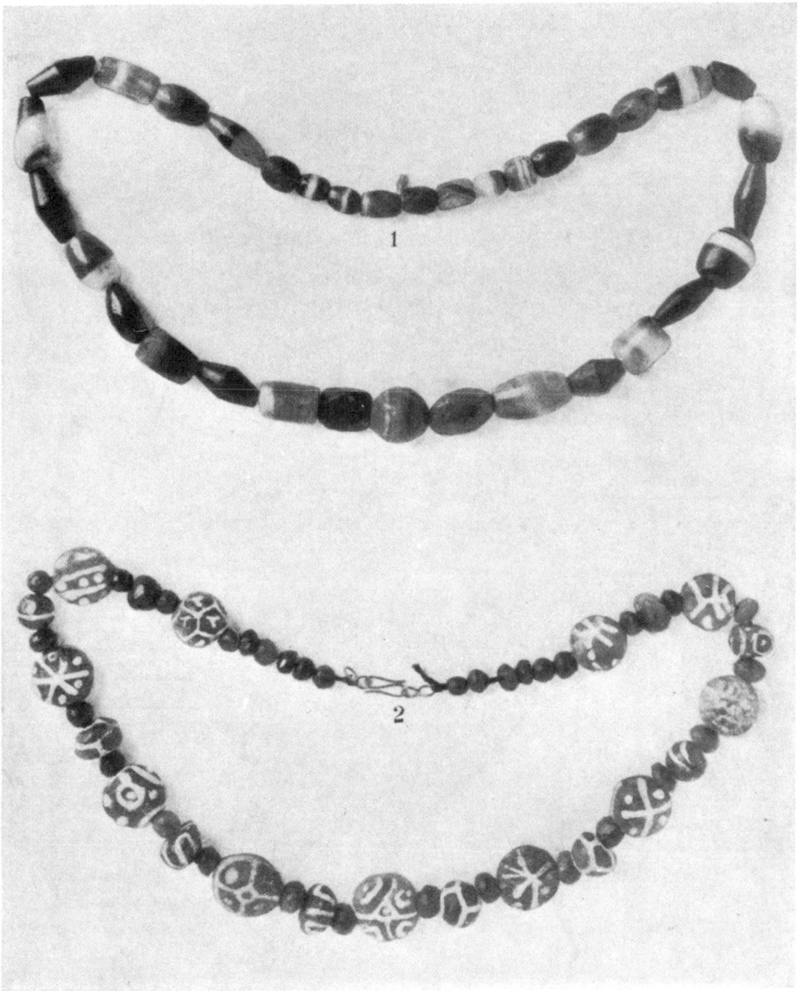

1. A necklace of banded and variegated agates, onyx, carnelians and sards. First Century A.D.
2. Beads of carnelian artificially marked for "good luck." The marking is produced by an application of potash and soda. Ancient Persian.

Scorpio.
October 23 to November 21.
The Beryl.

Through Scorpio this gem so fair
Is that which every one should wear,
Or tears of sad repentance bear,—
The beryl.

Sagittarius.
November 21 to December 21.
The Topaz.

Who first comes to this world below
Under Sagittarius should know
That their true gem should ever show
A topaz.

Capricorn.
December 21 to January 21.
The Ruby.

Those who live in Capricorn
No trouble shall their brows adorn
If they this glowing gem have worn,
The ruby.

An old Spanish list of the gems of the zodiacal signs differs from those given above, and probably represents Arab tradition:[12]

Aries—Crystal	Libra—Jasper
Taurus—Ruby and diamond	Scorpio—Garnet
Gemini—Sapphire	Sagittarius—Emerald
Cancer—Agate and beryl	Capricorn—Chalcedony
Leo—Topaz	Aquarius—Amethyst
Virgo—Magnet	Pisces—

Of planetary stones[13] there is assigned to the sun the jacinth and the chrysolite, when this latter name

[12] Morales, "De las virtudes y propiedades marvillosas de las piedras preciosas," Valladolid, 1604, fols. 15a, 15b.

[13] Rantzau, "Tractatus de genethliacorum thematum judiciis," Francofurti, 1633, pp. 46–55.

was applied to the yellow Brazilian chrysoberyl, while the moon controls the beryl, the rock-crystal and also the pearl. To the share of Venus fall the sapphire and carbuncle as well as coral and pearl; usually the emerald is the stone of Venus. Mars lays claim to the diamond, jacinth, and ruby, the last-named stone according with the ruddy hue of our neighbor planet. Under the control of Jupiter are placed the emerald, sapphire, amethyst, and turquoise, so that this planet has the richest assortment of gems; it will be remarked that the celestial sapphire unites the influence of Venus and Jupiter, the two especially propitious planets. Lastly, far-away Saturn must be content with all dark, black, and brittle stones; there was, indeed, little inducement to wear a Saturnian stone, for the influence of this cold and distant planet was always regarded as baleful.

The planetary controls of precious stones as given in the Lapidario of Alfonso X, according to "Chaldaic" tradition, show that the same stone was influenced in many or most cases by more than one of the "seven planets" (including the Sun and Moon). Thus the diamond, belonging to the first degree of the sign Taurus, was dominated by both Saturn and the Sun; the emerald was controlled by Jupiter, and also by Mercury and by Venus. The red jargoon was influenced by Mars, the yellow variety by Jupiter and the white jargoon by Venus. The carnelian received virtue from the Sun and from Venus. The ruby, although more especially a sun-stone, came as well under the influence of the Planet of Love. Coral belonged both to Venus and to the moon, while lapis-lazuli and chalcedony only owed allegiance to Venus; this planet also lent virtue to the beryl.[14]

---

[14] Lapidario del Rey D. Alfonso X; codice original, Madrid, 1881, fols. 101–109.

Among the Mohammedans, six of the seven heavens were supposed to be made of precious substances: the first was of emerald; the second, of white silver; the third, of large white pearls; the fourth, of ruby; the fifth, of red gold; and the sixth, of jacinth. The seventh and highest heaven, however, was of shining light.[15] Here we have the three precious colored stones, emerald, ruby, and sapphire (jacinth), to which is added the pearl.

The scarcity of the diamond in early times, and its comparative lack of brilliancy before the invention of rose and brilliant cutting, account for the absence of this king of gems.

Rabelais,[16] describing the temple of the oracle of the "Dive Bouteille," says that of its seven columns the first was of sapphire; the second, of jacinth; the third, of "dyamant"; the fourth, of the "male" balas-ruby; the fifth, of emerald, "more brilliant and glistening than were those which were set in place of eyes in the marble lion stretched before the tomb of King Hermias"; the sixth column was of agate, and the seventh of transparent selenite, " with a splendor like that of Hymettian honey, and within appeared the moon in form and motion such as she is in the heavens, full and new, waxing and waning." We are then told that these stones were attributed to the seven planets by the Chaldæans, as follows:

Sapphire	Saturn
Jacinth	Jupiter
Diamond	Sun
Ruby	Mars

[15] Lane, "Arabian Society in the Middle Ages," ed. by Stanley Lane-Poole, London, 1883, p. 98.

[16] Pantagruel, liv. v, chap. xlii, Paris, 1833, p. 341.

Emerald	Venus
Agate	Mercury
Selenite	Moon

Some of these attributions differ from those usually made and may represent another tradition.

PLANETARY INFLUENCES OF STONES [17]

Jasper	Venus and Mercury.
Sapphire	Jupiter and Mercury.
Emerald	Venus and Mercury.
Chalcedony	Jupiter, Mercury, and Saturn.
Sardonyx	Saturn and Mars.
Chrysolite	Mercury and Venus.
Beryl	Venus and Mars.
Topaz	Saturn and Mars.
Chrysoprase	Mercury and Venus.
Jacinth	Mars and Jupiter.
Amethyst	Mars and Jupiter.
Pearl	Venus and Mercury.
Carbuncle	Mars and Venus.
Diamond	Jupiter.
Agate	Venus and Mars.
Alectoria	Sun.
Turquoise	Venus and Mercury.
Chelidon	Jupiter.
Aetites	Sun.
Dionesia	Saturn.
Hematite	Mercury.
Lapis-lazuli	Venus.
Armena	Mercury and Venus.
Garnet	Sun.
Amber	Sun.
Jet	Saturn.
Lyncurius	Sun.
Crystal	Moon and Mars.
Bezoar	Jupiter.
Armenia.	Jupiter.

[17] Morales, "De las Piedras Preciosas," Valladolid, 1604.

Selenite	Moon.
Magnet	Mars.
Judaica, Hegolite or Cogolite	Mercury.
Iris	Jupiter.
Halcyon	Saturn and Mars.
Asbestus	Saturn.
Sarcophagus	Moon.
Arabian, white	Moon.
Arabian, green	Jupiter
Hyena	Sun.
Androdamas	Moon.
Pyrites:	
Copper-colored	Sun, Venus.
Gold-colored	Sun.
Silver-colored	Moon.
Tin-colored	Moon, Saturn.
Ash-colored	Jupiter.
Calatia	Moon.
Stalactite	Venus.
Thenarcus	Sun.
Carnelian	Jupiter, Mars, Venus.
Opal	Sun, Mercury.

Fixed stars associated with precious stones:[18]

Diamond. Caput Algol 18° of Taurus.
Crystal. The Pleiades 24° of Taurus.
Ruby, carbuncle. Aldebaran 3° of Gemini; also the Hyades.
Sapphire. The Goat 15° of Gemini.
Beryl. Sirius 10° of Cancer.
Garnet. Heart of Lion 23° of Leo.
Magnet. Tail of the Great Bear 8° of Scorpio; also the Pole Star.
Topaz. Right and left wing of Raven 8° of Libra.
Emerald and Jasper. Spica Virginis 17° of Libra.
Amethyst. Scorpion 3° of Sagittarius.
Chrysolite. Tortoise 8° of Capricorn.

[18] Morales, "De las piedras preciosas," Valladolid, 1604, pp. 16a–16b.

Chalcedony. Tail of Capricorn 15° of Aquarius.
Jacinth. Shoulder of Equis Major 18° of Pisces.
Pearl. Umbilicus Andromedæ 20° of Aries.
Sardonyx. Same as Topaz.

Images and virtues of the constellations as engraved on gems: [19]

URSA MAJOR, URSA MINOR, AND DRACO. Both bears are represented in the folds of a serpent, the Great Bear in the upper and the Lesser Bear in the lower folds. In almost all the signs. Nature: Ursa Major, Mars and Venus. Ursa Minor; Saturn. Draco: Saturn and Mars. Renders the wearer wise, cautious, versatile, and powerful.

The boundary lines of the various signs are carried up to the pole, and any constellation that is within these lines is considered to belong to the respective sign; thus, every constellation belongs to one or more signs.

CORONA BOREALIS. A royal crown, with many stars; sometimes the crowned head of a king. Sign: Sagittarius. Nature: Venus and Mercury. Engraved on the stone of one who is fitted for honors and knowledge, it gives him great favor with kings.

HERCULES. A man with knees bent, holding a club in his hand and killing a lion; sometimes a man with a lion's skin in his hand or on his shoulder and holding a club. Sign: Scorpio. Nature: Venus and Mercury. Engraved on a stone that brings victory, like the agate, it renders the wearer victorious in all conflicts in the field.

CYGNUS. A swan with outstretched wings and curved neck. In the North. Nature: Venus and Mercury. Renders the wearer popular, increases knowledge, and augments wealth. Cures gout, paralysis, and fever.

---

[19] Camilli Leonardi, "Speculum Lapidum," Venetia, 1502, f. liv–lvi.

CEPHEUS. A man girt with a sword and holding his hands and arms extended. Sign: Aries. Nature: Saturn and Jupiter. Causes pleasant visions if placed beneath the head of a sleeping person.

CASSIOPEIA. A woman seated in a chair and with hands extended in the form of a cross; sometimes with a triangle on her head. Sign: Taurus. Nature: Saturn and Venus. Restores the sickly, worn body to health, gives quiet and calm after labor and procures pleasant and tranquil sleep.

ANDROMEDA. A young girl with dishevelled hair, and hands hanging down. Sign: Taurus. Nature: Venus. Reconciles husband and wife, strengthens love, and protects the human body from many diseases.

PERSEUS. A man holding a sword in his right hand and the Gorgon's head in his left. Sign: Taurus. Nature: Saturn and Venus. Guards the wearer from misfortune and protects, not only the wearer but the place where it may be, from lightning and tempest. Dissolves enchantments.

SERPENS. A man in the folds of a serpent and holding its head in his right hand and its tail in his left. Sign: Taurus. Nature: Saturn and Venus. Antidote to poisons and to the bites of venomous creatures.

AQUILA. A flying eagle with an arrow beneath his feet. Sign: Cancer. Nature: Jupiter and Mercury; the arrow, however, is of Mars and Venus. Preserves former honors, adds new ones, and helps to victory.

PISCES or DELPHINUS. Figured in relief (?) Sign: Aquarius. Nature: Saturn and Mars. If this engraved gem be attached to nets it causes them to be filled with fish, and it renders the wearer fortunate in fishing.

PEGASUS. Some represent the half of a winged horse; others the whole figure and without a bridle. Sign: Aries. Nature: Mars and Jupiter. Gives victory in the field, and makes the wearer swift, cautious, and bold.

CETUS. Figure of a large fish with curved tail and capacious gullet. Sign: Taurus. Nature: Saturn. Renders the wearer fortunate on the sea and makes him prudent and agreeable. It also restores lost articles.

ORION. With or without armor, man holding a sword or a scythe in his hand. Sign: Gemini. Nature: Jupiter, Saturn, and Mars. Gives the wearer victory over his enemies.

NAVIS. A ship with prow curved back and spread sails; sometimes with and sometimes without oars. Sign: Leo. Nature: Saturn and Jupiter. Renders the wearer fortunate in his undertakings; he runs no risk on sea or water, neither can he be injured by water.

CANIS MAJOR. Figure of a dog for coursing hares, with a curved tail. Sign: Cancer. Nature: Venus. Cures lunacy, insanity, and demoniacal possession.

LEPUS. Figure of a hare with ears pricked up and the feet represented as though in swift motion. Sign: Gemini. Nature: Saturn and Mercury. Cures frenzy and protects from the wiles of demons. The wearer cannot be hurt by a malignant spirit.

CENTAUR. Half-figure of a bull, bearing a man on whose left shoulder rests a lance, from which depends a hare. In his right hand the man holds a small, supine animal with a vessel attached to it. Sign: Libra. Nature: Jupiter and Mars. Gives constancy and perpetual health.

CANIS MINOR. Figure of a dog, sitting. Sign: Cancer. Nature: Jupiter. Guards from dropsy, pestilence, and the bites of dogs.

SACRARIUS TURUBULUS (ARA). An altar with burning incense. Sign: Sagittarius. Nature: Venus and Mercury. Gives the wearer power to recognize spirits, to converse with them, and to command them; also confers chastity.

HYDRA. A serpent, having an urn at its head and a raven at its tail. Sign: Cancer. Nature: Saturn and Venus. Gives riches and all good gifts to the wearer and makes him cautious and prudent.

CORONA AUSTRALIS. An imperial crown. Sign: Libra. Nature: Saturn and Mars. Augments wealth and makes the wearer gay and happy.

AURIGA. A man in a chariot, bearing a goat on his left shoulder. Sign: Gemini. Nature: Mercury. Makes the wearer successful in hunting.

VEXILLUM. A flag flying from the extremity of a lance Sign: Scorpion. Gives skill in war and confers victory in the field.

FIGURES OF THE PLANETS

SATURN. An old man holding a curved scythe in his hand and with a not very heavy beard. Engraved on a stone of the nature of Saturn, it renders the wearer powerful and augments his power continually.

JUPITER. A seated figure, sometimes in a chariot, holding a staff in one hand and a spear in the other. It renders the wearer fortunate, especially if engraved on a Kabratis stone, and he easily gains what he wishes, especially from priests. He will be raised to honors and dignities.

MARS. Represented sometimes with a banner and sometimes with a lance or other weapon. He is, indeed, always armed and at times mounted on a horse. Gives victory, boldness in war, and success in everything, especially if engraved on an appropriate stone.

SUN. Sometimes as the solar disk with rays, sometimes as a man in a chariot, and this occasionally is surrounded by the signs of the zodiac. Renders the wearer powerful and a victor; this gem is prized by hunters.

VENUS Many forms, among them that of a woman with a voluminous dress and a stole, holding a laurel in her hand. Gives skill in handling affairs and usually brings them to a successful issue; removes the fear of drowning.

MERCURY. Figure of a slender man, usually with a beautiful beard, but sometimes without. He has winged feet and holds the caduceus. Increases knowledge and confers eloquence. It aids merchants, enabling them to acquire wealth.

MOON. Various forms. Sometimes as a crescent, sometimes as a young woman in a chariot and holding a quiver, and at others as a woman with a quiver and following the chase with dogs. Aids the fortunes of those who are sent on an embassy, and enables them to acquire wealth and honor thereby. Is said to confer speed and facility in undertakings and a happy issue.[20]

When Hudibras attacked and overcame the sorcerer Sidrophal, he rifled the latter's pockets of all his mystic treasures. Among these were

Several constellation stones,
Engraved in planetary hours,
That over mortals had strange powers,
To make them thrive in law or trade,
And stab or poison to evade,
In wit and wisdom to improve,
And be victorious in love.[21]

These manifold influences exerted by the stars and planets through the medium of the gems, not only concerned those actually present in a material form, but also those that were seen in dreams, and interpretations of such dreams are given by old writers.

Many Oneirocritica, or "dream-books," were written or compiled in the early centuries of our era, one of the most noted being the work of Artemidorus, who flourished in the second century A.D. Every object seen in a dream was given a special meaning, and it is interesting to note that Artemidorus believed dreams of rings or other ornaments, as well as of precious stones, to be of favorable significance only for women. Such dreams indicated marriage for unmarried women, and the birth

---

[20] Camilli Leonardi, "Speculum Lapidum," Venetia, 1502, f. liii.

[21] Butler, "Hudibras," Part II, Canto III, 11, 1096–1103.

STATUETTE KNOWN AS THAT OF SAINTE FOY, IN THE ABBEY-CHURCH AT CONQUES, DEPT. AVEYRON, FRANCE.

It is studded with precious and semi-precious stones and engraved gems, dating from various epochs, the pious offerings of those whose prayers have been answered. The figure is 85 cm. ($33\frac{1}{2}$ inches) high and is of gold in a core of wood. Probably of the tenth century. Two of the four crystal balls adorning the seat are said to replace golden doves. Rock-crystal was especially dedicated to the moon.

of children for those already married. If a woman was both wife and mother when she saw sparkling jewels in her dream, then the vision portended the acquisition of great wealth. Artemidorus here sagely remarks that women are by nature devoted to riches and passionately fond of ornaments. For men, on the other hand, to dream of jewels was an ill omen; probably because it foreshadowed the necessity of buying them for a good friend or a faithful wife.[22]

Another of these dream-books, probably composed in the eighth century A.D., appears under the name of Achametis and is of Arabic origin. Many of the interpretations in this book are referred to a Hindu source, and among these are visions of crowns that appear to kings. Such a dream, in itself, usually portended increased power and success for the sovereign, but this depended upon the color and character of the jewels which adorned the crown. For example, we read that if the gems were red and of the kind known as lychnites (carbuncles or rubies), the dream indicated that the king would have great joy and good fortune and would be more feared by his enemies than before; but if he saw blue gems in the crown, it was a bad omen, foreshadowing the loss of part of his kingdom. If the stones were of a light green hue (the color of the leek), the king would gain a great name in the world, both by his good faith and by the greatness of his kingdom; for, the writer adds, "this color in precious stones is universally accepted as signifying good-faith and religious devotion to God." [23]

---

[22] Artemidori Daldiani et Achametis Sereimi Oneirocritica, ed. Regaltius, Lutetiæ, 1603, pp. 86, 87.

[23] Ibid., p. 228.

There is signified by dreaming of

Agates	A journey.
Amber	A voyage.
Amethysts	Freedom from harm.
Aquamarines	New friends.
Beryls	Happiness in store.
Bloodstones	Distressing news.
Carbuncles	Acquirement of wisdom.
Carnelians	Impending misfortune.
Cat's-eyes	Treachery.
Chalcedony	Friends rejoined.
Chrysoberyls	A time of need.
Chrysolites	Necessary caution.
Coral	Recovery from illness.
Crystal	Freedom from enemies.
Diamonds	Victory over enemies.
Emeralds	Much to look forward to.
Garnets	The solution of a mystery.
Heliotropes	Long life.
Hyacinths	A heavy storm.
Jacinths	Success.
Jasper	Love returned.
Jet	Sorrow.
Lapis-lazuli	Faithful love.
Moonstones	Impending danger.
Moss-agates	An unsuccessful journey.
Onyx	A happy marriage.
Opals	Great possessions.
Pearls	Faithful friends.
Porphyry	Death.
Rubies	Unexpected guests.
Sapphires	Escape from danger.
Sardonyx	Love of friends.
Topaz	No harm shall befall.
Tourmalines	An accident.
Turquoises	Prosperity.

If precious stones be so combined in a ring, or other jewel that the initial letters of their names spell words significant of a tender sentiment or implying good fortune, or else the name of someone dear to the giver of the jewel, this is also supposed to strengthen their astral or planetary influence and to render them more potent charms. In the following examples the gems in the first column are the more expensive, those in the second column being comparatively inexpensive ones.

ACROSTICS FORMED WITH STONES

In France and England, during the 18th century, rings, bracelets, brooches, etc., were often set with gems the first letters of which, combined, formed a motto or expressed a sentiment. The following is a list of those that may be used in this way. The choice of stones afforded here brings these pretty devices within the reach of all.

FAITH

Fire-opal.	Feldspar.
Alexandrite.	Amethyst.
Iolite.	Idocrase.
Tourmaline.	Topaz.
Hyacinth.	Heliotrope.

HOPE

Hyacinth.	Hematite.
Opal.	Olivine.
Pearl.	Pyrope.
Emerald.	Essonite.

CHARITY

Cat's-eye.	Carbuncle.
Hyacinth.	Hematite.
Aquamarine.	Amethyst.
Ruby.	Rose quartz.
Iolite.	Idocrase.
Tourmaline.	Topaz.
Yellow sapphire.	Yu (Jade).

GOOD LUCK

Golden beryl.	Garnet.
Opal.	Onyx.
Olivine.	Obsidian.
Diamond.	Dendrite.
Lapis-lazuli.	Labradorite.
Uralian emerald.	Unio pearl.
Cat's-eye.	Carnelian.
Kunzite.	Krokidolite

FOREVER

Fire-opal.	Flèches d'amour
Opal.	Onyx.
Ruby.	Rutile.
Emerald.	Essonite.
Vermeille.	Verd antique.
Essonite.	Epidote.
Rubellite.	Rose quartz.

REGARD

Ruby.	Rubellite.
Emerald.	Essonite.
Garnet.	Garnet.
Alexandrite.	Amethyst.
Ruby.	Rock-crystal.
Diamond.	Demantcid.

ZES

Greek, meaning "Mayest thou live."

Zircon.	Zonochlorite.
Emerald.	Essonite.
Sapphire.	Sard.

MIZPAH

Moonstone.	Moldavite.
Indicolite.	Idocrase.
Zircon.	Zonochlorite.
Peridot.	Pyrope.
Asteria.	Aquamarine.
Hyacinth.	Hematite.

FRIENDSHIP

Flèches d'amour	Feldspar.
Ruby.	Rock crystal.
Indicolite.	Idocrase.
Emerald.	Epidote.
Nephrite.	Nicolo.
Diamond.	Diopside.
Sapphire.	Sard.
Hyacinth.	Hematite.
Iolite.	Idocrase.
Pearl.	Pyrite.

DEAREST

Diamond.	Demantoid.
Emerald.	Essonite.
Alexandrite.	Amethyst.
Ruby.	Rubellite.
Essonite.	Epidote.
Sapphire.	Spinel.
Turquoise.	Topaz.

SOUVENIR

Sapphire.	Sunstone.
Opal.	Onyx.
Uralian emerald.	Utahlite.
Vermeille.	Verd antique.
Emerald.	Epidote.
Nephrite.	Nephrite.
Iolite.	Indicolite.
Ruby.	Rock-crystal.

BONHEUR

Beryl.	Bloodstone.
Opal.	Onyx.
Nephrite.	Nephrite.
Hyacinth.	Hematite.
Emerald.	Essonite.
Uralian emerald.	Utahlite.
Ruby.	Rhodonite.

AMITIÉ

Alexandrite.	Almandine.
Moonstone.	Moonstone.
Indicolite.	Indicolite.
Tourmaline.	Topaz.
Idocrase.	Idocrase.
Emerald.	Essonite.

LOVE ME

Lapis-lazuli.	Labrador spar.
Opal.	Onyx.
Vermeille.	Verd antique.
Emerald.	Essonite.
Moonstone.	Moonstone.
Essonite.	Epidote.

A E I

Greek, meaning "forever," "eternity."

Alexandrite.	Almandine.
Emerald.	Essonite.
Indicolite.	Idocrase.

An attractive engagement ring can be formed of a central diamond from which extend the rays of a five-pointed star. Between the rays are set the stones emblematic of the zodiacal sign, of the guardian angel of the month, of the planet control of the hour and also the two stones indicating the initial letter of the two Christian names. This ring is in the form of the mystic Pentagon, the grand symbol of constancy and durability, since the number five is composed of three, which signifies creative power, and two, which typifies the balance, that is, stability.

As, according to the old fancy, the influences due to the light emanations from the planets or fixed stars, or from the combination of the stars in a zodiacal sign, would have a peculiar and more or less intimate connection with the fate of one country rather than of another, an attempt is here made to give a charcteristic stone for each country. In the case of the United States the various gemstones found within the boundaries of each of the States of the Union are given. That this special influence was exceptionally potent in regard to those born in the countries in question was also taught and hence a national gem would have a greater talismanic power than any other for the natives of each separate country. For those who may feel a certain degree of sympathy for time-honored fancies, and who may perhaps also have a trace of superstition hidden away in some part of their consciousness, one of our State gems would have a similar significance.

GEMS OF COUNTRIES

Alaska	Garnet
Algiers	Coral
Arabia	Pearl
Austria-Hungary	Opal
Belgium	Crystal
Bohemia	Garnet
Bokhara	Lapis-lazuli
Bolivia	Lapis-lazuli
Brazil	Tourmaline (Brazilian emerald)
Burma	Ruby
Canada	Sodalite
Ceylon	Cat's-eye
Chili	Lapis-lazuli
China	Jade
Congo	Dioptase
Denmark	Agate
Egypt	Peridot
England	Diamond
France	Pearl
Germany	Amber
German West Africa	Diamond
Greece	Sapphire
Holland	Diamond
Hungary	Opal
India	Pearl
Ireland	Precious serpentine (Connemara)
Italy	Coral
Japan	Rock-crystal
Korea	Abalone pearl

Madagascar	Morganite
Mexico	Obsidian
Morocco	Coral
New England	Tourmaline
New South Wales	Opal
New Zealand	Jade
Norway-Sweden	Carnelian
Panama	Agate
Persia	Turquoise
Peru	Emerald
Philippines	Pearl
Portugal	Chrysoberyl
Roumania	Amber
Russia	Rhodonite
Sandwich Islands	Olivine
Scotland	Cairngorm (smoky quartz)
Servia	Coral
Siam	Ruby
Sicily	Amber
South Africa	Diamond
Spain	Emerald
Switzerland	Rock-crystal
Turkestan	Jade
Turkey	Turquoise
United States	Sapphire
Uruguay	Amethyst

UNITED STATES STONES

Precious, semi-precious, or gem stones are found in nearly every State of the Union. The most important are enumerated below:

Alabama	Beryl, blue and yellow; smoky quartz.
Arizona	Agatized wood, azur-malachite, turquoise, garnet, peridot.
Arkansas	Rock-crystal, smoky quartz, agate, diamond, novaculite.
California	Agate, benitoite, californite, diamond, gold quartz, tourmaline, abalone pearl, chrysoprase, kunzite, morganite.

Colorado .... Beryl, aquamarine, phenacite, garnet, amethyst, agate, gold quartz, pyrite.
Connecticut .... Beryl, yellow and green; rose quartz, tourmaline.
Delaware .... Pearl.
Florida .... Chalcedony, conch pearl.
Georgia .... Ruby, beryl, amethyst, gold quartz, garnet.
Idaho .... Opal, agate, obsidian.
Illinois .... Fluorite, pearl.
Indian Territory .... Obsidian, pearl.
Indiana .... Pearl.
Iowa .... Fossil coral, pearl, chalcedony.
Kansas .... Chalcedony.
Kentucky .... Pearl.
Louisiana .... Chalcedony.
Maine .... Tourmaline, beryl, rose quartz, pearl, topaz, amazonite, smoky quartz, rock-crystal.
Maryland .... Beryl, clam-pearl.
Massachusetts ...... Beryl.
Michigan .... Agate, hematite.
Minnesota .... Chlorastrolite, thomsonite, agate
Mississippi .... Pearl, chalcedony.
Missouri .... Pearl, fluorite, pyrite.
Montana .... Sapphire, beryl, smoky quartz, agate, amethyst, agatized wood, obsidian.
Nebraska .... Chalcedony, pearl.
Nevada .... Gold quartz, rock-crystal.
New Hampshire .... Beryl, rock-crystal, garnet.
New Jersey .... Fowlerite, willemite, prehnite, smoky quartz, agate, pearl.
New Mexico .... Turquoise, garnet, obsidian, peridot, rock-crystal.
New York .... Beryl, brown tourmaline, rose quartz, fresh-water pearl, clam-pearl, chondrodite.
North Carolina ..... Aquamarine, beryl, emerald, almandite garnet, rhodolite, pyrope garnet, diamond, cyanite, hiddenite, amethyst, ruby, sapphire, smoky quartz, rock-crystal, rutile.
North Dakota ...... Chalcedony, agate.
Ohio .... Fossil coral, chalcedony.
Oregon .... Agate, obsidian, hydrolite.

Pennsylvania.......Amethyst, beryl, sunstone, moonstone, amazonite, almandite garnet, pyrope garnet, rutile.
Rhode Island.......Hornblende in quartz, amethyst, rock-crystal.
South Carolina.....Beryl, smoky quartz, rock-crystal.
South Dakota.......Quartzite, beryl, agate.
Tennessee..........Pearl.
Texas.............Beryl, pearl, tourmaline.
Utah...............Topaz, garnet.
Virginia............Amethyst, spessarite, garnet, beryl, moonstone, staurolite, allanite.
Vermont............Beryl, pearl.
Washington.........Pearl, agate.
West Virginia......Rock-crystal.
Wisconsin..........Agate, pearl.
Wyoming...........Moss-agate, agate.

XI

On the Therapeutic Use of Precious and Semi-Precious Stones

THE medicinal use of precious stones may be traced back to very ancient times. It has been conjectured that their employment for such purposes was introduced to Europe from India, whence many of the stones were derived. Nevertheless, the earliest evidence we have rather points to Egypt as the source, and, indeed, it appears that in early Egyptian times the chemical constituents of the stones were much more rationally considered than at a later period in Europe. The Ebers Papyrus, for instance, recommends the use of certain astringent substances, such as lapis-lazuli, as ingredients of eye-salves, and hematite, an iron oxide, was used for checking hemorrhages and for reducing inflammations. Little by little, however, superstition associated certain special virtues with the color and quality of precious stones, and their virtues were thought to be greatly enhanced by engraving on them the image of some god, or of some object symbolizing certain of the activities of nature. Later still, the science of astrology, most highly developed in Assyria and Babylonia, was brought into combination with the various superstitions above indicated, so that the image was believed to have much greater efficacy if the engraving were executed when the sun was in a certain constellation or when the moon or some one of the planets was in the ascendant at the time.

If we exclude certain fragmentary notices in Egyptian literature—notably the statements in the Ebers Papyrus—and the very uncertain sources in Hindu literature, the earliest authority for this branch of the subject is the Natural History of Pliny. In this connection, however, it is only just to call attention to a fact which has been often ignored—namely, that Pliny himself had very little faith in the teachings of the "magi," as he calls them, in regard to the superstitious use of gems for the prevention or cure of diseases; indeed, he seems to have been almost as sceptical in his attitude as many modern writers, for certain quite recent authorities still credit amber and a few other mineral substances with therapeutic effects other than those which can be explained by the known action of their chemical constituents. Still, Pliny yielded so far to the taste of his time as to preserve for us many of the statements of earlier writers on the subject, naming them in most cases and so enabling us to form some idea of the character of this pseudo-science in the Roman world in the first century of our era. With the gradual decay of ancient learning, the less valuable elements of popular belief came more and more into the foreground, and the old superstitions were freely copied by successive authors, each of whom felt called upon to add something new on his own account. This explains much of the confusion that reigns in regard to the attribution of special virtues to the different stones, for the wider the reading of the author the greater became the number of virtues attributed to each separate stone, until, at last, we might almost say that each and every precious stone could be used for the cure of all diseases. Nevertheless, it is comparatively easy to see that either the color or constitution of the stone originally indicated its use for this or that disease.

A distinction is often made between the talismanic

INSCRIPTION ON A SMALL PIECE OF LIMESTONE, IN CURSIVE EGYPTIAN WRITING.

It dates from about 1600 B.C., the period of the Ebers Papyrus, and gives directions for preparing certain remedies from precious stones. While the interpretation of this text offers considerable difficulty, one version finds in it the statement that lapis-lazuli—the "sapphire" of the ancients—was used for the wealthy, and malachite for those of limited means. Professor Oefele conjectures that the disease to be treated was hysteria. Munch Collection, Metropolitan Museum of Art, New York.

qualities of precious stones for the cure or prevention of disease and the properly medicinal use of them as mineral substances. In the former case the effect was attained by merely wearing them on the person, while in the latter case they were reduced to a powder, which was dissolved as far as possible in water or some other liquid and then taken internally. As, however, the end to be attained is the same whether the stone be worn or taken internally as a powder or liquid, it seems more logical to treat of both these methods of therapeutic use together, reserving for the chapter on the talismanic use of gems only their employment to avert misfortunes other than those caused by disease, and their influence in the procuring of wealth, honors, and happiness for their wearers.

The belief in the curative properties of precious stones was at one time universal among all those to whom gems were known. When we read to-day of the various ills that were supposed to be cured by the use of these gems, we find it difficult to understand what process of thought could have suggested the idea of employing such ineffectual remedies. It is true that the constituents of certain stones can be absorbed by the human body and have a definite effect upon it, but the greater part of the elements are so combined that they cannot be assimilated, and they pass through the system without producing any apparent effect.

In ancient and medieval times, however, other than chemical agencies were supposed to be efficient in the cure of diseases, and the primitive animistic conception of the cause of illness, and hence of the therapeutics of disease, long held sway among those who practised the medical art. Remedies were prized because of their rarity, and also because it was believed that certain spiritual or planetary influences had aided in their production and were latent in them. Besides this, the symbolism of color played a very important part in recommending the use of particular stones for special diseases. This may be noted in the case of the red or reddish stones, such as the ruby, spinel, garnet, carnelian, bloodstone, etc. These were thought to be sovereign remedies for hemorrhages of all kinds, as well as for all inflammatory diseases; they were also believed to exercise a calming influence and to remove anger and discord. The red hue of these stones was supposed to indicate their fitness for such use, upon the principle *similia similibus curantur*. In the same way yellow stones were prescribed for the cure of bilious disorders, for jaundice in all its forms and for other diseases of the liver.

The use of green stones to relieve diseases of the eye was evidently suggested by the beneficial influence exerted by this color upon the sight. The verdant emerald represented the beautiful green fields, upon which the tired eye rests so willingly, and which exert such a soothing influence upon the sight when it has been unduly strained or fatigued. One of the earliest, probably the very earliest reference in Greek writings to the therapeutic value of gems, appears in the works of Theophrastus, who wrote in the third century before Christ. Here we are told of the beneficial effect exercised by the emerald upon the eyes.

The sapphire, the lapis-lazuli, and other blue stones, with a hue resembling the blue of the heavens, were believed to exert a tonic influence, and were supposed to counteract the wiles of the spirits of darkness and procure the aid and favor of the spirits of light and wisdom. These gems were usually looked upon as emblems of chastity, and for this reason the sapphire came to be

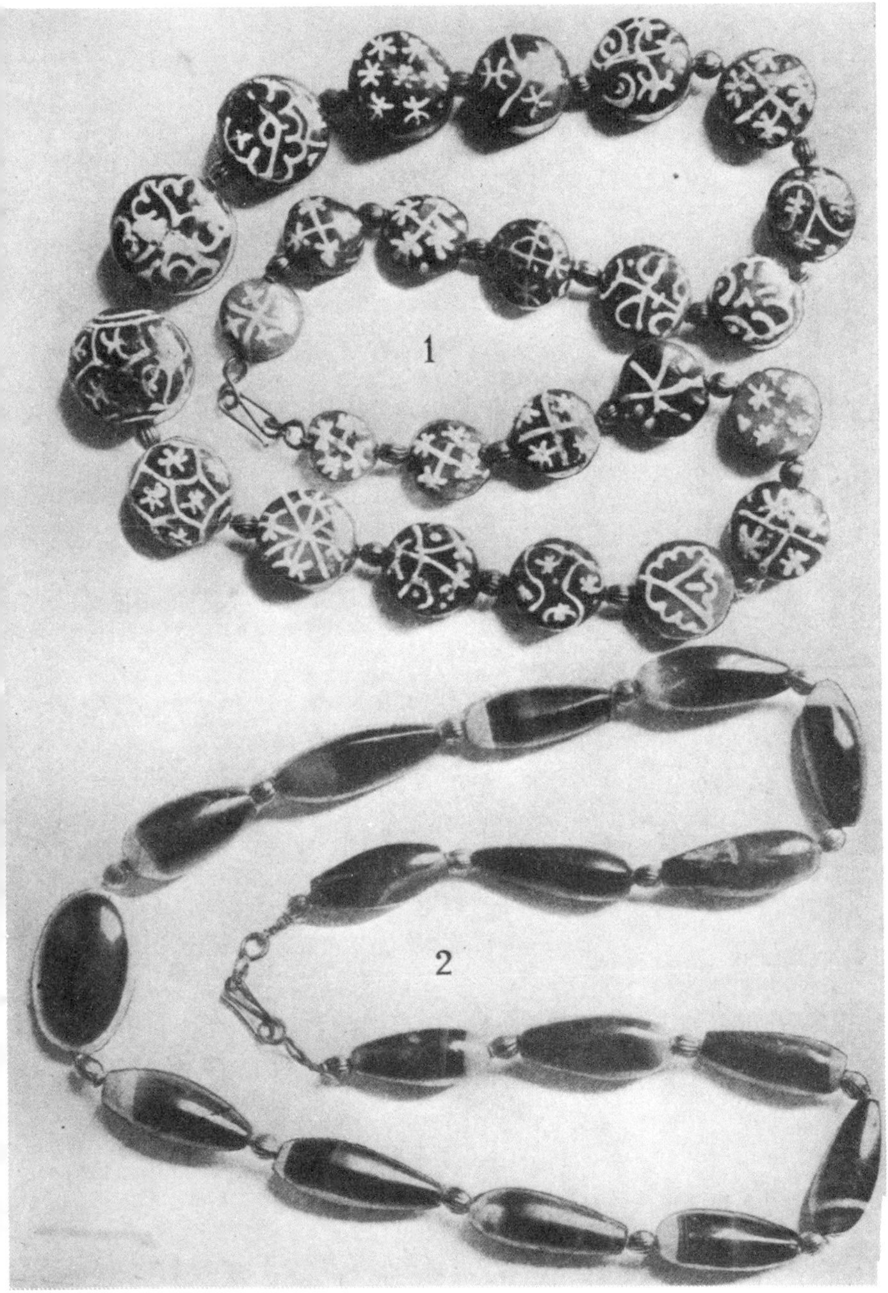

1. Necklace of carnelian beads. Persian. The decoration is made with carbonate of lime and fixed by firing. Charms against the Evil Eye.
2. Necklace of onyx beads. Early Christian.

regarded as especially appropriate for use in ecclesiastical rings. Among purple stones, the amethyst is particularly noteworthy. The well-known belief that this gem counteracted the effects of undue indulgence in intoxicating beverages is indicated by its name, derived from μεθύω — "to be intoxicated," and the privative α, the name thus signifying the "sobering" gem. It is not unlikely that a fancied resemblance between the prevailing hue of these stones and that of certain kinds of wine first gave rise to the name and to the idea of the peculiar virtues of the amethyst.

We have mentioned only a few of the more obvious analogies suggested by the color of gems, and we might be tempted to cite many others were it not that symbolism is always treacherous ground, since there is practically no limit to the correspondences that may be found between sensuous impressions and ideas.

One great difficulty which besets any one who is trying to find a clue to guide him through the labyrinth of the medical affinities of gems is the fact that there was, from an early period, a tendency to attribute the virtues of one gem to another, probably owing to the commercial instinct which urged the dealer to praise his wares in every possible way, so that no part of his stock should fail to find a purchaser. This tendency is especially marked in the old Hindu Lapidaries, wherein it is almost impossible to find any differentiation of the stones in respect to their curative or talismanic virtues. Only the condition and perfection of the gems are made the criterion of their worth. Any given stone, if perfect, was a source of all blessings to the wearer and possessed all remedial powers, while a defective stone, or one lacking the proper lustre or color, was destined to be a source of untold misfortune to the owner.

The European writers on the medical properties of precious stones were influenced by quite different considerations; their chief aim was to represent each stone, regarded simply as a mineral substance, as being the abode of the greatest possible number of curative properties. Indeed, many of the most highly recommended electuaries contained all kinds of stones, as though the effect to be produced did not depend upon the qualities of any single stone, or class of stones, but rather upon the quantity used. In Arnobio's "Tesoro delle Gioie,"[1] we have a receipt for the composition of "the most noble electuary of jacinth." This contains jacinth, emerald, sapphire, topaz, garnet, pearl, ruby, white and red coral, and amber, as well as many animal and vegetable substances, in all, thirty-four ingredients. It would indeed seem that a good dose of such a mixture should have provided a cure for "all the ills that flesh is heir to," by the simple and effective means of removing the unhappy patient to a better world.

Treating of the metallic affinities of precious stones, Paracelsus (1493–1541) affirmed that the emerald was a copper stone; the carbuncle and the jasper were golden stones; the ruby and the chalcedony, silver stones. The "white sapphire" (corundum) was a stone of Jupiter, while the jacinth was a mercurial stone. Powdered jacinth mixed with an equal quantity of laudanum was recommended as a remedy for fevers resulting from "putrefaction of the air or water." This illustrates the custom of combining an inefficacious material, such as the powder of a precious stone, with another possessing genuine remedial virtue, the name of the stone appealing to the popular superstitions regarding its thera-

[1] Venice, 1602, p. 254.

peutic powers, and thus rendering the preparation more acceptable.[2]

It is related by Plutarch that when Pericles was dying of the plague, he showed to one of his friends, who was visiting him, an amulet suspended from his neck. This had been given to Pericles by the women of his household, and Plutarch cites the instance as a proof that even the strongest minds will at certain times yield to the influence of superstition.[3]

There were sceptics in ancient times who put no faith in the popular superstitions as to the curative powers of precious stones. Eusebius (ca. 264–ca. 349), in his oration on the Emperor Constantine the Great (272–337), says:[4]

> He held that the varieties of stones so greatly admired were useless and ineffective things. They possessed no other qualities than their natural ones, and hence no efficacy to hold evils aloof; for what power can such things have either to cure disease or to avert death? Nevertheless, although he well knew this, he was in no wise opposed to their use simply as ornaments by his subjects.

The Middle High German didactic poem on precious stones, composed by Volmar, or Volamar, about 1250, appears to have been written as a rejoinder to a satirical poem, the work of a writer called the "Stricker" (rascal). What chiefly aroused Volmar's wrath was the fact that this irreverent personage dared to assert that a piece of

---

[2] "The Hermetic and Alchemical writings of Aureolus Philippus Theophrastus Bombast of Hohenheim, called Paracelsus the Great," trans. by Arthur Edward Waite, London, 1894, Vol. I, pp. 14, 225, Vol. II, p. 218.

[3] Plutarchi, "Vitæ," ed. Sinteris, Lipsiæ, 1884, p. 339; Pericles, 38.

[4] Eusebii Pamphili, "De laudibus Constantini," cap. v; in Eusebii, "Opera Omnia," ed. Migne, Parisiis, 1857, cols. 1337, 1340; Patrologiæ Græcæ, vol. xx.

colored glass set in a ring looked just as well and possessed the same virtues as a genuine precious stone of the same color. Volmar does not mince matters, and roundly declares that whoever should kill the man who wrote thus would do no sinful act. While we can scarcely recommend such drastic action, we must admit that we feel a little sympathy with the medieval champion of genuine stones against imitations.

A most interesting item recording one phase of a great tyrant's character is reported by Sir Jerome Horsey, who was entrusted with messages to and from Elizabeth of England and Ivan the Terrible of Russia. He gives, in his "Travels," a graphic recital of an interview with Ivan just before the latter's death in 1584. We retain the archaic spelling as it is reproduced in the Hakluyt publication from the original manuscript. Writing of Ivan, Horsey says:[5]

Carried every daye in his chair into his treasure. One daye the prince beckoned me to follow. I strode emonge the rest venturously, and heard him call for som precious stones and jewells. Told the prince and nobles present before and aboute him the virtue of such and such, which I observed, and do pray I may a littell degress to declare for my own memorie sake.

" The load-stone you all know hath great and hidden vertue, without which the seas that compas the world ar not navigable, nor the bounds nor circles of the earth cannot be knowen. Mahomett, the Percians proffit, his tombe of steell hangs in their Repatta at Darbent most miraculously "—Caused the waiters to bringe a chaine of nedells towched by his load-stone, hanged all one by the other.—" This faire currell (coral) and this faire turcas you see; take in your hand; of his natur arr orient coullers; put them on my hand and arm. I am poisoned with disease: you see they shewe their virtue by the chainge of their pure culler into pall: declares my death. Reach owt my staff roiall; an unicorns horn garnished with verie fare diomondes, rubies,

[5] The Travels of Sir Jerome Horsey, Hakluyt Society, London, 1856, pp. 199, 200.

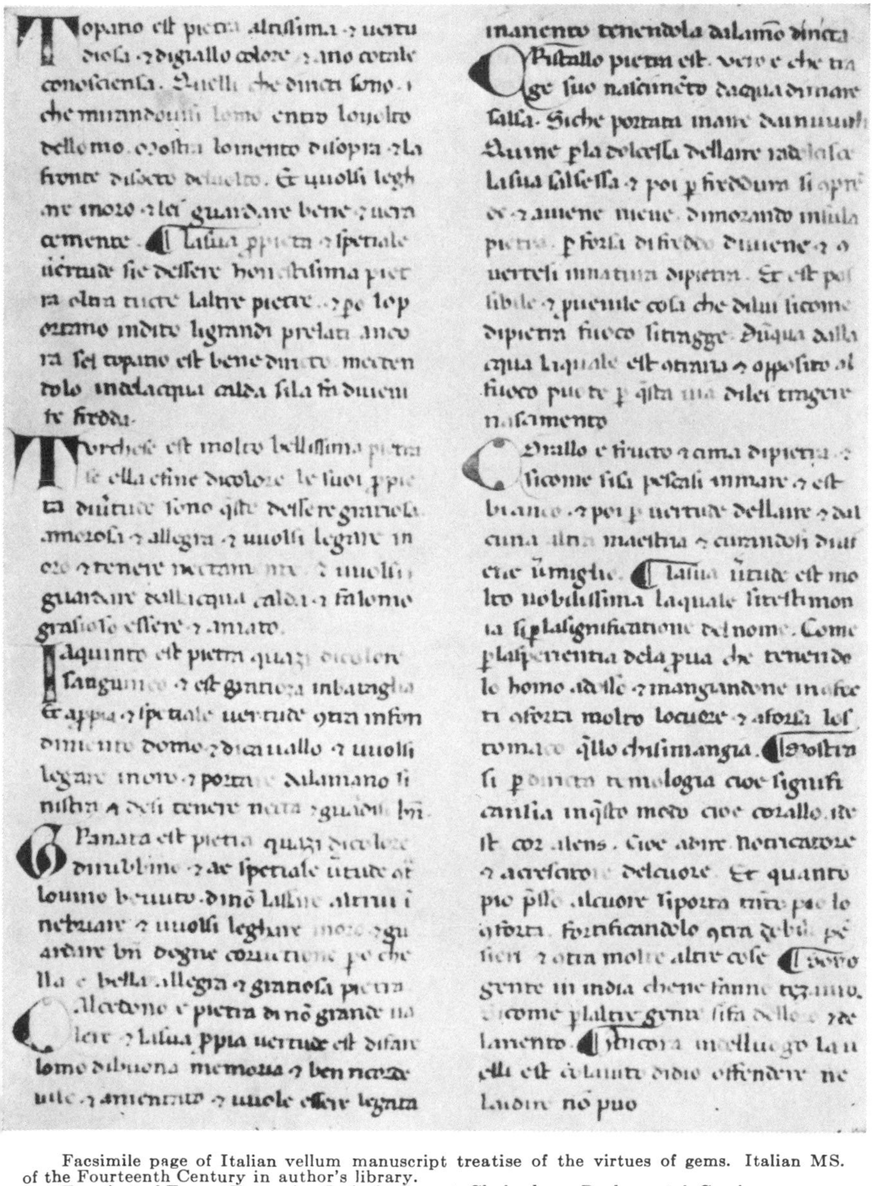

Facsimile page of Italian vellum manuscript treatise of the virtues of gems. Italian MS. of the Fourteenth Century in author's library.
Treating of Topaz, Turquoise, Jacinth, Garnet, Chalcedony, Rock-crystal, Coral.

saphiers, emeralls and other precious stones that ar rich in vallew; cost 70 thousand marckes sterlinge of David Gower from the fowlkers of Ousborghe.[6] Seek owt for som spiders." Caused his phiziccians, Johannes Lloff, to scrape a circle thereof upon the tabell; putt within it one spider and so one other and died, and some other without that ran alive apace from it.—"It is too late, it will not preserve me. Behold these precious stones. This diomond is the orients richest and most precious of all other. I never affected it; yt restreyns furie and luxurie and abstinacie and chasticie; the least parcell of it in powder will poysen a horse geaven to drinck, much more a man." Poynts at the ruby. "Oh! this is most comfortable to the hart, braine, vigar and memorie of man, clarifies congelled and corrupt bloud."—Then at the emerald.—"The natur of the reyn-bowe; this precious stone is an enemye to uncleanness. The saphier I greatlie delight in; yt preserves and increaseth courage, joies the hart, pleasinge to all the vitall sensis, precious and verie soveraigne for the eys, clears the sight, takes awaye bloudshott and streingthens the mussells and strings thereof."—Then takes the onex in hand.—"All these are Gods wonderfull guifts, secreats in natur, and yet revells [reveals] them to mans use and contemplacion, as frendes to grace and vertue and enymies to vice. I fainte, carie me awaye till an other tyme."

Some believed that when precious stones were worn to relieve or prevent disease, it was important that the different stones should be worn on different parts of the body. According to one authority, the jacinth should be worn on the neck; the diamond, on the left arm; the sapphire, on the ring-finger; the emerald, or the jacinth, on the index-finger; and the ruby or turquoise, on either the index-finger or the little finger.[7] There is, however, little reason to assume that these rules were generally known and observed.

That precious stones not only appealed to the eye by

[6] The Fuggers of Augsburg, the jeweller bankers of the 15th and 16th centuries.

[7] Wolffii, "Curiosus amuletorum scrutator," Francofurti et Lipsiæ, 1692, p. 363; citing Rodolphus Goclenius (De peste, p. 70).

their beautiful colors, but also possessed a fragrant odor, was one of the many fanciful ideas regarding them. If we could believe the following circumstantial account, this was once experimentally proved:[8]

When precious stones are to be used in medicine, they must be pulverized until they are reduced to a powder so fine that it will not grate under the teeth, or, in the words of Galen, this powder must be as impalpable "as that which is blown into the eyes." Since this trituration is not usually operated with sufficient care by the apothecaries, I begged a medical student, who was lodging with me, to pass an entire month in grinding some of these stones. I gave him emeralds, jacinths, sapphires, rubies, and pearls, an ounce of each kind. As these stones were rough and whole, he first crushed them a little in a well-polished iron mortar, using a pestle of the same metal; afterward he employed a pestle and mortar of glass, devoting several hours each day to this work. At the end of about three weeks, his room, which was rather large, became redolent with a perfume, agreeable both from its variety and sweetness. This odor, which much resembled that of March violets, lingered in the room for more than three days. There was nothing in the room to produce it, so that it certainly proceeded from the powder of precious stones.

Diamond

Of the many medicinal virtues attributed to the diamond, one of the most noteworthy is that of an antidote for poisons. Strangely enough, the belief in its efficacy in this respect was coupled with the idea that the stone in itself was a deadly poison. The origin of this latter fancy must be sought in the tradition that the place wherein the diamonds were generated—"in the land where it is six months day and six months night"—was guarded by venomous creatures who, in passing over the stones, were wounded by the sharp points of the crystals,

[8] Olaus Borrichius, in the Collection Académique, Paris, 1757, tome iv, p. 338.

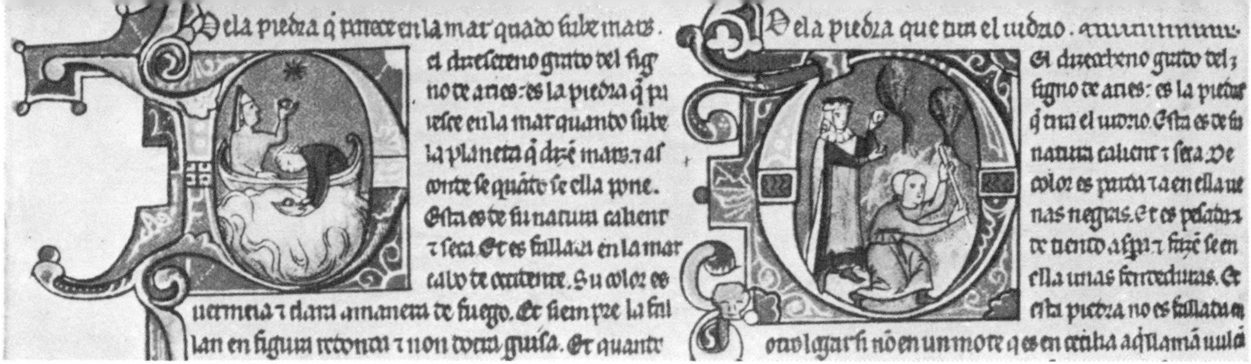

INITIALS FROM THE LAPIDARIO DE ALFONSO X (XIII CENTURY).

Codice Original (fol. 6), published in Madrid, 1881. That on the left figures "the stone found in the sea when the planet Mars rises"; that on the right, "the stone that attracts glass." Author's library.

INITIALS FROM THE LAPIDARIO DE ALFONSO X.

Codice Original (fol. 4), published in Madrid, 1881. On the left, "the stone that recoils from milk"; on the right, pearl-fishers

and thus embued the stones with some of their venom.[9] The attribution of curative properties in case of poisoning arose from association of ideas. The Lapidario of Alfonso X recommends the diamond for diseases of the bladder; it adds, however, that this stone should be used only in desperate cases.

The diamond was also believed to afford protection from plague or pestilence, and a proof of its powers in this direction was found in the fact that the plague first attacked the poorer classes, sparing the rich, who could afford to adorn themselves with diamonds. Naturally, in common with other precious stones, this brilliant gem was supposed to cure many diseases. Marbodus[10] tells us that it was even a cure for insanity.

In the Babylonian Talmud we read of a marvellous precious stone belonging to Abraham. This was perhaps a diamond, or possibly a pearl; the accounts vary, and the same word is often used to designate "precious stone" and "pearl." The following version represents it to be a diamond:[11]

R. Simeon, ben Johanan said: "A diamond was hanging on Abraham's neck, and when a sick man looked upon it he was cured. And when Abraham passed away, the Lord sealed it in the planet of the sun."

The Hindus believed that it was extremely dangerous to use diamonds of inferior quality for curative purposes, as they would not only fail to remedy the disease for which they were prescribed, but might cause lameness,

---

[9] Lapidario del Rey D. Alfonso X, Codice Original, Madrid, 1881, f. xi.

[10] "De lapidibus," Friburgi, 1531, f. 8.

[11] New edition of the Babylonian Talmud, ed. and trans. by Michael L. Rodkinson, vol. v (xiii), Baba Barat, New York, 1902, p. 53. See also Beer, "Leben Abraham's," Leipzig, 1859, p. 79.

jaundice, pleurisy, and even leprosy. As to the use of diamonds of good quality, very explicit directions are given. On some day regarded as auspicious for the operation, the stone was to be dipped in the juice of the *kantakára* (*Solanum jaquiri*) and subjected for a whole night to the heat of a fire made by dried pieces of the dung of a cow or of a buffalo. In the morning it was to be immersed in cow's urine and again subjected to fire. These processes were to be repeated for seven days, at the end of which term the diamond could be regarded as purified. After this the stone was to be buried in a paste of certain leguminous seeds mixed with asafœtida and rock salt. Herein it was to be heated twenty-one successive times, when it would be reduced to ashes. If these ashes were then dissolved in some liquid, the potion would "conduce to longevity, general development of the body, strength, energy, beauty of complexion, and happiness," giving an adamantine strength to the limbs.[12]

An Austrian nobleman, who for a long time had not been able to sleep without having terrible dreams, was immediately cured by wearing a small diamond set in gold on his arm, so that the stone came in contact with his skin.[13]

The fact that in this case, as in many others, the stone was required to touch the skin, proves that the effect supposed to be produced was not altogether magical, but in the nature of a physical emanation from the stone to the body of the wearer.

We are told that when Pope Clement VII was seized by his last illness, in 1534, his physicians resorted to

[12] Surindro Mohun Tagore, "Mani Málá," Pt. I, Calcutta, 1879, pp. 137, 139, 141.

[13] Andrea Spigello, "De semitert."; cited in Gimma, "Della Storia naturale delle gemme," Napoli, 1730, vol. i, p. 208.

powders composed of various precious stones. In the space of fourteen days they are asserted to have given the pope forty thousand ducats' worth of these stones, a single dose costing as much as three thousand ducats. The most costly remedy of all was a diamond administered to him at Marseilles. Unfortunately, this lavish expenditure was of no avail; indeed, according to our modern science, the remedies might have sufficed to end the pope's life, without the help of his disease.[14]

The old fancy that the diamond grew dark in the presence of poison is explained by the Italian physician Gonelli as caused by minute and tenuous particles which emanated from the poison, impinged upon the surface of the diamond, and, unable to penetrate its dense mass, accumulated on the surface, thus producing a superficial discoloration. The diamond, being a cold substance, may have condensed moisture from the body, and the one suffering from the poison may have emitted exudations. But this elaborate explanation of a phenomenon which never existed except in the imagination of those who related it is characteristic of Gonelli, who was always ready to elucidate in some similar way any of the marvels recounted in regard to precious stones.[15]

Emerald

The emerald was employed as an antidote for poisons and for poisoned wounds, as well as against demoniacal possession.[16] If worn on the neck it was said to cure the

[14] Raumer, "Historisches Taschenbuch," I Ser., vol. vi, Leipzig, 1835, p. 370.

[15] Josephi Gonelli, "Thesaurus philosophicus, seu de gemmis," Neapoli, 1702, pp. 76, 77.

[16] Lapidario del Rey D. Alfonso X, Codice Original, Madrid, 1881, f. xv.

"semitertian" fever and epilepsy.[17] The use of the emerald to rest and relieve the eye is the only remedial use of a precious stone mentioned by Theophrastus in his treatise on gems, written in the third century B.C. Alluding to its powers as an antidote for poisons, Rueus asserts [18] that if the weight of eighty barley-corns of its powder were given to one dying from the effects of poison, the dose would save his life. The Arabs prized emeralds highly for this purpose, and Abenzoar states that, having once taken a poisonous herb, he placed an emerald in his mouth and applied another to his stomach, whereupon he was entirely cured.[19]

A certain cure for dysentery also was to wear an emerald suspended so that it touched the abdomen and to place another emerald in the mouth. Michaele Paschali, a learned Spanish physician of the sixteenth century, declared that he had effected a cure of the disease by means of the emerald in the case of Juan de Mendoza, a Spanish grandee, and Wolfgang Gabelchover, of Calw, in Würtemberg, writing in 1603, asserts that he had often tested the virtues of the emerald in cases of dysentery and with invariable success.[20]

It speaks not a little for the beauty of the emerald that so good a judge of precious stones as Pliny should have pronounced this gem to be the only one that delighted the eye without fatiguing it, adding that when the vision was wearied by gazing intently at other objects, it gained renewed strength by viewing an emerald. So general in the early centuries of our era was the persuasion that

[17] Marbodus, l. c., f. 48.

[18] Rueus, l. c., p. 36.

[19] Morales, "De las piedras preciosas," Valladolid, 1604, f. 101.

[20] Andreæ Bacci, "De gemmis et lapidibus pretiosis," Francofurti, 1603, pp. 63, 64 (annotation of Gabelchover to his Latin version).

the pure green hue of emeralds aided the eyesight, that gem engravers are said to have kept some of them on their work-tables, so as to be able to look at the stones from time to time and thus relieve the eye-strain caused by close application to their delicate task.[21]

Psellus says that a cataplasm made of emeralds was of help to those suffering from leprosy; he adds that if pulverized and taken in water they would check hemorrhages.[22] They were especially commended for use as amulets to be hung on the necks of children, as they were believed to ward off and prevent epilepsy. If, however, the violence of the disease was such that it could not be overcome by the stone, the latter would break.[23] Hermes Trismegistus says the emerald cures ophthalmia and hemorrhages. The great Hermes must have had a special preference for this stone, since his treatise on chemistry (*peri chemeias*) is said to have been found inscribed on an emerald.[24]

By the Hindu physicians of the thirteenth century the emerald was considered to be a good laxative. It cured dysentery, diminished the secretion of bile, and stimulated the appetite. In short, it promoted bodily health and destroyed demoniacal influences. In the curious phrase of the school the emerald was "cold and sweet."[25]

Teifashi (1242 A.D.) believed that the emerald was a

---

[21] Plinii, "Naturalis historia," lib. xxxvii, cap. 16.

[22] Psellus, "De lapidum virtutibus," Lug. Bat., 1745, p. 32.

[23] Johannis Braunii, "De Vestitu sacerd. Heb.," Amstel., 1680, p. 659.

[24] From an old book the title-page of which reads: "In hoc volumine de Alchemia," etc., Norimberghe, 1541, p. 363.

[25] Garbe, "Die indische Mineralien; Naharari's Rajanighantu, Varga xiii," Leipzig, 1882, p. 76.

cure for hæmoptysis and for dysentery if it were worn over the liver of the person affected; to cure gastric troubles, the stone was to be laid upon the stomach. Furthermore, the wearer was protected from the attacks of venomous creatures, and evil spirits were driven from the place where emeralds were kept.[26] The direction to place the stone on the affected part, a recommendation often met with in the treatises on the therapeutic use of ornamental stones, shows that these were believed to send forth emanations of subtle power. Probably enough, the brilliant play of reflected light which proceeds from many of these gems suggested the idea that they radiated a certain curative energy. This theory need not surprise us, for, although it is altogether fanciful in the case of the diamond, ruby, emerald, etc., the newly discovered substance, radium, really possesses the active properties ascribed by old writers to precious stones.

Jade

A stone the therapeutic quality of which was specialized is the jade or nephrite. Strange to say, although there are very few places where this mineral can now be obtained,—the chief sources of supply being the province of Khotan in Turkistan and New Zealand,—in prehistoric times the stone must have been found in many different localities, since axe-heads and other artefacts of jade have been discovered in many lands both of the old and new world.

When the Spaniards discovered and explored the southern part of the American continent, they came across numerous native ornaments and amulets made of jade (jadeite) and brought many of these with them to

[26] Teifashi, "Fior di pensieri sulle pietre preziose," Ital. trans. by Antonio Raineri, Firenzi, 1818, p. 20.

Europe. The name jade is derived from the Spanish designation, *piedra de hijada,* meaning literally "stone of the flank," which is said to have been bestowed on the stone because the Indians used it for all diseases of the kidneys. The name nephrite owes its origin to the same idea. In ancient times jade appears to have been looked upon as a great aid in parturition, and many ingenious conjectures have been advanced as to the connection between this belief and the form of some of the prehistoric objects made of this material. Whether the Spaniards really learned from the Indians that the stone was especially adapted to cure renal diseases, or whether they only suggested this special and peculiar virtue in order to give an enhanced value to their jade ornaments, is a question not easily answered.

An early notice of jade as a remedial agent appears in Sir Walter Raleigh's account of his travels in Guiana. Treating of a people of "Amazons" said to dwell in the interior of the country, Raleigh says:[27]

> These Amazones have likewise great store of these plates of golde, which they recover by exchange, chiefly for a kinde of greene stone, which the Spaniards call Piedras Hijadas, and we use for spleene stones and for the disease of the stone we also esteeme them: of these I saw divers in Guiana, and commonly every King or *Casique* hath one, which theire wives for the most part weare, and they esteeme them as great jewels.

By the middle of the seventeenth century the curative powers of jade for the various forms of calculi was very generally admitted. A singular instance is offered us in one of Voiture's letters. He was a great sufferer from "the stone" and he had received, from a Mademoiselle Paulet, a beautiful jade bracelet. Gratefully acknowl-

---

[27] "The Discovery of the Large, Rich, and Beautiful Empire of Guiana," London, 1848, p. 29, Hakluyt Pub. Originally published in 1596.

edging the receipt of this peculiar gift, he expresses himself in the following frank way, a mixture of indelicacy and gallantry that seems strange to us: "If the stones you have given me do not break mine, they will at least make me bear my sufferings patiently; and it seems to me that I ought not to complain of my colic, since it has procured me this happiness." The name used for jade by Voiture, "*l'éjade*," supplied a missing link in the derivation of our name jade from the Spanish *hijada*. When the lady's gift was received by Voiture, some friends chanced to be present, and they were disposed to regard it as a token of love until he assured them that it was only a remedy. It appears that Mlle. Paulet was a fellow sufferer, and, alluding to this, Voiture writes: "On this occasion the jade had for you an effect you did not expect from it, and its virtue defended your own." [28]

Renal calculi and poetry do not seem to have much in common, but the following lines freely rendered from an old Italian poem on the subject by Ciri de Pers show that even this unpromising theme is susceptible of poetic treatment: [29]

"Other white stones serve to mark happy days,
But mine do mark days full of pain and gloom.
To build a palace, or a temple fair,
Stones should be used; but mine do serve
To wreck the fleshly temple of my soul.

. . . .

Well do I know that Death doth whet his glaive
Upon these stones, and that the marble white
That grows in me is there to form my tomb."

[28] Lettres de Voiture, ed. by Octave Uzanne, Paris, 1880, vol. i, p. 66, Letter XXIII.

[29] Josephi Gonnelli, Thesaurus philosophicus, seu de gemmis," Neapoli, 1702, pp. 157, 158.

As jade was and still is the most favored stone in China, although never found within the boundaries of China proper, it was very naturally accorded wonderful medical virtues. An old Chinese encyclopedia, the work of Li She Chan, and presented by him to the emperor Wan Lih of the Ming dynasty, in 1596, contains many interesting notices of jade. When reduced to a powder of the size of rice grains it strengthened the lungs, the heart, and the vocal organs, and prolonged life, more especially if gold and silver were added to the jade powder. Another, and certainly a pleasanter way of absorbing this precious mineral, was to drink what was enthusiastically called the "divine liquor of jade." To concoct this elixir equal parts of jade, rice, and dew-water were put into a copper pot and boiled, the resultant liquid being carefully filtered. This mixture was said to strengthen the muscles and make them supple, to harden the bones, to calm the mind, to enrich the flesh, and to purify the blood. Whoever took it for a long space of time ceased to suffer from either heat or cold and no longer felt either hunger or thirst.

Galen (b. ca. 130 A.D.) wrote thus of the green jasper:[30]

> Some have testified to a virtue in certain stones, and this is true of the green jasper, that is to say, this stone aids the stomach and navel by contact. And some, therefore, set the stone in rings and engrave on it a dragon surrounded by rays, according to what King Nechepsos has transmitted to posterity in the fourteenth book (of his works). Indeed, I myself have thoroughly tested this stone, for I hung a necklace composed of them about my neck so that they touched the navel, and I received not less benefit from them than I would had they borne the engraving of which Nechepsos wrote.

---

[30] Claudii Galeni, "De simplic. medicament., etc.," lib. ix, cap. 19. "Opera Omnia," ed. C. G. Kühn, Lipsiæ, 1826, vol. xii, p. 207. See also Duffield Osborne, "Engraved Gems," New York, 1912, pp. 138, 139.

Ruby

Sanskrit medical literature as represented by Naharari, a physician of Cashmere, who wrote in the thirteenth century, finds in the ruby a valuable remedy for flatulency and biliousness. Moreover, aside from these special uses, an elixir of great potency could be made from rubies by those who properly understood the employment of precious stones in the compounding of medicines.[31] This famous "ruby elixir" may have had little in common with the stone except its color, as such remedies were generally said to have been made by some secret and mysterious process, in the course of which all material evidence of the presence of any precious stone or stones completely disappeared.

Sapphire

One of the earliest specimens of English literature, William Langley's "Vision of William concerning Piers the Plowman" (written about 1377), contains a mention of the sapphire as a cure for disease:[32]

> I looked on my left half as þe lady me taughte
> And was war of a woman wortheli yeclothed,
> Purfiled with pelure[33] þe finest vpon erthe,
> Y-crowned with a corone þe kyng hath none better.
> Fetislich[34] hir fyngres were fretted[35] with gold wyre,

[31] Garbe, "Die indische Mineralien"; Naharari's "Rajanighantu," Varga XIII, Leipzig, 1882, p. 70.

[32] The Vision of William Concerning Piers the Plowman, by William Langley (or Langland). Ed. by Rev. Walter W. Skeat, Oxford, 1881, p. 16. Passus II, lines 8–15.

[33] Trimmed with fur.

[34] Handsomely.

[35] Adorned.

And þere-on red rubyes as red as any glede,[36]
And diamants of derrest pris, and double manere safferes,
Orientales and ewages [37] enuenymes [38] to destroye.

Among the rich gifts offered at the shrine of St. Erkinwald, in Old Saint Paul's, was a sapphire given in 1391 by Richard Preston, "a citizen and grocer of London." He stipulated that the stone should be kept at the shrine for the cure of diseases of the eyes, and that proclamation should be made of its remedial virtues. St. Erkinwald was the son of Offa, King of the East Saxons, and was converted to Christianity by Melittus, the first bishop of London. In 675 A.D. he himself became bishop of London, being the third to attain that rank after the death of Melittus. His body was interred in the cathedral, and his shrine, which was richly embellished during the reign of Edward III (1327–1377), received many valuable donations.[39]

The usefulness of the sapphire as an eyestone for the removal of all impurities or foreign bodies from the eye is noted by Albertus Magnus, who writes that he had seen it employed for this purpose. He adds that when a sapphire was used in this way it should be dipped in cold water both before and after the operation.[40] This was probably not so much to make the stone colder to the touch as to cleanse it, certainly a very necessary proceeding when the same stone was used by many persons suffering from contagious diseases of the eyes.

---

[36] Burning coal.

[37] Aquamarines.

[38] Poisons.

[39] Dugdale, "History of Saint Paul's Cathedral in London," London, 1818, vol. i, pp. 15, 16. First edition published in 1658.

[40] Alberti Magni, "Opera omnia," ed. Borgnet, Paris, 1890, vol. v, p. 44.

Richard Preston's sapphire appears to have been only one of a class regarded as having special virtue to cure diseased eyes, as is shown by the existence of various other similar sapphires in different parts of Europe. It is not very easy to determine the precise reason—if there be one—which rendered any single sapphire more useful than another in this respect. An entry in the inventory of Charles V notes "an oval Oriental sapphire for touching the eyes, set in a band of gold."[41] Possibly the fact that a particular gem of this kind was used remedially, and was not set for wear as an ornament, may have been the only cause for a belief in its special virtue.

That the sapphire should have been regarded as especially valuable for the cure of eye diseases serves to illustrate the wide-reaching and persistent influence of Egyptian thought, and the curious transformations through which an originally reasonable idea may pass in the course of time. We have already noted that the sapphire of the ancients was our lapis-lazuli, and in the Ebers Papyrus lapis-lazuli is given as one of the ingredients of an eye-wash. This ingredient is believed to have originally been the oxide of copper sometimes called lapis Armenus, a material possessing marked astringent properties, and which might be used to advantage in certain morbid conditions of the eye. Lapis-lazuli, another blue stone, was later substituted because of its greater intrinsic value, its similarity of color rendering it equally efficacious according to primitive ideas on this subject. When, however, in medieval times, the name sapphire came to signify the blue corundum gem known to us by this designation, the special curative virtues of the lapis-lazuli were transferred to this still more valuable stone.

[41] Labarte, "Inventaire du mobilier de Charles V," Paris, 1879, p. 308, No. 2937.

The proper method of applying a sapphire to cure plague boils is given at some length by Von Helmont. A gem of a fine, deep color was to be selected and rubbed gently and slowly around the pestilential tumor. During and immediately after this operation, the patient would feel but little alleviation; but a good while after the removal of the stone, favorable symptoms would appear, provided the malady were not too far advanced. This Von Helmont attributes to a magnetic force in the sapphire by means of which the absent gem continued to extract "the pestilential virulency and contagious poyson from the infected part."[42]

Topaz

The use of a topaz to cure dimness of vision is strongly recommended by St. Hildegard. To attain the desired end the stone was to be placed in wine and left there for three days and three nights. When retiring to sleep, the patient should rub his eyes with the moistened topaz, so that this moisture lightly touched the eyeball. After the stone had been removed, the wine could be used for five days.[43]

A Roman physician of the fifteenth century was reputed to have wrought many wonderful cures of those stricken by the plague, through touching the plague sores with a topaz which had belonged to two popes, Clement VI and Gregory II. The fact that this particular topaz had been in the hands of two supreme pontiffs must have added much to the faith reposed in the curative powers

---

[42] "A Ternary of Paradoxes, written originally by Joh. Bapt. Van Helmont and translated, illustrated, and amplicated by Walter Charleton," London, 1650, p. 17.

[43] S. Hildegardae, "Opera omnia," in Pat. Lat. ed. by J. P. Migne, vol. cxcvii, Parisiis, 1855, col. 1255.

of the stone by those upon whom it was used, and this faith may really have helped to hasten their recovery.[44]

Bloodstone

A historical instance of the use of the bloodstone to check a hemorrhage is recorded in the case of Giorgio Vasari (1514–1578), the author of the lives of the Italian painters of the Renaissance period. On a certain occasion, when the painter Luca Signorelli (1439–1521) was placing one of his pictures in a church at Arezzo, Vasari, who was present, was seized with a violent hemorrhage and fainted away. Without a moment's hesitation, Signorelli took from his pocket a bloodstone amulet and slipped it down between Vasari's shoulder-blades. The hemorrhage is said to have ceased immediately.[45]

The bloodstone was used as a remedy by the Indians of New Spain, and Monardes notes that they often cut the material into the shape of hearts. This seems a very appropriate form for an object used to check hemorrhages. The best effect was attained when the stone was first dipped in cold water and then held by the patient in his right hand. Of course the application of any cold object would serve to congeal the blood, but the connection with the heart vanishes in the direction to place the stone in the *right* hand. Monardes states that both Spaniards and Indians used the bloodstone in this way.[46]

The Franciscan friar Bernardino de Sahagun, a missionary to the Mexican Indians, shortly after the Spanish Conquest, writes that in 1576 he cured many natives who

---

[44] Arnobio, "Il tesoro delle gioie," Venice, 1602, p. 21.

[45] Bellucci, "Il feticismo in Italia," Perugia, 1907, p. 91, note.

[46] Monardes, Semplicium medicamentorum ex novo orbe delatorum historia (Latin version by Clusius), Antverpiæ, 1579, p. 51.

were at the point of death from hemorrhage, a result of the plague, by causing them to hold in the hand a piece of bloodstone. By this means he claims to have saved many lives.[47]

Robert Boyle, in his "Essay about the Origin and Virtues of Gems" (London, 1672, pp. 177–78), tells of a gentleman of his acquaintance who was "of a complexion extraordinary sanguin," and was much afflicted with bleeding of the nose. A gentlewoman sent to him a bloodstone, directing him to wear it suspended from his neck, and from the time he put it on he was no longer troubled with his malady. It recurred, however, if he removed the stone. When Boyle objected that this might be a result of imagination, his friend disposed of his objection by relating the instance of a woman to whom the stone had been applied when she was unconscious from loss of blood. Nevertheless, as soon as it touched her, the flow of blood was checked. Boyle states that this stone did not seem to him to resemble a true bloodstone. It may have been that the cold of the stone congealed the blood, or that the flow was checked by exhaustion.

---

[47] Sahagun, "Historia general de las cosas de Nueva España, vol. iii. Mexico, 1830, pp. 300, 301; lib. xi, cap. viii.

Index

D

I

J

K

A CATALOG OF SELECTED

DOVER BOOKS

IN ALL FIELDS OF INTEREST

STICKLEY CRAFTSMAN FURNITURE CATALOGS, Gustav Stickley and L. & J. G. Stickley. Beautiful, functional furniture in two authentic catalogs from 1910. 594 illustrations, including 277 photos, show settles, rockers, armchairs, reclining chairs, bookcases, desks, tables. 183pp. 6½ x 9¼. 23838-5 Pa. $11.95

AMERICAN LOCOMOTIVES IN HISTORIC PHOTOGRAPHS: 1858 to 1949, Ron Ziel (ed.). A rare collection of 126 meticulously detailed official photographs, called "builder portraits," of American locomotives that majestically chronicle the rise of steam locomotive power in America. Introduction. Detailed captions. xi + 129pp. 9 x 12. 27393-8 Pa. $13.95

AMERICA'S LIGHTHOUSES: An Illustrated History, Francis Ross Holland, Jr. Delightfully written, profusely illustrated fact-filled survey of over 200 American lighthouses since 1716. History, anecdotes, technological advances, more. 240pp. 8 x 10¾. 25576-X Pa. $12.95

TOWARDS A NEW ARCHITECTURE, Le Corbusier. Pioneering manifesto by founder of "International School." Technical and aesthetic theories, views of industry, economics, relation of form to function, "mass-production split" and much more. Profusely illustrated. 320pp. 6⅛ x 9¼. (USO) 25023-7 Pa. $9.95

HOW THE OTHER HALF LIVES, Jacob Riis. Famous journalistic record, exposing poverty and degradation of New York slums around 1900, by major social reformer. 100 striking and influential photographs. 233pp. 10 x 7⅞. 22012-5 Pa. $11.95

FRUIT KEY AND TWIG KEY TO TREES AND SHRUBS, William M. Harlow. One of the handiest and most widely used identification aids. Fruit key covers 120 deciduous and evergreen species; twig key 160 deciduous species. Easily used. Over 300 photographs. 126pp. 5⅜ x 8½. 20511-8 Pa. $3.95

COMMON BIRD SONGS, Dr. Donald J. Borror. Songs of 60 most common U.S. birds: robins, sparrows, cardinals, bluejays, finches, more–arranged in order of increasing complexity. Up to 9 variations of songs of each species. Cassette and manual 99911-4 $8.95

ORCHIDS AS HOUSE PLANTS, Rebecca Tyson Northen. Grow cattleyas and many other kinds of orchids–in a window, in a case, or under artificial light. 63 illustrations. 148pp. 5⅜ x 8½. 23261-1 Pa. $5.95

MONSTER MAZES, Dave Phillips. Masterful mazes at four levels of difficulty. Avoid deadly perils and evil creatures to find magical treasures. Solutions for all 32 exciting illustrated puzzles. 48pp. 8¼ x 11. 26005-4 Pa. $2.95

MOZART'S DON GIOVANNI (DOVER OPERA LIBRETTO SERIES), Wolfgang Amadeus Mozart. Introduced and translated by Ellen H. Bleiler. Standard Italian libretto, with complete English translation. Convenient and thoroughly portable–an ideal companion for reading along with a recording or the performance itself. Introduction. List of characters. Plot summary. 121pp. 5¼ x 8½. 24944-1 Pa. $3.95

TECHNICAL MANUAL AND DICTIONARY OF CLASSICAL BALLET, Gail Grant. Defines, explains, comments on steps, movements, poses and concepts. 15-page pictorial section. Basic book for student, viewer. 127pp. 5⅜ x 8½. 21843-0 Pa. $4.95

BRASS INSTRUMENTS: Their History and Development, Anthony Baines. Authoritative, updated survey of the evolution of trumpets, trombones, bugles, cornets, French horns, tubas and other brass wind instruments. Over 140 illustrations and 48 music examples. Corrected and updated by author. New preface. Bibliography. 320pp. 5⅜ x 8½. 27574-4 Pa. $9.95

HOLLYWOOD GLAMOR PORTRAITS, John Kobal (ed.). 145 photos from 1926-49. Harlow, Gable, Bogart, Bacall; 94 stars in all. Full background on photographers, technical aspects. 160pp. 8⅜ x 11¼. 23352-9 Pa. $12.95

MAX AND MORITZ, Wilhelm Busch. Great humor classic in both German and English. Also 10 other works: "Cat and Mouse," "Plisch and Plumm," etc. 216pp. 5⅜ x 8½. 20181-3 Pa. $6.95

THE RAVEN AND OTHER FAVORITE POEMS, Edgar Allan Poe. Over 40 of the author's most memorable poems: "The Bells," "Ulalume," "Israfel," "To Helen," "The Conqueror Worm," "Eldorado," "Annabel Lee," many more. Alphabetic lists of titles and first lines. 64pp. $5\frac{3}{16}$ x 8¼. 26685-0 Pa. $1.00

PERSONAL MEMOIRS OF U. S. GRANT, Ulysses Simpson Grant. Intelligent, deeply moving firsthand account of Civil War campaigns, considered by many the finest military memoirs ever written. Includes letters, historic photographs, maps and more. 528pp. 6⅛ x 9¼. 28587-1 Pa. $12.95

AMULETS AND SUPERSTITIONS, E. A. Wallis Budge. Comprehensive discourse on origin, powers of amulets in many ancient cultures: Arab, Persian Babylonian, Assyrian, Egyptian, Gnostic, Hebrew, Phoenician, Syriac, etc. Covers cross, swastika, crucifix, seals, rings, stones, etc. 584pp. 5⅜ x 8½. 23573-4 Pa. $12.95

RUSSIAN STORIES/PYCCKNE PACCKA3bl: A Dual-Language Book, edited by Gleb Struve. Twelve tales by such masters as Chekhov, Tolstoy, Dostoevsky, Pushkin, others. Excellent word-for-word English translations on facing pages, plus teaching and study aids, Russian/English vocabulary, biographical/critical introductions, more. 416pp. 5⅜ x 8½. 26244-8 Pa. $9.95

PHILADELPHIA THEN AND NOW: 60 Sites Photographed in the Past and Present, Kenneth Finkel and Susan Oyama. Rare photographs of City Hall, Logan Square, Independence Hall, Betsy Ross House, other landmarks juxtaposed with contemporary views. Captures changing face of historic city. Introduction. Captions. 128pp. 8¼ x 11. 25790-8 Pa. $9.95

AIA ARCHITECTURAL GUIDE TO NASSAU AND SUFFOLK COUNTIES, LONG ISLAND, The American Institute of Architects, Long Island Chapter, and the Society for the Preservation of Long Island Antiquities. Comprehensive, well-researched and generously illustrated volume brings to life over three centuries of Long Island's great architectural heritage. More than 240 photographs with authoritative, extensively detailed captions. 176pp. 8¼ x 11. 26946-9 Pa. $14.95

NORTH AMERICAN INDIAN LIFE: Customs and Traditions of 23 Tribes, Elsie Clews Parsons (ed.). 27 fictionalized essays by noted anthropologists examine religion, customs, government, additional facets of life among the Winnebago, Crow, Zuni, Eskimo, other tribes. 480pp. 6⅛ x 9¼. 27377-6 Pa. $10.95

FRANK LLOYD WRIGHT'S HOLLYHOCK HOUSE, Donald Hoffmann. Lavishly illustrated, carefully documented study of one of Wright's most controversial residential designs. Over 120 photographs, floor plans, elevations, etc. Detailed perceptive text by noted Wright scholar. Index. 128pp. 9¼ x 10¾. 27133-1 Pa. $11.95

THE MALE AND FEMALE FIGURE IN MOTION: 60 Classic Photographic Sequences, Eadweard Muybridge. 60 true-action photographs of men and women walking, running, climbing, bending, turning, etc., reproduced from rare 19th-century masterpiece. vi + 121pp. 9 x 12. 24745-7 Pa. $10.95

1001 QUESTIONS ANSWERED ABOUT THE SEASHORE, N. J. Berrill and Jacquelyn Berrill. Queries answered about dolphins, sea snails, sponges, starfish, fishes, shore birds, many others. Covers appearance, breeding, growth, feeding, much more. 305pp. 5¼ x 8¼. 23366-9 Pa. $8.95

GUIDE TO OWL WATCHING IN NORTH AMERICA, Donald S. Heintzelman. Superb guide offers complete data and descriptions of 19 species: barn owl, screech owl, snowy owl, many more. Expert coverage of owl-watching equipment, conservation, migrations and invasions, etc. Guide to observing sites. 84 illustrations. xiii + 193pp. 5⅜ x 8½. 27344-X Pa. $8.95

MEDICINAL AND OTHER USES OF NORTH AMERICAN PLANTS: A Historical Survey with Special Reference to the Eastern Indian Tribes, Charlotte Erichsen-Brown. Chronological historical citations document 500 years of usage of plants, trees, shrubs native to eastern Canada, northeastern U.S. Also complete identifying information. 343 illustrations. 544pp. 6½ x 9¼. 25951-X Pa. $12.95

STORYBOOK MAZES, Dave Phillips. 23 stories and mazes on two-page spreads: Wizard of Oz, Treasure Island, Robin Hood, etc. Solutions. 64pp. 8¼ x 11. 23628-5 Pa. $2.95

NEGRO FOLK MUSIC, U.S.A., Harold Courlander. Noted folklorist's scholarly yet readable analysis of rich and varied musical tradition. Includes authentic versions of over 40 folk songs. Valuable bibliography and discography. xi + 324pp. 5⅜ x 8½. 27350-4 Pa. $9.95

MOVIE-STAR PORTRAITS OF THE FORTIES, John Kobal (ed.). 163 glamor, studio photos of 106 stars of the 1940s: Rita Hayworth, Ava Gardner, Marlon Brando, Clark Gable, many more. 176pp. 8⅜ x 11¼. 23546-7 Pa. $12.95

BENCHLEY LOST AND FOUND, Robert Benchley. Finest humor from early 30s, about pet peeves, child psychologists, post office and others. Mostly unavailable elsewhere. 73 illustrations by Peter Arno and others. 183pp. 5⅜ x 8½. 22410-4 Pa. $6.95

YEKL and THE IMPORTED BRIDEGROOM AND OTHER STORIES OF YIDDISH NEW YORK, Abraham Cahan. Film Hester Street based on Yekl (1896). Novel, other stories among first about Jewish immigrants on N.Y.'s East Side. 240pp. 5⅜ x 8½. 22427-9 Pa. $6.95

SELECTED POEMS, Walt Whitman. Generous sampling from *Leaves of Grass*. Twenty-four poems include "I Hear America Singing," "Song of the Open Road," "I Sing the Body Electric," "When Lilacs Last in the Dooryard Bloom'd," "O Captain! My Captain!"–all reprinted from an authoritative edition. Lists of titles and first lines. 128pp. 5$\frac{3}{16}$ x 8¼. 26878-0 Pa. $1.00

THE BEST TALES OF HOFFMANN, E. T. A. Hoffmann. 10 of Hoffmann's most important stories: "Nutcracker and the King of Mice," "The Golden Flowerpot," etc. 458pp. 5⅜ x 8½. 21793-0 Pa. $9.95

FROM FETISH TO GOD IN ANCIENT EGYPT, E. A. Wallis Budge. Rich detailed survey of Egyptian conception of "God" and gods, magic, cult of animals, Osiris, more. Also, superb English translations of hymns and legends. 240 illustrations. 545pp. 5⅜ x 8½. 25803-3 Pa. $13.95

FRENCH STORIES/CONTES FRANÇAIS: A Dual-Language Book, Wallace Fowlie. Ten stories by French masters, Voltaire to Camus: "Micromegas" by Voltaire; "The Atheist's Mass" by Balzac; "Minuet" by de Maupassant; "The Guest" by Camus, six more. Excellent English translations on facing pages. Also French-English vocabulary list, exercises, more. 352pp. 5⅜ x 8½. 26443-2 Pa. $9.95

CHICAGO AT THE TURN OF THE CENTURY IN PHOTOGRAPHS: 122 Historic Views from the Collections of the Chicago Historical Society, Larry A. Viskochil. Rare large-format prints offer detailed views of City Hall, State Street, the Loop, Hull House, Union Station, many other landmarks, circa 1904-1913. Introduction. Captions. Maps. 144pp. 9⅜ x 12¼. 24656-6 Pa. $12.95

OLD BROOKLYN IN EARLY PHOTOGRAPHS, 1865-1929, William Lee Younger. Luna Park, Gravesend race track, construction of Grand Army Plaza, moving of Hotel Brighton, etc. 157 previously unpublished photographs. 165pp. 8⅞ x 11¾. 23587-4 Pa. $13.95

THE MYTHS OF THE NORTH AMERICAN INDIANS, Lewis Spence. Rich anthology of the myths and legends of the Algonquins, Iroquois, Pawnees and Sioux, prefaced by an extensive historical and ethnological commentary. 36 illustrations. 480pp. 5⅜ x 8½. 25967-6 Pa. $10.95

AN ENCYCLOPEDIA OF BATTLES: Accounts of Over 1,560 Battles from 1479 B.C. to the Present, David Eggenberger. Essential details of every major battle in recorded history from the first battle of Megiddo in 1479 B.C. to Grenada in 1984. List of Battle Maps. New Appendix covering the years 1967-1984. Index. 99 illustrations. 544pp. 6½ x 9¼. 24913-1 Pa. $16.95

SAILING ALONE AROUND THE WORLD, Captain Joshua Slocum. First man to sail around the world, alone, in small boat. One of great feats of seamanship told in delightful manner. 67 illustrations. 294pp. 5⅜ x 8½. 20326-3 Pa. $6.95

ANARCHISM AND OTHER ESSAYS, Emma Goldman. Powerful, penetrating, prophetic essays on direct action, role of minorities, prison reform, puritan hypocrisy, violence, etc. 271pp. 5⅜ x 8½. 22484-8 Pa. $7.95

MYTHS OF THE HINDUS AND BUDDHISTS, Ananda K. Coomaraswamy and Sister Nivedita. Great stories of the epics; deeds of Krishna, Shiva, taken from puranas, Vedas, folk tales; etc. 32 illustrations. 400pp. 5⅜ x 8½. 21759-0 Pa. $12.95

BEYOND PSYCHOLOGY, Otto Rank. Fear of death, desire of immortality, nature of sexuality, social organization, creativity, according to Rankian system. 291pp. 5⅜ x 8½. 20485-5 Pa. $8.95

A THEOLOGICO-POLITICAL TREATISE, Benedict Spinoza. Also contains unfinished Political Treatise. Great classic on religious liberty, theory of government on common consent. R. Elwes translation. Total of 421pp. 5⅜ x 8½. 20249-6 Pa. $9.95

PERSPECTIVE FOR ARTISTS, Rex Vicat Cole. Depth, perspective of sky and sea, shadows, much more, not usually covered. 391 diagrams, 81 reproductions of drawings and paintings. 279pp. 5⅜ x 8½. 22487-2 Pa. $7.95

DRAWING THE LIVING FIGURE, Joseph Sheppard. Innovative approach to artistic anatomy focuses on specifics of surface anatomy, rather than muscles and bones. Over 170 drawings of live models in front, back and side views, and in widely varying poses. Accompanying diagrams. 177 illustrations. Introduction. Index. 144pp. 8⅜ x11¼. 26723-7 Pa. $8.95

GOTHIC AND OLD ENGLISH ALPHABETS: 100 Complete Fonts, Dan X. Solo. Add power, elegance to posters, signs, other graphics with 100 stunning copyright-free alphabets: Blackstone, Dolbey, Germania, 97 more–including many lower-case, numerals, punctuation marks. 104pp. 8⅛ x 11. 24695-7 Pa. $8.95

HOW TO DO BEADWORK, Mary White. Fundamental book on craft from simple projects to five-bead chains and woven works. 106 illustrations. 142pp. 5⅜ x 8. 20697-1 Pa. $4.95

THE BOOK OF WOOD CARVING, Charles Marshall Sayers. Finest book for beginners discusses fundamentals and offers 34 designs. "Absolutely first rate . . . well thought out and well executed."–E. J. Tangerman. 118pp. 7¾ x 10⅝. 23654-4 Pa. $6.95

ILLUSTRATED CATALOG OF CIVIL WAR MILITARY GOODS: Union Army Weapons, Insignia, Uniform Accessories, and Other Equipment, Schuyler, Hartley, and Graham. Rare, profusely illustrated 1846 catalog includes Union Army uniform and dress regulations, arms and ammunition, coats, insignia, flags, swords, rifles, etc. 226 illustrations. 160pp. 9 x 12. 24939-5 Pa. $10.95

WOMEN'S FASHIONS OF THE EARLY 1900s: An Unabridged Republication of "New York Fashions, 1909," National Cloak & Suit Co. Rare catalog of mail-order fashions documents women's and children's clothing styles shortly after the turn of the century. Captions offer full descriptions, prices. Invaluable resource for fashion, costume historians. Approximately 725 illustrations. 128pp. 8⅜ x 11¼. 27276-1 Pa. $11.95

THE 1912 AND 1915 GUSTAV STICKLEY FURNITURE CATALOGS, Gustav Stickley. With over 200 detailed illustrations and descriptions, these two catalogs are essential reading and reference materials and identification guides for Stickley furniture. Captions cite materials, dimensions and prices. 112pp. 6½ x 9¼. 26676-1 Pa. $9.95

EARLY AMERICAN LOCOMOTIVES, John H. White, Jr. Finest locomotive engravings from early 19th century: historical (1804–74), main-line (after 1870), special, foreign, etc. 147 plates. 142pp. 11⅜ x 8¼. 22772-3 Pa. $10.95

THE TALL SHIPS OF TODAY IN PHOTOGRAPHS, Frank O. Braynard. Lavishly illustrated tribute to nearly 100 majestic contemporary sailing vessels: Amerigo Vespucci, Clearwater, Constitution, Eagle, Mayflower, Sea Cloud, Victory, many more. Authoritative captions provide statistics, background on each ship. 190 black-and-white photographs and illustrations. Introduction. 128pp. 8⅞ x 11¾. 27163-3 Pa. $14.95

PIANO TUNING, J. Cree Fischer. Clearest, best book for beginner, amateur. Simple repairs, raising dropped notes, tuning by easy method of flattened fifths. No previous skills needed. 4 illustrations. 201pp. 5⅜ x 8½. 23267-0 Pa. $6.95

A SOURCE BOOK IN THEATRICAL HISTORY, A. M. Nagler. Contemporary observers on acting, directing, make-up, costuming, stage props, machinery, scene design, from Ancient Greece to Chekhov. 611pp. 5⅜ x 8½. 20515-0 Pa. $12.95

THE COMPLETE NONSENSE OF EDWARD LEAR, Edward Lear. All nonsense limericks, zany alphabets, Owl and Pussycat, songs, nonsense botany, etc., illustrated by Lear. Total of 320pp. 5⅜ x 8½. (USO) 20167-8 Pa. $7.95

VICTORIAN PARLOUR POETRY: An Annotated Anthology, Michael R. Turner. 117 gems by Longfellow, Tennyson, Browning, many lesser-known poets. "The Village Blacksmith," "Curfew Must Not Ring Tonight," "Only a Baby Small," dozens more, often difficult to find elsewhere. Index of poets, titles, first lines. xxiii + 325pp. 5⅜ x 8¼. 27044-0 Pa. $8.95

DUBLINERS, James Joyce. Fifteen stories offer vivid, tightly focused observations of the lives of Dublin's poorer classes. At least one, "The Dead," is considered a masterpiece. Reprinted complete and unabridged from standard edition. 160pp. $5\frac{3}{16}$ x 8¼. 26870-5 Pa. $1.00

THE HAUNTED MONASTERY and THE CHINESE MAZE MURDERS, Robert van Gulik. Two full novels by van Gulik, set in 7th-century China, continue adventures of Judge Dee and his companions. An evil Taoist monastery, seemingly supernatural events; overgrown topiary maze hides strange crimes. 27 illustrations. 328pp. 5⅜ x 8½. 23502-5 Pa. $8.95

THE BOOK OF THE SACRED MAGIC OF ABRAMELIN THE MAGE, translated by S. MacGregor Mathers. Medieval manuscript of ceremonial magic. Basic document in Aleister Crowley, Golden Dawn groups. 268pp. 5⅜ x 8½. 23211-5 Pa. $9.95

NEW RUSSIAN-ENGLISH AND ENGLISH-RUSSIAN DICTIONARY, M. A. O'Brien. This is a remarkably handy Russian dictionary, containing a surprising amount of information, including over 70,000 entries. 366pp. 4½ x 6⅛. 20208-9 Pa. $9.95

HISTORIC HOMES OF THE AMERICAN PRESIDENTS, Second, Revised Edition, Irvin Haas. A traveler's guide to American Presidential homes, most open to the public, depicting and describing homes occupied by every American President from George Washington to George Bush. With visiting hours, admission charges, travel routes. 175 photographs. Index. 160pp. 8¼ x 11. 26751-2 Pa. $11.95

NEW YORK IN THE FORTIES, Andreas Feininger. 162 brilliant photographs by the well-known photographer, formerly with *Life* magazine. Commuters, shoppers, Times Square at night, much else from city at its peak. Captions by John von Hartz. 181pp. 9¼ x 10¾. 23585-8 Pa. $12.95

INDIAN SIGN LANGUAGE, William Tomkins. Over 525 signs developed by Sioux and other tribes. Written instructions and diagrams. Also 290 pictographs. 111pp. 6⅛ x 9¼. 22029-X Pa. $3.95

CATALOG OF DOVER BOOKS

ANATOMY: A Complete Guide for Artists, Joseph Sheppard. A master of figure drawing shows artists how to render human anatomy convincingly. Over 460 illustrations. 224pp. 8⅜ x 11¼. 27279-6 Pa. $11.95

MEDIEVAL CALLIGRAPHY: Its History and Technique, Marc Drogin. Spirited history, comprehensive instruction manual covers 13 styles (ca. 4th century thru 15th). Excellent photographs; directions for duplicating medieval techniques with modern tools. 224pp. 8⅜ x 11¼. 26142-5 Pa. $12.95

DRIED FLOWERS: How to Prepare Them, Sarah Whitlock and Martha Rankin. Complete instructions on how to use silica gel, meal and borax, perlite aggregate, sand and borax, glycerine and water to create attractive permanent flower arrangements. 12 illustrations. 32pp. 5⅜ x 8½. 21802-3 Pa. $1.00

EASY-TO-MAKE BIRD FEEDERS FOR WOODWORKERS, Scott D. Campbell. Detailed, simple-to-use guide for designing, constructing, caring for and using feeders. Text, illustrations for 12 classic and contemporary designs. 96pp. 5⅜ x 8½. 25847-5 Pa. $3.95

SCOTTISH WONDER TALES FROM MYTH AND LEGEND, Donald A. Mackenzie. 16 lively tales tell of giants rumbling down mountainsides, of a magic wand that turns stone pillars into warriors, of gods and goddesses, evil hags, powerful forces and more. 240pp. 5⅜ x 8½. 29677-6 Pa. $6.95

THE HISTORY OF UNDERCLOTHES, C. Willett Cunnington and Phyllis Cunnington. Fascinating, well-documented survey covering six centuries of English undergarments, enhanced with over 100 illustrations: 12th-century laced-up bodice, footed long drawers (1795), 19th-century bustles, 19th-century corsets for men, Victorian "bust improvers," much more. 272pp. 5⅝ x 8¼. 27124-2 Pa. $9.95

ARTS AND CRAFTS FURNITURE: The Complete Brooks Catalog of 1912, Brooks Manufacturing Co. Photos and detailed descriptions of more than 150 now very collectible furniture designs from the Arts and Crafts movement depict davenports, settees, buffets, desks, tables, chairs, bedsteads, dressers and more, all built of solid, quarter-sawed oak. Invaluable for students and enthusiasts of antiques, Americana and the decorative arts. 80pp. 6½ x 9¼. 27471-3 Pa. $8.95

HOW WE INVENTED THE AIRPLANE: An Illustrated History, Orville Wright. Fascinating firsthand account covers early experiments, construction of planes and motors, first flights, much more. Introduction and commentary by Fred C. Kelly. 76 photographs. 96pp. 8¼ x 11. 25662-6 Pa. $8.95

THE ARTS OF THE SAILOR: Knotting, Splicing and Ropework, Hervey Garrett Smith. Indispensable shipboard reference covers tools, basic knots and useful hitches; handsewing and canvas work, more. Over 100 illustrations. Delightful reading for sea lovers. 256pp. 5⅜ x 8½. 26440-8 Pa. $7.95

FRANK LLOYD WRIGHT'S FALLINGWATER: The House and Its History, Second, Revised Edition, Donald Hoffmann. A total revision–both in text and illustrations–of the standard document on Fallingwater, the boldest, most personal architectural statement of Wright's mature years, updated with valuable new material from the recently opened Frank Lloyd Wright Archives. "Fascinating"–*The New York Times*. 116 illustrations. 128pp. 9¼ x 10¾. 27430-6 Pa. $12.95

PHOTOGRAPHIC SKETCHBOOK OF THE CIVIL WAR, Alexander Gardner. 100 photos taken on field during the Civil War. Famous shots of Manassas Harper's Ferry, Lincoln, Richmond, slave pens, etc. 244pp. 10⅝ x 8¼. 22731-6 Pa. $9.95

FIVE ACRES AND INDEPENDENCE, Maurice G. Kains. Great back-to-the-land classic explains basics of self-sufficient farming. The one book to get. 95 illustrations. 397pp. 5⅜ x 8½. 20974-1 Pa. $7.95

SONGS OF EASTERN BIRDS, Dr. Donald J. Borror. Songs and calls of 60 species most common to eastern U.S.: warblers, woodpeckers, flycatchers, thrushes, larks, many more in high-quality recording. Cassette and manual 99912-2 $9.95

A MODERN HERBAL, Margaret Grieve. Much the fullest, most exact, most useful compilation of herbal material. Gigantic alphabetical encyclopedia, from aconite to zedoary, gives botanical information, medical properties, folklore, economic uses, much else. Indispensable to serious reader. 161 illustrations. 888pp. 6½ x 9¼. 2-vol. set. (USO) Vol. I: 22798-7 Pa. $9.95
Vol. II: 22799-5 Pa. $9.95

HIDDEN TREASURE MAZE BOOK, Dave Phillips. Solve 34 challenging mazes accompanied by heroic tales of adventure. Evil dragons, people-eating plants, bloodthirsty giants, many more dangerous adversaries lurk at every twist and turn. 34 mazes, stories, solutions. 48pp. 8¼ x 11. 24566-7 Pa. $2.95

LETTERS OF W. A. MOZART, Wolfgang A. Mozart. Remarkable letters show bawdy wit, humor, imagination, musical insights, contemporary musical world; includes some letters from Leopold Mozart. 276pp. 5⅜ x 8½. 22859-2 Pa. $7.95

BASIC PRINCIPLES OF CLASSICAL BALLET, Agrippina Vaganova. Great Russian theoretician, teacher explains methods for teaching classical ballet. 118 illustrations. 175pp. 5⅜ x 8½. 22036-2 Pa. $5.95

THE JUMPING FROG, Mark Twain. Revenge edition. The original story of The Celebrated Jumping Frog of Calaveras County, a hapless French translation, and Twain's hilarious "retranslation" from the French. 12 illustrations. 66pp. 5⅜ x 8½. 22686-7 Pa. $3.95

BEST REMEMBERED POEMS, Martin Gardner (ed.). The 126 poems in this superb collection of 19th- and 20th-century British and American verse range from Shelley's "To a Skylark" to the impassioned "Renascence" of Edna St. Vincent Millay and to Edward Lear's whimsical "The Owl and the Pussycat." 224pp. 5⅜ x 8½. 27165-X Pa. $5.95

COMPLETE SONNETS, William Shakespeare. Over 150 exquisite poems deal with love, friendship, the tyranny of time, beauty's evanescence, death and other themes in language of remarkable power, precision and beauty. Glossary of archaic terms. 80pp. 5³⁄₁₆ x 8¼. 26686-9 Pa. $1.00

BODIES IN A BOOKSHOP, R. T. Campbell. Challenging mystery of blackmail and murder with ingenious plot and superbly drawn characters. In the best tradition of British suspense fiction. 192pp. 5⅜ x 8½. 24720-1 Pa. $6.95

THE WIT AND HUMOR OF OSCAR WILDE, Alvin Redman (ed.). More than 1,000 ripostes, paradoxes, wisecracks: Work is the curse of the drinking classes; I can resist everything except temptation; etc. 258pp. 5⅜ x 8½. 20602-5 Pa. $5.95

SHAKESPEARE LEXICON AND QUOTATION DICTIONARY, Alexander Schmidt. Full definitions, locations, shades of meaning in every word in plays and poems. More than 50,000 exact quotations. 1,485pp. 6½ x 9¼. 2-vol. set.
Vol. 1: 22726-X Pa. $17.95
Vol. 2: 22727-8 Pa. $17.95

SELECTED POEMS, Emily Dickinson. Over 100 best-known, best-loved poems by one of America's foremost poets, reprinted from authoritative early editions. No comparable edition at this price. Index of first lines. 64pp. 5$\frac{3}{16}$ x 8¼.
26466-1 Pa. $1.00

CELEBRATED CASES OF JUDGE DEE (DEE GOONG AN), translated by Robert van Gulik. Authentic 18th-century Chinese detective novel; Dee and associates solve three interlocked cases. Led to van Gulik's own stories with same characters. Extensive introduction. 9 illustrations. 237pp. 5⅜ x 8½. 23337-5 Pa. $7.95

THE MALLEUS MALEFICARUM OF KRAMER AND SPRENGER, translated by Montague Summers. Full text of most important witchhunter's "bible," used by both Catholics and Protestants. 278pp. 6⅝ x 10. 22802-9 Pa. $12.95

SPANISH STORIES/CUENTOS ESPAÑOLES: A Dual-Language Book, Angel Flores (ed.). Unique format offers 13 great stories in Spanish by Cervantes, Borges, others. Faithful English translations on facing pages. 352pp. 5⅜ x 8½.
25399-6 Pa. $8.95

THE CHICAGO WORLD'S FAIR OF 1893: A Photographic Record, Stanley Appelbaum (ed.). 128 rare photos show 200 buildings, Beaux-Arts architecture, Midway, original Ferris Wheel, Edison's kinetoscope, more. Architectural emphasis; full text. 116pp. 8¼ x 11. 23990-X Pa. $9.95

OLD QUEENS, N.Y., IN EARLY PHOTOGRAPHS, Vincent F. Seyfried and William Asadorian. Over 160 rare photographs of Maspeth, Jamaica, Jackson Heights, and other areas. Vintage views of DeWitt Clinton mansion, 1939 World's Fair and more. Captions. 192pp. 8⅞ x 11. 26358-4 Pa. $12.95

CAPTURED BY THE INDIANS: 15 Firsthand Accounts, 1750-1870, Frederick Drimmer. Astounding true historical accounts of grisly torture, bloody conflicts, relentless pursuits, miraculous escapes and more, by people who lived to tell the tale. 384pp. 5⅜ x 8½. 24901-8 Pa. $8.95

THE WORLD'S GREAT SPEECHES, Lewis Copeland and Lawrence W. Lamm (eds.). Vast collection of 278 speeches of Greeks to 1970. Powerful and effective models; unique look at history. 842pp. 5⅜ x 8½. 20468-5 Pa. $14.95

THE BOOK OF THE SWORD, Sir Richard F. Burton. Great Victorian scholar/adventurer's eloquent, erudite history of the "queen of weapons"–from prehistory to early Roman Empire. Evolution and development of early swords, variations (sabre, broadsword, cutlass, scimitar, etc.), much more. 336pp. 6⅛ x 9¼.
25434-8 Pa. $9.95

AUTOBIOGRAPHY: The Story of My Experiments with Truth, Mohandas K. Gandhi. Boyhood, legal studies, purification, the growth of the Satyagraha (nonviolent protest) movement. Critical, inspiring work of the man responsible for the freedom of India. 480pp. 5⅜ x 8½. (USO) 24593-4 Pa. $8.95

CELTIC MYTHS AND LEGENDS, T. W. Rolleston. Masterful retelling of Irish and Welsh stories and tales. Cuchulain, King Arthur, Deirdre, the Grail, many more. First paperback edition. 58 full-page illustrations. 512pp. 5⅜ x 8½. 26507-2 Pa. $9.95

THE PRINCIPLES OF PSYCHOLOGY, William James. Famous long course complete, unabridged. Stream of thought, time perception, memory, experimental methods; great work decades ahead of its time. 94 figures. 1,391pp. 5⅜ x 8½. 2-vol. set.
Vol. I: 20381-6 Pa. $13.95
Vol. II: 20382-4 Pa. $14.95

THE WORLD AS WILL AND REPRESENTATION, Arthur Schopenhauer. Definitive English translation of Schopenhauer's life work, correcting more than 1,000 errors, omissions in earlier translations. Translated by E. F. J. Payne. Total of 1,269pp. 5⅜ x 8½. 2-vol. set. Vol. 1: 21761-2 Pa. $12.95
Vol. 2: 21762-0 Pa. $12.95

MAGIC AND MYSTERY IN TIBET, Madame Alexandra David-Neel. Experiences among lamas, magicians, sages, sorcerers, Bonpa wizards. A true psychic discovery. 32 illustrations. 321pp. 5⅜ x 8½. (USO) 22682-4 Pa. $9.95

THE EGYPTIAN BOOK OF THE DEAD, E. A. Wallis Budge. Complete reproduction of Ani's papyrus, finest ever found. Full hieroglyphic text, interlinear transliteration, word-for-word translation, smooth translation. 533pp. 6½ x 9¼.
21866-X Pa. $11.95

MATHEMATICS FOR THE NONMATHEMATICIAN, Morris Kline. Detailed, college-level treatment of mathematics in cultural and historical context, with numerous exercises. Recommended Reading Lists. Tables. Numerous figures. 641pp. 5⅜ x 8½.
24823-2 Pa. $11.95

THEORY OF WING SECTIONS: Including a Summary of Airfoil Data, Ira H. Abbott and A. E. von Doenhoff. Concise compilation of subsonic aerodynamic characteristics of NACA wing sections, plus description of theory. 350pp. of tables. 693pp. 5⅜ x 8½. 60586-8 Pa. $14.95

THE RIME OF THE ANCIENT MARINER, Gustave Doré, S. T. Coleridge. Doré's finest work; 34 plates capture moods, subtleties of poem. Flawless full-size reproductions printed on facing pages with authoritative text of poem. "Beautiful. Simply beautiful."–*Publisher's Weekly*. 77pp. 9¼ x 12. 22305-1 Pa. $7.95

NORTH AMERICAN INDIAN DESIGNS FOR ARTISTS AND CRAFTSPEOPLE, Eva Wilson. Over 360 authentic copyright-free designs adapted from Navajo blankets, Hopi pottery, Sioux buffalo hides, more. Geometrics, symbolic figures, plant and animal motifs, etc. 128pp. 8⅜ x 11. (EUK) 25341-4 Pa. $8.95

SCULPTURE: Principles and Practice, Louis Slobodkin. Step-by-step approach to clay, plaster, metals, stone; classical and modern. 253 drawings, photos. 255pp. 8⅛ x 11.
22960-2 Pa. $11.95